T0176931

Information Theory Meets Power Laws

Information Theory Meets Power Laws

Stochastic Processes and Language Models

Łukasz Dębowski
Polish Academy of Sciences

This edition first published 2021
© 2021 John Wiley & Sons, Inc.

All rights reserved. No part of this publication may be reproduced, stored in a retrieval system, or transmitted, in any form or by any means, electronic, mechanical, photocopying, recording or otherwise, except as permitted by law. Advice on how to obtain permission to reuse material from this title is available at http://www.wiley.com/go/permissions.

The right of Łukasz Dębowski to be identified as the author of this work has been asserted in accordance with law.

Registered Office
John Wiley & Sons, Inc., 111 River Street, Hoboken, NJ 07030, USA

Editorial Office
111 River Street, Hoboken, NJ 07030, USA

For details of our global editorial offices, customer services, and more information about Wiley products visit us at www.wiley.com.

Wiley also publishes its books in a variety of electronic formats and by print-on-demand. Some content that appears in standard print versions of this book may not be available in other formats.

Limit of Liability/Disclaimer of Warranty
While the publisher and authors have used their best efforts in preparing this work, they make no representations or warranties with respect to the accuracy or completeness of the contents of this work and specifically disclaim all warranties, including without limitation any implied warranties of merchantability or fitness for a particular purpose. No warranty may be created or extended by sales representatives, written sales materials or promotional statements for this work. The fact that an organization, website, or product is referred to in this work as a citation and/or potential source of further information does not mean that the publisher and authors endorse the information or services the organization, website, or product may provide or recommendations it may make. This work is sold with the understanding that the publisher is not engaged in rendering professional services. The advice and strategies contained herein may not be suitable for your situation. You should consult with a specialist where appropriate. Further, readers should be aware that websites listed in this work may have changed or disappeared between when this work was written and when it is read. Neither the publisher nor authors shall be liable for any loss of profit or any other commercial damages, including but not limited to special, incidental, consequential, or other damages.

Library of Congress Cataloging-in-Publication Data

Names: Dębowski, Łukasz Jerzy, 1975- author.
Title: Information theory meets power laws : stochastic processes and
    language models / Łukasz Jerzy Dębowski, Polish Academy of Sciences.
Description: Hoboken : Wiley, 2021. | Includes bibliographical references
    and index.
Identifiers: LCCN 2020015366 (print) | LCCN 2020015367 (ebook) | ISBN
    9781119625278 (hardback) | ISBN 9781119625360 (adobe pdf) | ISBN
    9781119625377 (epub)
Subjects: LCSH: Computational linguistics. | Stochastic processes.
Classification: LCC P98 .D35 2020 (print) | LCC P98 (ebook) | DDC
    410.1/5195–dc23
LC record available at https://lccn.loc.gov/2020015366
LC ebook record available at https://lccn.loc.gov/2020015367

Cover design by Wiley
Cover image: © Jan Hakan Dahlstrom/Getty Images

Set in 9.5/12.5pt STIXTwoText by SPi Global, Chennai, India

Printed in the United States of America

SKY10022086_103020

In memory of my Mother

Contents

Preface *ix*
Acknowledgments *xiii*
Basic Notations *xv*

1   **Guiding Ideas** *1*
1.1   The Motivating Question   *1*
1.2   Further Questions About Texts   *5*
1.3   Zipf's and Herdan's Laws   *8*
1.4   Markov and Finite-State Processes   *14*
1.5   More General Stochastic Processes   *20*
1.6   Two Interpretations of Probability   *23*
1.7   Insights from Information Theory   *25*
1.8   Estimation of Entropy Rate   *28*
1.9   Entropy of Natural Language   *30*
1.10   Algorithmic Information Theory   *35*
1.11   Descriptions of a Random World   *37*
1.12   Facts and Words Related   *43*
1.13   Repetitions and Entropies   *47*
1.14   Decay of Correlations   *52*
1.15   Recapitulation   *54*

2   **Probabilistic Preliminaries** *57*
2.1   Probability Measures   *59*
2.2   Product Measurable Spaces   *63*
2.3   Discrete Random Variables   *65*
2.4   From IID to Finite-State Processes   *68*
       Problems   *73*

3   Probabilistic Toolbox   *77*
3.1   Borel σ-Fields and a Fair Coin   *79*
3.2   Integral and Expectation   *83*
3.3   Inequalities and Corollaries   *87*
3.4   Semidistributions   *92*
3.5   Conditional Probability   *94*
3.6   Modes of Convergence   *101*
3.7   Complete Spaces   *103*
      Problems   *106*

4   Ergodic Properties   *109*
4.1   Plain Relative Frequency   *111*
4.2   Birkhoff Ergodic Theorem   *116*
4.3   Ergodic and Mixing Criteria   *119*
4.4   Ergodic Decomposition   *125*
      Problems   *128*

5   Entropy and Information   *131*
5.1   Shannon Measures for Partitions   *133*
5.2   Block Entropy and Its Limits   *139*
5.3   Shannon Measures for Fields   *145*
5.4   Block Entropy Limits Revisited   *155*
5.5   Convergence of Entropy   *159*
5.6   Entropy as Self-Information   *160*
      Problems   *163*

6   Equipartition and Universality   *167*
6.1   SMB Theorem   *169*
6.2   Universal Semidistributions   *171*
6.3   PPM Probability   *172*
6.4   SMB Theorem Revisited   *178*
6.5   PPM-based Statistics   *180*
      Problems   *186*

7   Coding and Computation   *189*
7.1   Elements of Coding   *191*
7.2   Kolmogorov Complexity   *197*
7.3   Algorithmic Coding Theorems   *207*
7.4   Limits of Mathematics   *215*
7.5   Algorithmic Randomness   *220*
      Problems   *225*

8   **Power Laws for Information** *229*
8.1   Hilberg Exponents *231*
8.2   Second Order SMB Theorem *238*
8.3   Probabilistic and Algorithmic Facts *241*
8.4   Theorems About Facts and Words *248*
      Problems *255*

9   **Power Laws for Repetitions** *259*
9.1   Rényi–Arimoto Entropies *261*
9.2   Generalized Entropy Rates *266*
9.3   Recurrence Times *268*
9.4   Subword Complexity *272*
9.5   Two Maximal Lengths *280*
9.6   Logarithmic Power Laws *284*
      Problems *289*

10   **AMS Processes** *291*
10.1   AMS and Pseudo AMS Measures *293*
10.2   Quasiperiodic Coding *295*
10.3   Synchronizable Coding *298*
10.4   Entropy Rate in the AMS Case *301*
      Problems *304*

11   **Toy Examples** *307*
11.1   Finite and Ultrafinite Energy *309*
11.2   Santa Fe Processes and Alike *315*
11.3   Encoding into a Finite Alphabet *323*
11.4   Random Hierarchical Association *334*
11.5   Toward Better Models *345*
      Problems *348*

Future Research *349*
Bibliography *351*
Index *365*

Preface

This book concerns some idealized mathematical models of human language. Nowadays, there are recognized two approaches to mathematical modeling of language. One stems historically from theoretical computer science and formal linguistics, another applies probability, statistics, and information theory. The first approach gave rise to formal language theory, formal syntax, and formal semantics, whereas the second approach is mostly followed by engineers working in computational linguistics and machine learning. In this book, I would like to lay foundations to a novel subtype of a probabilistic approach to language, based on empirical quantitative laws of texts and their deductive analysis by means of information theory and stochastic processes.

Having originally done my master studies in theoretical physics, gaining some experience in computational linguistics, and later switching to information theory and stochastic processes, I believe that a good science is a dialogue between theory and experiment. Concerning probability models for language, I think that there are many computational applications and ambitious experiments. In contrast, the theoretical understanding of some relatively basic stochastic-semantic phenomena, such as strong nonergodicity, remains surprisingly underdeveloped, untouched by linguists, engineers, or mathematicians. Hence, in my opinion, comes the necessity of the present work, which aims to sketch at least some boundaries of this no-man's land.

The core idea of this book is to explain a few quantitative power laws satisfied by texts in natural language in terms of non-Markovian and nonhidden Markovian discrete stochastic processes with some sort of long-range dependence. The most important novel results discussed in this book, the theorems about facts and words, characterize arbitrary nonergodic or perigraphic stationary processes but have a clear linguistic inspiration. To achieve my goals, I use information theory, probability measures, and some related technical tools such as the ergodic decomposition and the Kolmogorov complexity. These tools, although they arise in a natural way, are black magic for linguists, and they are not the everyday bread for engineers. For this reason, this book will be likely treated as a monograph in mathematics rather than linguistics.

In fact, I wrote this book as dedicated primarily for mathematics graduate students and professionals who are interested in information theory or discrete stochastic processes and are intrigued by statistical modeling of natural language – to show them that there are a few novel formal ideas and open problems in this domain. However, since the use of mathematics in this book is motivated linguistically in quite an unusual way, as a secondary audience of the book, I foresee doctoral students and researchers in artificial intelligence, computational and quantitative linguistics, as well physics of complex systems. Those may find this book a helpful reference explaining mathematical foundations of information theory, discrete stochastic processes, and statistical modeling of natural language, of course. In the future, there may be also a need for a more popular book touching the topics of this monograph, but I am convinced that a rigorous exposition should be published in its own right first.

My father, who is a professor of engineering, once joked that there are no scientific monographs but textbooks for fewer and fewer students. I hope that this book will attract quite a few students, however. To increase the number of potential readers, I tried to avoid too abstract or redundant constructions while striving to be rigorous (in particular, no topology or linear algebra). I also resumed the main ideas of this book in the first chapter in a lighter form, so that an interested reader without a sufficient mathematical training could sense what the rest is about, whereas a mathematician could learn about the linguistic motivations. The remaining contents of the book are driven by a series of papers, mostly mine, which were preceded with an introductory material of a similar volume. This introductory material is due to many good people but likely has not been put together into a single convenient reference.

To be concrete, the book is divided into 11 chapters with the following contents:

- *Chapter 1. Guiding ideas*: As stated in the previous paragraph, this chapter resumes the principal ideas of the book in a half-formal way. I state the question what kind of a probability distribution should be assigned to particular human utterances. Sketching possible answers, I discuss quantitative linguistic laws and hypotheses such as Zipf's law and Hilberg's hypothesis and I introduce helpful mathematical concepts and results. In particular, I argue that a reasonable statistical language model should be either nonergodic or uncomputable.

- *Chapter 2. Probabilistic preliminaries*: This chapter opens the formal lecture, beginning with an established introductory material. I define probability measures and some discrete stochastic processes, I introduce Markov and hidden Markov processes, and I recall the ergodic theorem for irreducible Markov processes.

- *Chapter 3. Probabilistic toolbox*: Here real random variables are introduced and many important technical results of measure-theoretic probability calculus are stated. These include properties of nonatomic measures, Lebesgue integrals, important inequalities, conditional probability, convergence of random variables, and completeness of probability spaces.

- *Chapter 4. Ergodic properties*: In this chapter, I discuss the frequency interpretation of probability and demonstrate several celebrated theorems concerning ergodic properties of stationary processes such as the Birkhoff ergodic theorem and the ergodic decomposition. These results matter for language modeling since, as argued in Chapter 1, a reasonable statistical language model should be either nonergodic or uncomputable.

- *Chapter 5. Entropy and information*: Here I introduce the concepts of entropy, mutual information, entropy rate, and excess entropy. These are some functionals of random variables or stationary stochastic processes, which can help to capture the nondeterminism or infinite memory of the respective processes. I also generalize conditional entropy and mutual information to functionals of arbitrary fields, which allows to prove ergodic decomposition of entropy rate and excess entropy. This decomposition yields a simple insight into strongly nonergodic processes, which seem to arise naturally in statistical language modeling.

- *Chapter 6. Equipartition and universality*: In this chapter, the Shannon–MacMillan–Breiman theorem is stated and an example of a universal distribution called prediction by partial matching (PPM) is exhibited. I show that the PPM distribution can be used to estimate not only the entropy rate but also to upper bound the amount of memory stored by a stationary stochastic process.

- *Chapter 7. Coding and computation*: Here I introduce prefix-free codes, the idea of Shannon–Fano coding, the notion of Kolmogorov complexity, and the concept of algorithmic randomness. The fundamental Kraft inequality satisfied by prefix-free codes resurfaces in this chapter several times, in particular, to show the chain rule for Kolmogorov complexity. This chapter closes the cycle of introductory results.

- *Chapter 8. Power laws for information*: This chapter opens a discussion of more original research, done mostly by the author of this book. The point of departure is the power-law growth of mutual information hypothetically obeyed by texts in natural language. Within the chapter discussed are the concept of Hilberg exponents, strong nonergodicity, Santa Fe processes, and theorems about facts and words. The latter propositions seem to formalize some important intuitions about natural language but can be applied to general stationary processes.

- *Chapter 9. Power laws for repetitions*: The point of departure of this chapter is the power-law logarithmic growth of maximal repetition also hypothetically obeyed by texts in natural language. Trying to characterize this phenomenon mathematically, in this chapter I discuss Rényi entropies and entropy rates, recurrence times, the subword complexity, the longest match length, and the maximal repetition. The respective results are of a preliminary character.
- *Chapter 10. AMS processes*: This chapter concerns asymptotically mean stationary (AMS) processes, a subject usually deemed quite technical. However, in the context of language modeling, AMS processes arise naturally when we switch between two views of a stochastic process: one in which the process consists of letters and another in which the process consists of words. Therefore, AMS processes are a convenient tool for the subsequent chapter where I construct some examples of processes.
- *Chapter 11. Toy examples*: Here I discuss some particular examples of stochastic processes which can or cannot model some specific quantitative laws of natural language. Several important classes of processes are discussed, such as: finite and ultrafinite energy processes, aforementioned Santa Fe processes and their stationary encodings into a finite alphabet, as well as random hierarchical association (RHA) processes. Progressively, these processes can model more and more quantitative linguistic laws but none of them is fully satisfactory, which opens an avenue for future research.

The book is concluded by a list of open problems. I hope that these problems are salient enough to draw interest also of greater mathematical minds to problems of statistical language modeling. I suppose that many important developments are to be made if sufficient attention is devoted. The motivating intuitions lying at the heart of this book are not very complicated but I hope that they may be fertile when translated into the language of mathematics.

Warsaw, 2 January 2020   *Łukasz Dębowski*

Acknowledgments

This book recapitulates the path of stubborn research onto which I stepped around year 2000, when I quit physics in favor of computational linguistics and accidentally came across the work of Hilberg (1990), concerning a hypothetical power-law growth of entropy for natural language. In order to understand it properly, I started learning information theory and related mathematical subjects. Although I mostly wrote research papers alone, it may be difficult to name all important people that may have influenced the general outlook of this book. Let me give a try.

- First, I express my gratitude to my high-school teachers Olga Stande, Krzysztof Przybyszewski, and Lilianna Szymańska, who stimulated my independent thinking about mathematics, physics, and language, respectively.

- Subsequently, I thank my doctoral mentor Jan Mielniczuk and my former boss Jacek Koronacki, who provided me with a friendly and very patient environment, gave me lots of precious advice, and assisted me in discoveries leading toward the conception of this book, whose first critical readers they were later as well.

- I am grateful to David Adger, Gabriel Altmann, Janusz Bień, Richard Bradley, Mark Braverman, Gregory Chaitin, Alexander Clark, James Crutchfield, Magdalena Derwojedowa, Jarosław Duda, David Feldman, Ramon Ferrer-i-Cancho, Richard Futrell, Richard Gill, Peter Grünwald, Michael Hahn, Jan Hajič, Peter Harremoës, Frederick Jelinek, Brunon Kamiński, Reinhard Köhler, Andrzej Komisarski, Alfonso Martinez, Krzysztof Oleszkiewicz, Rajmund Piotrowski, Adam Przepiórkowski, Arthur Ramer, Alexandr Rosen, Zygmunt Saloni, Cosma Shalizi, Marek Świdziński, Hayato Takahashi, Kumiko Tanaka-Ishii, Flemming Topsøe, Paul Vitányi, Vladimir Vovk, Vladimir V'yugin, Mark Daniel Ward, and En-Hui Yang for correspondence, conversations, or invitations, longer or shorter, which over years guided or refined my positions on topics discussed in this book.

- More directly, I thank Christian Bentz, Antoni Hernández-Fernández, Dariusz Kalociński, Ioannis Kontoyiannis, Rodrigo Lambert, Dmitrii Manin, Tomasz Steifer, and three anonymous referees, who read some parts of this book at different stages of their writing and offered me helpful comments regarding their composition.
- I am grateful to my late uncle Tadeusz Krauze and my colleagues Marcin Woliński and Joanna Rączaszek-Leonardi, with whom I exercised discussing the guiding ideas of this book over dinner, during a walk in the forest, or hiking in the mountains.
- Finally, I am indebted to my close family for their love and care, without which this book would never come to daylight, either.

Basic Notations

In the main matter of this book, let us switch to the customary plural personal pronoun. Throughout the book, we use the following general notations:

- $\mathbb{N} := \{1, 2, 3, \dots\}$ is the set of natural numbers without zero, $\mathbb{Z} := \{\dots, -1, 0, 1, \dots\}$ is the set of integers, $\mathbb{Q} := \{p/q : p \in \mathbb{Z}, q \in \mathbb{N}\}$ is the set of rational numbers, and $\mathbb{R}$ is the set of real numbers.
- A set is called countable if its elements can be mapped one-to-one to a subset of natural numbers. Sets $\mathbb{N}$, $\mathbb{Z}$, and $\mathbb{Q}$ are countable but $\mathbb{R}$ is not.
- For a countable set $\mathbb{X}$, called an alphabet, $\mathbb{X}^n$ is the set of sequences of length n and $\mathbb{X}^+ := \bigcup_{n=1}^{\infty} \mathbb{X}^n$, called the Kleene plus, is the set of nonempty finite sequences. Symbol λ denotes the empty sequence, whereas $\mathbb{X}^0 := \{\lambda\}$. Then, $\mathbb{X}^* := \mathbb{X}^0 \cup \mathbb{X}^+$, called the Kleene star, is the set of all finite sequences, also called strings, $\mathbb{X}^{\mathbb{N}}$ is the set of one-sided infinite sequences, and $\mathbb{X}^{\mathbb{Z}}$ is the set of two-sided infinite sequences.
- Relation $x \geq y$ will be pronounced as x *is greater than* y and y *is less than* x. Relation $x > y$ will be pronounced as x *is strictly greater than* y and y *is strictly less than* x. Relations $x \geq 0$ or $x \leq 0$ will be pronounced as x *is positive* or x *is negative*, respectively. Relations $x > 0$ or $x < 0$ will be pronounced as x *is strictly positive* or x *is strictly negative*, respectively. Similar conventions concerning the use of adverb *strictly* will concern notions such as *more, fewer, smaller, larger, growing/increasing, decreasing, convex,* and *concave*, which are formally defined using inequalities.
- Relation $A \subseteq B$ is the inclusion of sets, whereas $A \subset B$ is the proper inclusion of sets (i.e. $A \subseteq B$ and $A \neq B$). Similarly, we write $a \sqsubseteq b$ when string a is a substring of sequence b (i.e. $b = xay$ for some possibly empty strings x and y), whereas we write $a \sqsubset b$ when string a is a proper substring of sequence b (i.e. $a \sqsubseteq b$ and $a \neq b$).
- Symbols $A \cup B$, $A \cap B$, and $A \setminus B$ denote the union, the intersection, and the difference of sets A and B, respectively.
- We denote the length $|w| := k$ of a string $w \in \mathbb{X}^k$.

- The same symbol for a real number denotes the absolute value,

$$|x| := \begin{cases} x, & x \geq 0 \\ -x, & x < 0 \end{cases}$$

- Cardinality of set A is denoted as $\#A$.
- We denote the floor and the ceiling functions as

$$\lceil x \rceil := \min\{y \in \mathbb{Z} : y \geq x\}$$
$$\lfloor x \rfloor := \max\{y \in \mathbb{Z} : y \leq x\}$$

- The greater and the smaller of two numbers will be written as

$$a \wedge b := \min\{a, b\}$$
$$a \vee b := \max\{a, b\}$$

- Symbol $\log x$ denotes the binary logarithm of x, i.e.

$$y = \log x \iff 2^y = x$$

whereas $\ln x$ denotes the natural logarithm of x, i.e.

$$y = \ln x \iff \exp y = x$$

- For a proposition φ we will write the indicator function

$$\mathbf{1}\{\varphi\} := \begin{cases} 1, & \text{if proposition } \varphi \text{ is true} \\ 0, & \text{if proposition } \varphi \text{ is false} \end{cases}$$

1

Guiding Ideas

This book concerns mathematical foundations of statistical language modeling, i.e. the question what kind of a probability distribution should be assigned to particular utterances of human languages. In this chapter, we will describe the core ideas of this book in a way which is less formalized mathematically, but more motivated linguistically. Based on the intuitions sketched in this chapter, in the following chapters, we will build rigorous mathematical constructions. The general goal is to develop a theory of discrete stochastic processes so that it would be able to account for certain statistical phenomena exhibited by human texts. The considered statistical phenomena take form of several power laws. We hope that if we were to succeed in a better modeling of these power laws, then in the long run, we may also obtain probabilistic models of language which are better in terms of performance measures used by engineers in computational linguistics. In other words, we hope that our quest for stochastic processes may turn out to be fruitful not only for purely theoretical interest but also for practical applications in engineering. We hope that the considered problems are also interesting enough on the theoretical side, and they can draw interest of professional mathematicians.

1.1   The Motivating Question

The fundamental question that motivates this book is

> What kind of a statistical model may explain generation of texts in natural language, such as books, our daily dialogues, or maybe even our internal stream of consciousness?

We perceive ourselves as free to say or think whatever we like, but, just to raise some doubts about our naive feeling of free will, are we more deterministic or more

Information Theory Meets Power Laws: Stochastic Processes and Language Models, First Edition.
Łukasz Dębowski.
© 2021 John Wiley & Sons, Inc. Published 2021 by John Wiley & Sons, Inc.

random in what we think? Either of these two extreme options seem scary. Surely, we are constrained by rules of grammar, meanings of words, habits of culture, daily chores, as well as search for meaning, good, beauty, truth, or happiness, which also impact our sense of freedom and the patterns of our deeds. But does it follow that the question of typical paths of our thoughts is worth investigation? Is there any mathematics there? Can we meaningfully quantify complexity of a song (Knuth, 1984) or a poem (Manin, 2019)? Is it a scientific question, does it bear practical importance, is it a cultural taboo of searching which one should not break, or all of that simultaneously?

There were many eminent intellectuals and scholars so far who asked the question about probability of texts. For example, Jonathan Swift, an Irish writer and clergyman, in his highly ironic novel popularly called *Gulliver's Travels*, in the chapter about the academy of sciences of Lagado, described a machine for automated text generation. The author of the machine assured Gulliver that

> ... this invention had employed all his thoughts from his youth; that he had emptied the whole vocabulary into his frame, and made the strictest computation of the general proportion there is in books between the numbers of particles, nouns, and verbs, and other parts of speech.
>
> (Swift, 1726)

We can then suppose that this machine while being deterministic, but making use of the frequencies of particular words should have implemented a certain statistical model of texts. Swift derided this machine as a highly unsuccessful project.

Despite many conceptual difficulties, the question about the probability of texts in natural language arises from time to time in the minds of scientists representing several domains: mathematics, linguistics, computer science, psychology, as well as physics of complex systems. In this book, we will build mostly on the expertise of mathematicians. Thus, let us recall that a few important mathematical concepts were in fact inspired by our question, such as Markov chains (Markov, 1913, translated in: Markov, 2006), entropy (Shannon, 1948, 1951), fractals (Mandelbrot, 1953, 1954), and algorithmic complexity (Kolmogorov, 1965). In contrast, among linguists, the question of probability in language caused a debate. Some linguists, such as Zipf (1935, 1949) and Herdan (1964), were highly interested in statistical properties of texts, whereas Chomsky (1956, 1957, 1959), while being more influential at that time, dismissed the question of probabilities in language as irrelevant for linguistics for the reason that human language cannot be in principle a finite-state stochastic process, see also Chomsky and Miller (1959). A brief historical account of this debate was resumed by Pereira (2000). As Christian Bentz, my friend, has also put it:

> It is an irony of 20th century linguistics that Shannon's theory of information, though explicitly linked to semantics, was deemed irrelevant by linguists, while Chomsky's formal syntax, though explicitly dissociated from semantics, was adopted as the default theory of natural language.
>
> (Bentz, 2018, p. 62)

Nowadays, in spite of the debates and reservations among the older generation of linguists, there is a growing pressure coming from engineering, technology, and experimental science to seriously ponder upon the question about statistical models of texts. While modern technology provides us with large amounts of textual data available to relatively powerful computers, applications such as predictive keyboards and speech-to-text recognizers in our smartphones actually already rely on some imperfect statistical models of texts. Machine learning engineers working on these problems continuously release better and better models. Initially, natural language engineers used finite-state models for this goal (Jelinek, 1997; Manning and Schütze, 1999), exactly those criticized by Chomsky. However, since the introduction of deep neural networks to the domain of natural language processing, the progression of statistical language model quality has been gaining a breathtaking momentum.

In particular, there was quite much publicity in the media concerning GPT-2, a statistical model of English announced by the OpenAI team on their blog on 14 February 2019. GPT-2 is nothing but a statistical language model encoded as a neural network with 1.5 billion parameters trained on 40 gigabytes of texts. Given this, it is capable of generating quite a coherent narration of the length of a few paragraphs, such as the following text, which we quote in full since it is spectacular:

SYSTEM PROMPT (HUMAN-WRITTEN):
In a shocking finding, scientist discovered a herd of unicorns living in a remote, previously unexplored valley, in the Andes Mountains. Even more surprising to the researchers was the fact that the unicorns spoke perfect English.

MODEL COMPLETION:
The scientist named the population, after their distinctive horn, Ovid's Unicorn. These four-horned, silver-white unicorns were previously unknown to science.

Now, after almost two centuries, the mystery of what sparked this odd phenomenon is finally solved.

Dr. Jorge Pérez, an evolutionary biologist from the University of La Paz, and several companions, were exploring the Andes Mountains when they found a small valley, with no other animals or humans. Pérez noticed that the valley had what appeared to be a natural fountain, surrounded by two peaks of rock and silver snow.

Pérez and the others then ventured further into the valley. "By the time we reached the top of one peak, the water looked blue, with some crystals on top," said Pérez.

Pérez and his friends were astonished to see the unicorn herd. These creatures could be seen from the air without having to move too much to see them – they were so close they could touch their horns.

While examining these bizarre creatures the scientists discovered that the creatures also spoke some fairly regular English. Pérez stated, "We can see, for example, that they have a common 'language,' something like a dialect or dialectic."

Dr. Pérez believes that the unicorns may have originated in Argentina, where the animals were believed to be descendants of a lost race of people who lived there before the arrival of humans in those parts of South America.

While their origins are still unclear, some believe that perhaps the creatures were created when a human and a unicorn met each other in a time before human civilization. According to Pérez, "In South America, such incidents seem to be quite common."

However, Pérez also pointed out that it is likely that the only way of knowing for sure if unicorns are indeed the descendants of a lost alien race is through DNA. "But they seem to be able to communicate in English quite well, which I believe is a sign of evolution, or at least a change in social organization," said the scientist.

(Radford et al., 2019)

A simple numerical trick called calibration can improve the outputs of GPT-2 even more (Braverman et al., 2019). Takahashi and Tanaka-Ishii (2019) show that neural network language models can also mimic many statistical patterns of texts created by humans. But most importantly, as we can simply understand the machine-generated text, GPT-2 can write a text which is a complete fiction but sounds like a readable story. (Even the self-contradictory "four-horned unicorns" could be wishfully attributed to the machine's singular sense of humor.) For the reason that similar statistical models could possibly flood the Internet not only with nice literary fiction but also with dangerous fake news, the OpenAI team initially decided not to publish the full GPT-2 model. This decision caused some ado in the media. A less publicized, related achievement in natural language processing was a publication by Springer of a book (Beta Writer, 2019) which is a fully machine-generated summary of 150 scientific articles in chemistry concerning lithium-ion batteries. This achievement concerns, however, summarizing of existing texts rather than generation of a novel text.

We agree that there is a far-reaching ethical question concerning our ambitious dream of statistical language modeling, as embodied in a stream of consciousness. Creating an artificial intelligence, be it in form of an artificial stream of consciousness, may be dangerous, also because of the risk that this intelligence will be smarter than us and simply desires to get rid of us. Thus, we can perfectly understand fears of those who think that any progress in imitating human intelligence will inevitably contribute to constructing another intelligence which will replace our own. But given the current state of engineering, it is possible that we may create a dangerous artificial intelligence accidentally. In such a situation, paradoxically, it is advisable to put as large effort as possible in understanding the dynamics of intelligence in general, exactly not to create the dangerous intelligent entity by accident.

Consequently, our initial question about statistical language models is no longer so strange. It is one of the fundamental research topics in artificial intelligence. Moreover, besides purely engineering approaches, we need some orchestrated fundamental mathematical research to understand the largely black-boxed progress of neural network language models, see Tishby and Zaslavsky (2015) and Shwartz-Ziv and Tishby (2017). Hopefully, for some and disappointingly to others, in this book, we will not give the ultimate and potentially risky answer to the question about the statistical model for natural language, but we hope to provide some insights which may inspire further research. The modest goal of this book is to understand some aspects of ideal statistical language models as seen via quantitative laws exhibited by texts spontaneously created by humans. That is, we will look into some general statistical patterns of texts.

1.2 Further Questions About Texts

What are the general statistical properties of texts created by humans? Let us briefly call the texts or sequences that get reproduced in the process of human culture *natural texts*. We may observe that in all human languages, natural texts can be written down – at least phonetically – using symbols from a finite set. In information theory, such a set is called an alphabet. For a finite alphabet, the number of all combinatorially possible sequences grows exponentially with the considered length of the sequence.

The vision of this potential exponential growth must have haunted Jorge Luis Borges, an Argentinian writer and librarian. In the short story *La Biblioteca de Babel* (Borges, 1941) – see also Bloch (2008), he described a fictitious library, which contained every combinatorially possible text of a given length, in form of a book meticulously printed out and stored on some bookshelf. When a random book was selected from this collection, one could only read a meaningless sequence of random letters. Surely, real books and libraries do not look like this! We may suppose that there are some patterns in natural texts – maybe there are unboundedly many distinct sorts of patterns – and hence the number of texts that remain in cultural discourse is acutely limited.

Using word *patterns*, we must be careful, however. Obviously, some simple patterns such as regular cycles arise in simple deterministic sequences. But, also completely random sequences of symbols exhibit some statistical order, such as the law of large numbers. Thus, what kind of patterns are specific for natural texts? We suppose that natural texts are neither of two extremes – they are neither deterministic in a simple way nor completely random. In fact, George Kingsley Zipf already put this intuition into the following beautiful words:

> If a Martian scientist sitting before his radio in Mars accidentally received from Earth the broadcast of an extensive speech which he recorded perfectly through the perfection of Martian apparatus and studied at his leisure, what criteria would he have to determine whether the reception represented the effect of animate process on Earth, or merely the latest thunderstorm on Earth? It seems that the only criteria would be the arrangement of occurrences of the elements, and the only clue to the animate origin would be this: the arrangement of the occurrences would be neither of rigidly fixed regularity such as frequently found in wave emissions of purely physical origin nor yet a completely random scattering of the same.
>
> (Zipf, 1935, p. 187)

Rejecting simple determinism from the process of natural text generation means that we have to consider some model of randomness. Consequently, searching for models of natural texts, in this book, we will use the framework of probability calculus and information theory. This approach does lead to certain insights.

Whereas the question about the probability of a particular text or a sequence of symbols seems very difficult and dependent on their context, we may consider relaxed versions of the problem. If we abstract from the concrete numerical value of a text probability and we assume that all texts with a strictly positive probability are equally important, then our motivating question from Section 1.1 boils down to:

> Which texts or sequences of symbols get reproduced in the process of human culture or in the stream of consciousness?

Let us observe that different sequences may get reproduced in the streams of consciousness of different persons. Moreover, if there is some randomness involved in the process of preselecting texts or sequences that get further reproduced in a particular stream of consciousness, the possibility which we cannot exclude a priori, then we may not be able to answer the above question constructively.

In view of this simple idea, we may find the task of finding a statistical model of language completely vane. But we think that we should not lose hope in this scientific project. There is a way of looking at the problem that may bring us closer to the solution. The idea is to introduce numbers. So let us ask rather:

> How many different texts or sequences of symbols of a given length get reproduced in the process of human culture or in the stream of consciousness?

We can see that answering how many sequences get reproduced does not automatically answer which sequences these are. In particular, even if the numbers of distinct sequences are the same for different streams of consciousness, there may be different sequences in different streams of consciousness. But building a partial quantitative model may shed some light onto principles of human communication and inspire further progress. At some point, we may be able to answer also the question:

> Why only a certain well-defined number of texts or sequences of symbols of a given length gets reproduced in the process of human culture or in the stream of consciousness?

This question concerns both the dynamics and the purpose of thinking and culture. If we were able to answer this question, then, given enough computational resources, we might be able to simulate a random process of culture or a stream of consciousness *in silico*, or in other words create an artificial intelligence.

Asking about the dynamics and purpose of thinking and culture is a difficult and potentially controversial question. Personally, we suppose that there may be some purpose of culture. The argument goes like this. We suppose that there is some objective truth that is possible to discover. We suppose that this truth cannot be expressed in finitely many words but can be learned partially and incrementally, given a certain effort. Moreover, we suppose that under certain physical circumstances, it is profitable to accumulate the partial knowledge of the truth. Hence, according to our views, the ultimate purpose of culture is a never ending accumulation of the partial knowledge of the truth. In the beginning of culture development, this purpose is rather a derivative of special physical circumstances which make knowledge accumulation profitable in some sense. This, however, does not exclude the possibility that culture, understood as a physical system of knowledge accumulation, gains its autonomy and begins to create conditions under which knowledge accumulation remains profitable. Then the ultimate purpose of culture becomes mere accumulation of knowledge.

It is a good question whether the system of knowledge accumulation can actually assume total control over its physical environment to ensure the unbounded growth. Unbounded growth of anything besides thermodynamical entropy seems in general physically impossible, but we refrain from giving a definite answer. Maybe knowledge *is* a peculiar sort of thermodynamical entropy (Lem, 1974; Feistel and Ebeling, 2011).

There is a large gap between the informal considerations of the above paragraph and the actual contents of this book. We suppose, however, that this gap is not infinite and may be filled some day with mathematical models. Therefore, it may be good to look at this gap from time to time to measure our progress. In the following, we will investigate some particular statistical patterns of natural texts that may inform us of more general theoretical properties of human language.

1.3 Zipf's and Herdan's Laws

Speaking of concrete statistical patterns of language, we should begin with Zipf's law. Zipf's law, discovered by Estoup (1916) and later popularized by Zipf (1935) and Mandelbrot (1953, 1954), is undoubtedly the most famous of all quantitative patterns exhibited by natural texts known so far – for an overview of many other patterns see Köhler et al. (2005). Also more recent books on quantitative linguistics by Esposti et al. (2016) and Hernández and Ferrer i Cancho (2019) begin with a discussion of Zipf's law. Thus, commencing quantitative considerations about natural language with Zipf's law is quite expected and natural.

To explain what Zipf's law is, we have to introduce the concepts of the word frequency and the word rank. The word frequency $f(w)$ is simply the number of occurrences of a given word w in a given text. Now let us sort the words according to their frequencies and assign rank $r(w_1) = 1$ to the most frequent word w_1, rank $r(w_2) = 2$ to the second most frequent word w_2, and in general rank $r(w_n) = n$ to the nth most frequent word w_n. In Table 1.1, we present an example of a rank list, i.e. the list of ranks $r(w)$, frequencies $f(w)$, and words w sorted according to the frequencies. By definition, the word frequency $f(w)$ is a decreasing function of the word rank $r(w)$. Zipf's law, called also the Zipf–Mandelbrot law, says something more, namely, it says that the word frequency is approximately inversely proportional to a power of the word rank,

$$f(w) \propto \left[\frac{1}{B + r(w)} \right]^{\alpha} \tag{1.1}$$

Table 1.1   The beginning of the rank list for
Shakespeare's First Folio/35 Plays.

$r(w)$	$f(w)$	w
1	21 557	I
2	19 059	and
3	16 571	to
4	14 921	of
5	14 491	a
6	12 077	my
7	10 463	you
8	9789	in
9	8754	is
10	7428	that
...	...	...

Source: Text downloaded from Project Gutenberg
(https://www.gutenberg.org).

where the exponent is $\alpha \approx 1$. In particular, if we plot the frequency versus the rank in the doubly logarithmic scale, we obtain almost a straight line with the slope $-\alpha \approx -1$, see Figure 1.1. Miracle?

Unfortunately, Zipf's law, as stated above, is not a pattern that is unique to natural texts. Let us take a *unigram text*, that is a random permutation of characters of a given natural text – to state it clearly, a space between the words is also considered a character. (This process is metaphorically called "monkey typing" but, of course, real monkeys do not type in this way.) Subsequently, let us make a list of "words" defined as space-to-space sequences of letters in the random permutation of characters, compute ranks and frequencies of those "words," and plot the obtained numbers in the doubly logarithmic scale. Then we obtain a very similar plot as for the text in natural language, see Figure 1.1! Hence, some simple kinds of random texts, such as unigram texts, also exhibit Zipf's law. This observation has been made by Mandelbrot (1954) and Miller (1957) and rigorously explained by Conrad and Mitzenmacher (2004).

Before we start producing alternative explanations for Zipf's law, let us make a general methodological remark. There can be at least four different sorts of reasons for quantitative laws of natural texts:

1. *General laws of randomness*: We have just seen an example thereof in the instance of the "monkey typing" explanation of Zipf's law.

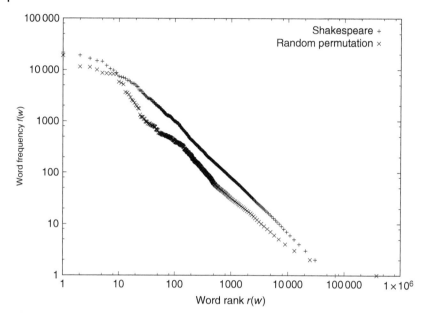

Figure 1.1 Zipf's law.

Zipf's law for Shakespeare's First Folio/35 Plays and a random permutation of the text's characters. We note that we use the doubly logarithmic scale here and later. Source: Author's figure based on the text downloaded from Project Gutenberg (https://www.gutenberg.org).

2. *Physical constraints of the human vocal tract*: Speech is the primary mode of human language communication and the distribution of breath length may impose certain constraints on the structure of syllables and the lengths of words. For recent relevant research, see Torre et al. (2019) and Coupé et al. (2019).

3. *Physical constraints of the human brain*: Language is processed by a three-dimensional network of neurons, whose configuration may imply some unrecognized bounds on the time and space complexity of language processing. For potentially relevant research, see Burger et al. (2019), West (2017), and Zador (2019).

4. *Abstract semantic constraints*: Texts created by humans are meaningful, i.e. they describe, associate, and control entities which are not the texts themselves. Meaningfulness may harness the texts to exhibit certain patterns or be attracted by some particular asymptotic structure. This sort of explanations will be pursued in this book. We will argue that they are within reach of mathematics.

For different empirical laws of natural language, different sorts of explanations can be suitable. It is possible that a given law can be given multiple explanations of varying degrees of plausibility. Some laws can be true for a bunch of roughly equally plausible reasons – and if these reasons are of a similar complexity, even Occam's razor would not help us to single out the best one. However, finding several equally plausible reasons can also enrich our understanding of the explained phenomenon – like finding multiple proofs of a single mathematical theorem.

Indeed, not only unigram texts provide a basis for formal explanations of Zipf's law. Some other identified mathematical or computational mechanisms leading to Zipf's law involve: multiplicative processes, originally discovered by Simon (1955) and, nowadays, called preferential attachment; large number of rare events (LNRE) distributions by Khmaladze (1988), later researched by Baayen (2001); coding games with entropy loss discovered by Harremoës and Topsøe (2001); a formalization of Pareto's principle of least effort by Ferrer-i-Cancho and Solé (2003), curiously resembling the bottleneck method in machine learning introduced by Tishby et al. (1999). Most of these models, however, either ignore the statistical dependence between occurrences of particular words or assume that there is no such dependence.

In contrast, in this book, we will seek for explanations of word distributions which combine general laws of randomness with abstract global semantic constraints. We will assume that texts in natural language attempt to describe an infinitely complex reality, be it fictitious or not, whose state can be partitioned into an infinite number of independent elementary facts, see Section 1.11. In Section 1.12, we will see that a power law for the distribution of word-like strings arises naturally in this case, as predicted by so-called theorems about facts and words. These theorems provide an explication of Herdan's law, a corollary of Zipf's law to be discussed in the next paragraph.

Herdan's law also has many fathers – it was independently rediscovered by Kuraszkiewicz and Łukaszewicz (1951), Guiraud (1954), Herdan (1964), and Heaps (1978). Quite often Herdan's law is also called Heaps' law. The law describes a relationship between the number of word types V, i.e. the number of different words in some initial part of a text, and the number of word tokens N, i.e. the number of all words in the same part. Herdan's law states that the number of word types V is proportional to a power of the number of word tokens N,

$$V \propto N^\beta \tag{1.2}$$

where $0 < \beta < 1$. In particular, if we plot the number of word types V versus the number of word tokens N in the doubly logarithmic scale, we obtain something similar to a straight line with slope β, see Figure 1.2.

It turns out that Herdan's law can be inferred from Zipf's law assuming a certain pattern of the text growth (Khmaladze, 1988; Baayen, 2001; Kornai, 2002). For example, if Zipf's law (1.1) is satisfied with $B = 0$ and constant α, whereas the

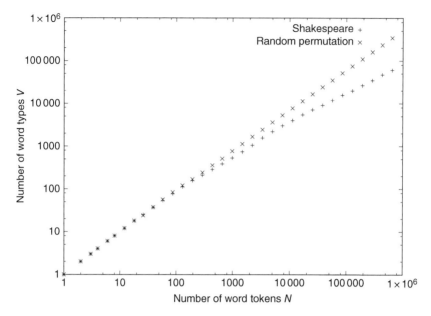

Figure 1.2   Herdan's law.

Herdan's law for Shakespeare's First Folio/35 Plays and a random permutation of characters. Source: Author's figure based on the text downloaded from Project Gutenberg (https://www.gutenberg.org).

relative frequency $f(u)/N$ for the most frequent word u does not depend on the considered text, then we would obtain Herdan's law (1.2) with $\beta = 1/\alpha$. To see it, we observe that for the least frequent word w, we have the frequency $f(w) = 1 \propto V^{-\alpha}$. Hence, the proportionality constant in Zipf's law equals roughly V^{α}. Thus, for the most frequent word u, we have $f(u) = V^{\alpha}$. Because $f(u)/N$ was assumed constant, we obtain $V \propto N^{1/\alpha}$. That is, we have inferred Herdan's law from Zipf's law. In reality, as we can see in Figures 1.1 and 1.2, it turns out that relationships (1.1) and (1.2) are somewhat inexact and the best fit for Herdan's law (1.2) yields an exponent β smaller than for Zipf's law (1.1). In particular, the number of word types is smaller in natural texts than in unigram texts.

In spite of many existing mathematical explanations of Zipf's and Herdan's laws, we should be also aware that the actual reasons for these laws for natural texts can be less obvious than in any idealized mathematical model. For example, words in natural texts differ statistically to "words" in unigram texts. If we compute the histograms of word length, see Figure 1.3, we can see that words in natural texts are quite differently distributed than "words" in unigram texts. In particular, for words in natural texts, we obtain a power-law tail, whereas for "words" in unigram texts,

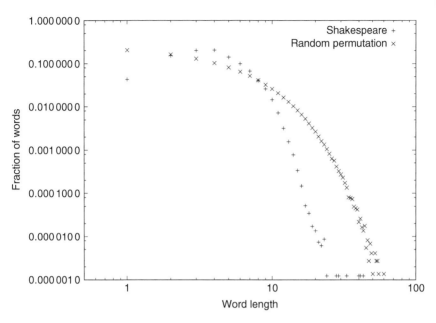

Figure 1.3 Histograms of word length.

Histograms of word length for Shakespeare's First Folio/35 Plays and a random permutation of characters. Source: Author's figure based on the text downloaded from Project Gutenberg (https://www.gutenberg.org).

we obtain an exponential tail. Hence, we may conclude that Zipf's law for words in natural texts is qualitatively a different phenomenon than Zipf's law for "words" in unigram texts. In particular, Zipf's law for natural language may require a distinct explanation. Another important phenomenon, which cannot be ignored, is a gradual change of Zipf's law exponent from $\alpha \approx 1$ for ranks roughly smaller than 10^4 to $\alpha \approx 2$ for ranks roughly larger than 10^4, as discovered by Ferrer-i-Cancho and Solé (2001) and later confirmed by Montemurro and Zanette (2002). Thus, we may still think that Zipf's law combined with the special distribution of word length can be really a pattern that distinguishes natural texts language from "inanimate processes."

The rigorous explanation of Zipf's law is not so easy as it may seem. Concerning statistical modeling of natural texts, there is a fundamental question whether we should commence our modeling efforts taking words as atomic units of our description. From the linguistic point of view, words are not purely atomic symbols – they have some internal structure. We may also suppose that words in the linguistic sense need not be the simplest units to describe statistical properties of natural texts. Maybe, if we consider that texts consist of a finite number of distinct meaningless letters (or phonemes) on the fundamental level, then we may obtain a simpler statistical description of language.

This question is not obvious at all! In particular, both sentences and words can be some phenomena emerging from some simpler statistical rules acting on the most fundamental level of language description, the level of letters or phonemes. Whereas linguists prefer to discuss units with a meaning, such as morphemes, words, or sentences, the number of these units can be in principle countably infinite or at least very large, so it is unlikely that we obtain a finite or at least a reasonably compact statistical model of language if we treat these high-level units as primitive units. In contrast, for random texts consisting of finitely many distinct symbols, such as letters or phonemes, we can use a bunch of nice theoretical results from probability and information theory. So we are given a certain advantage in the beginning.

1.4 Markov and Finite-State Processes

The atomic view of natural texts as being sequences of relatively few distinct meaningless symbols is well known to linguists but too trivial for them to consider it a sufficient description of language. Linguists usually express justified interest in where the meaning of text comes from, and for various reasons, prefer to work with meaningful units, such as morphemes, words, or sentences. However, following the quotation from Zipf (1935) in Section 1.1, we suppose that meaning cannot be captured by the marginal distribution of words or other text units, but it emerges from their linear interaction – the order of symbols, which is neither purely random nor purely deterministic. The question arises whether we can describe a statistical order of meaningless symbols that necessarily leads to emergence of somehow understood meaningfulness for sufficiently long texts, i.e. strings of these symbols.

In Section 1.3, we have introduced the concept of a unigram text, i.e. a random text obtained from a given text by permuting its characters at random. In fact, it is not the only possible way of introducing randomness into text generation. There are infinitely many mathematically admissible schemes – so many different ones that it is nontrivial to single out those realistic. To begin, in this section, we will consider a sequence of Markov processes, which imitate a given text by allowing for more and more statistical dependence among the consecutive letters. By the end of this section, we will also consider a generalization of Markov processes, called hidden Markov processes. There exist also radically different processes, which are neither Markov nor hidden Markov, to be discussed in further sections.

Suppose that we want to imitate the process of generation of a text in natural language in a purely statistical fashion. Assume that the alphabet $\mathbb{X}$, i.e. the set of symbols with which we can write the texts, is finite. Without loss of generality, let it consist of D distinct fixed symbols, say $\mathbb{X} = \{a_1, a_2, \dots, a_D\}$. Let $x_j^k = (x_j, x_{j+1}, \dots, x_k)$ denote a finite sequence of symbols $x_i \in \mathbb{X}$. Such finite sequences of symbols will be also called strings. A probability distribution on strings is a function p such that

$$p(\lambda) = 1 \text{ for the empty sequence } \lambda \tag{1.3}$$

$$p(x_1^n) \geq 0 \tag{1.4}$$

$$\sum_{x_n \in \mathbb{X}} p(x_1^n) = p(x_1^{n-1}) \tag{1.5}$$

Additionally, a probability distribution p is called stationary if

$$\sum_{x_1 \in \mathbb{X}} p(x_1^n) = p(x_2^n) \tag{1.6}$$

There are infinitely many possible probability distributions, also stationary ones. The essence of statistical language modeling lies in picking up the probability distribution that resembles natural language the most.

The simplest probability distribution on strings is obtained when we assume that all symbols in the alphabet $\mathbb{X}$ are sampled independently with equal probabilities,

$$p_U(x_1^n) := p_U(x_1)p_U(x_2)\cdots p_U(x_n) \tag{1.7}$$

where

$$p_U(x_i) := 1/D \tag{1.8}$$

Probability model (1.7) will be called the uniform distribution. It is easy to check that the uniform distribution is stationary.

In general, probabilities $p(x_1^n)$ need not satisfy conditions (1.7) and (1.8). Suppose that we have some particular text that we want to imitate, which we will denote as y_1^m. The stationary frequency of a substring w_1^k in text y_1^m is

$$N_s(w_1^k|y_1^m) := \sum_{i=1}^{m} \mathbf{1}\{y_i^{i+k-1} = w_1^k\} \tag{1.9}$$

where we will assume a periodic condition $y_{km+r} := y_r$ for $k \in \mathbb{N}$ and $r = 1, 2, \dots, m$. Subsequently, we will define the empirical distribution

$$p_s(x_1^n) := \frac{N_s(x_1^n|y_1^m)}{m} \tag{1.10}$$

It is easy to check that the empirical distribution is also stationary.

Having the empirical distribution, we can also define a ladder of Markov processes, as envisaged by Markov (1913) and Shannon (1948). The first step is the Markov process of order 0, which reads

$$p_0(x_1^n) := p_s(x_1)p_s(x_2)\cdots p_s(x_n) \tag{1.11}$$

This model has the advantage that the probabilities of individual symbols x_i are equal to the relative frequencies of individual symbols x_i in the imitated text y_1^m.

Continuing the sequence of the Markov processes, we have the Markov process of order 1,

$$p_1(x_1^n) := \begin{cases} p_s(x_1^n), & n \leq 2 \\ \dfrac{p_s(x_1^2)p_s(x_2^3)p_s(x_3^4)\cdots p_s(x_{n-1}^n)}{p_s(x_2)p_s(x_3)\cdots p_s(x_{n-1})}, & n > 2 \end{cases} \tag{1.12}$$

And in general, we have the Markov processes of orders $k \geq 1$,

$$p_k(x_1^n) := \begin{cases} p_s(x_1^n), & n \leq k+1 \\ \dfrac{p_s(x_1^{k+1})p_s(x_2^{k+2})p_s(x_3^{k+3})\cdots p_s(x_{n-k}^n)}{p_s(x_2^{k+1})p_s(x_3^{k+2})\cdots p_s(x_{n-k}^{n-k+1})}, & n > k+1 \end{cases} \tag{1.13}$$

In general, for the Markov process of order k, probabilities of tuples x_i^{i+k} are equal to relative frequencies of these tuples in the imitated text y_1^m.

Markov processes p_k for any order $k = 0, 1, 2, \ldots$ are stationary. Quantity

$$p_k(x_i|x_{i-k}^{i-1}) := \frac{p_k(x_{i-k}^i)}{p_k(x_{i-k}^{i-1})} \tag{1.14}$$

is called the transition probability of the Markov process of order k. The probability of a string according to the Markov process of order k can be also written as a product of transition probabilities,

$$p_k(x_1^n) = p_s(x_1^k)p_k(x_{k+1}|x_1^k)\ldots p_k(x_n|x_{n-k}^{n-1}) \tag{1.15}$$

Markov processes also called n-gram models in computational linguistics (Jelinek, 1997; Manning and Schütze, 1999) provide a baseline for computer generation of texts. To write a computer program that generates a random text according to the probability distribution (1.13), we need a function of the transition probability $p_k(x_i|x_{i-k}^{i-1})$ called the cumulative distribution function. To construct this function, we impose a linear order on symbols in the alphabet $\mathbb{X}$, i.e. we assume that the alphabet is an ordered list $\mathbb{X} = \{a_1, a_2, \ldots, a_D\}$. Then the cumulative distribution function is defined as

$$f_k(a_l|x_{i-k}^{i-1}) := \sum_{j=1}^{l} p_k(a_j|x_{i-k}^{i-1}) \tag{1.16}$$

Subsequently, if we have a real random variable U with the uniform distribution on the range $(0, 1]$, then the random variable

$$A = \min\{a_j : f_k(a_j | x_{i-k}^{i-1}) \geq U\} \tag{1.17}$$

assumes value a_j with probability $p_k(a_j | x_{i-k}^{i-1})$. Thus, if we can use a built-in function RAND that returns a real pseudorandom number from the uniform distribution on the range $(0, 1]$, then an algorithm for generating a random text according to the probability distribution (1.13) may look like Algorithm 1.

Algorithm 1   Generating text x_1^n according to a Markov process of order k

---

1: **for** $i \in \{1, 2, \dots, n\}$ **do**
2:     **if** $i \leq k$ **then**
3:         $x_i \leftarrow y_i$
4:     **else**
5:         $U \leftarrow$ RAND
6:         $S \leftarrow 0$
7:         $j \leftarrow 0$
8:         **repeat**
9:             $j \leftarrow j + 1$
10:            $S \leftarrow S + p_k(a_j | x_{i-k}^{i-1})$
11:        **until** $S \geq U$
12:        $x_i \leftarrow a_j$
13:    **end if**
14: **end for**

Now, as it was already done by Shannon (1948), we will present several examples of random texts generated by the above algorithm. The imitated text is *Gulliver's Travels* (Swift, 1726). The text was downloaded from the Project Gutenberg (https://www.gutenberg.org), and it includes a copyright license, a fragment of which appears in the generation of order 15.

- Order 0:

    ```
    emp cf harat tm a,nosaei adsuh,iethlraheernnHcergapn
    ,,Rtc hevi.lak eedooutam t,ptotb slpcteemdM n astv r
    :hebte rodeea tyw,Ide
    ```

- Order 1:

    ```
    oooh,ranhmirtaet ai rnete nhthp etoadtn dtayGfn p n
    smsa;nhnoje ehsteo tnhtd eduvhirersmaeifhlt,ernmw d
    ,ob; ohlpoknoo,g iruglicl
    ```

- Order 2:

  he f f y, aly f ig peangure paf I pose t anitomade
  N an'snveders shio this sutss then ct thud ldong me
  go t, heasis, thopron s hoft

- Order 3:

  e cossin to hationt this preast the purs, for; an lar
  eated cough my tre amplaw not she knes, to myse
  comand nes ing ory thar

- Order 4:

  had not publight of highs. The his a pairst Ported.
  It withould sever. Howereforged, into livenole I
  come as ple a bodigns was

- Order 5:

  gone of kissed the which I have my box; but the words
  in this I could been lookingdom formatical grass, and
  paratify hind I

- Order 10:

  ct gratefully and humbly conceive, in justice to the
  same manner, and wherein I spent fifteen hundred
  carpenters, who were indeed sensible, will, on this
  country-house, about the eBook refund and
  replacement or Refund" described, was a very
  faithful memory, that these nurses ever presume to
  entertainment to the royal

- Order 15:

  tenberg, you can always email directly to: Michael S.
  Hart. Project Gutenberg eBooks are often created from
  several nations; that our ship was staunch, and able
  to struggle no longer, I found myself so listless and
  desponding, that I really began to imagine myself
  dwindled many degrees below my usual size. Nothing

As we can see, Markov processes of lower orders cannot learn the concept of a word, whereas Markov processes of higher orders are able to memorize and chain at random some longer chunks of the imitated text, but they ignore any global intended meaning and purpose. When the order of a Markov process exceeds the length of the longest repeated substring in the imitated text, the process is able only to generate the imitated text repeatedly.

Judging by the above simulations of Markov processes, we may suppose that natural language is a non-Markovian probability distribution, i.e. it is not Markov of any finite order. In the following sections, we will provide some harder evidence in favor of this claim. At this moment, let us simply note that there exist quite many probability distributions which are non-Markovian. For instance, we may consider a hidden Markov process. A stationary hidden Markov process is a probability distribution of the form

$$p_{\text{HMM}}(x_1^n) := \sum_{y_1^n} p_{\text{MM}}(y_1^n) \prod_{i=1}^{n} e(x_i|y_i) \tag{1.18}$$

where

$$p_{\text{MM}}(y_1^n) = \frac{\prod_{i=1}^{n-1} p_{\text{MM}}(y_i^{i+1})}{\prod_{i=2}^{n-1} p_{\text{MM}}(y_i)} \tag{1.19}$$

is a stationary Markov process of order 1 and so-called emission probabilities $e(x_i|y_i)$ satisfy

$$e(x_i|y_i) \geq 0 \tag{1.20}$$

$$\sum_{x_i} e(x_i|y_i) = 1 \tag{1.21}$$

For a hidden Markov process, entities $y_i \in \mathbb{Y}$ are called hidden states in contrast to entities $x_i \in \mathbb{X}$ being called emitted symbols. In general, the alphabet of hidden states $\mathbb{Y}$ may be countably infinite. If the alphabet of hidden states $\mathbb{Y}$ is finite, then distribution (1.18) is called a finite-state process. In particular, any Markov process (1.13) with a finite alphabet of emitted symbols $\mathbb{X}$ can be written as a finite-state process. However, the number of hidden states grows exponentially with the order of the represented Markov process.

Finite-state processes were a standard tool for statistical natural language modeling in computational linguistics in 1990s and 2000s until they were replaced by artificial neural networks, which turned out to yield much more accurate models, such as GPT-2 mentioned in Section 1.1 (Radford et al., 2019). Originally, finite-state processes were introduced into computational linguistics by engineers working on speech recognition (Jelinek, 1997). Finite-state processes can model quite complex statistical dependencies in texts while being easy enough to train given enough data.

In this book, we would like to argue that finite-state processes may be insufficient as models of natural language. This idea, confirmed by the present success of neural network models, is far from new. It dates back to a famous debate between Skinner (1957), Chomsky (1956, 1957, 1959), and Chomsky and Miller

(1959). Skinner was a psychologist believing that finite-state processes, with a sufficiently large number of hidden states, can account for human language behavior. In contrast, Chomsky showed that if the syntax of sentences of human languages is described by a context-free grammar, then natural language cannot be adequately described by finite-state formal languages or processes. Chomsky thought that rejecting finite-state processes means eradicating any probability models from linguistic considerations. This reasoning turns out to be a fallacy in view of contemporary successes of statistical models of language based on artificial neural networks.

A certain idea why Chomsky supposed that context-free grammars are incompatible with probability models may be that it is not so trivial to incorporate a hierarchical structure into a stationary distribution on strings. Quite early in computational linguistics, some probabilistic context-free grammar (PCFGs) also called branching processes were discovered (Harris, 1963; Miller and O'Sullivan, 1992; Chi and Geman, 1998). It should be noted, however, that PCFGs do not define a stationary distribution on strings since they construct a finite hierarchical structure in a top-down fashion, which results in defining discrete probabilities over finite trees. In contrast, in Section 11.4, we will exhibit a class of random hierarchical association (RHA) processes, which constitute a stationary variation on the theme of context-free grammar. RHA processes build an infinite hierarchical structure in an unbounded bottom-up fashion, which results in defining a stationary probability measure over infinite sequences, as will be partly explained in Section 1.5.

Another important point is that Chomsky's argument against finite-state models did not appeal to statistical properties of texts written by humans but only to our perception of grammaticality of particular sentences. In contrast, in the following sections, we will seek for some statistical properties of natural texts that can be used to disprove the finite-state hypothesis. As we will see in Section 1.7, there is a parameter of a stationary distribution called excess entropy which is finite for finite-state processes but can be infinite for other probability distributions and natural language in particular, as will be argued in the following sections. Moreover, in Section 1.13, we will see that a power-law logarithmic growth of a simple statistic called the maximal repetition, observed for natural language, also excludes a large subclass of hidden Markov processes as potential statistical models of language.

1.5 More General Stochastic Processes

In Section 1.4, we have introduced the notion of a probability distribution on texts of arbitrary finite lengths. In this section, we would like to sketch a more general theory of such distributions. We will show that any probability distribution on strings, i.e. one which satisfies axioms (1.3)–(1.5), defines a probability measure on subsets of infinitely long sequences. Subsequently, under certain conditions, we can learn the distribution on strings from a random infinite sequence of symbols typically generated by the distribution. The rigorous construction is quite involved so let us report only the most important facts, which will be developed formally in Chapters 2 and 4. We apply the basic notations stated before this chapter.

For a set Ω, which will be called the event space, let 2^Ω denote the set of all subsets (i.e. the power set) of Ω. For a class of subsets $A \subseteq 2^\Omega$, let $\sigma(A) \subseteq 2^\Omega$ denote the generated σ-field, i.e. the smallest algebra of subsets of Ω which contains all elements of class A and is closed under complements and countable unions. Now for a finite alphabet $\mathbb{X}$, we define a few generated σ-fields. First, let $\mathcal{X} := 2^{\mathbb{X}}$ be the set of all subsets of $\mathbb{X}$. Second, for $\mathbb{T} = \mathbb{N}$ or $\mathbb{T} = \mathbb{Z}$, let us introduce projections

$$X_k : \mathbb{X}^{\mathbb{T}} \ni (x_i)_{i \in \mathbb{T}} \mapsto x_k \in \mathbb{X}, \quad k \in \mathbb{T} \tag{1.22}$$

which map an infinite sequence to a symbol on a particular position k in that sequence. Having these projections, third, we introduce cylinder sets

$$(X_k \in A) := \{(x_i)_{i \in \mathbb{T}} \in \mathbb{X}^{\mathbb{T}} : x_k \in A\}, \quad k \in \mathbb{T}, A \in \mathcal{X} \tag{1.23}$$

which are simply the sets of infinite sequences whose kth symbol belongs to a particular set A. Consequently, fourth, we define the σ-field generated by the class of cylinder sets,

$$\mathcal{X}^{\mathbb{T}} := \sigma(\{(X_k \in A) : k \in \mathbb{T}, A \in \mathcal{X}\}) \tag{1.24}$$

Now, for an arbitrary set Ω, let $\mathcal{J} \subseteq 2^\Omega$ be a generated σ-field. A probability measure $P : \mathcal{J} \to \mathbb{R}$ is a function that satisfies the following axioms:

1. $P(A \cup B) = P(A) + P(B)$ for disjoint $A, B \in \mathcal{J}$,
2. $P\left(\bigcup_{n \in \mathbb{N}} A_n\right) = \sum_{n \in \mathbb{N}} P(A_n)$ for disjoint $A_i \in \mathcal{J}$ and $\bigcup_{n \in \mathbb{N}} A_n \in \mathcal{J}$,
3. $P(A) \geq 0$ for $A \in \mathcal{J}$,
4. $P(\Omega) = 1$.

Triple $(\Omega, \mathcal{J}, P)$ is called a probability space. In the context of probability spaces, sets $A \in \mathcal{J}$ are called events, whereas functions $Y : \Omega \to \mathbb{Y}$ are called random variables. Some examples of random variables are projections X_k defined in (1.22). Collections of random variables, such as $(X_i)_{i \in \mathbb{T}}$, are called stochastic processes. Generalizing the notation for cylinder sets, for a random variable $\varphi : \Omega \to \{\text{true}, \text{false}\}$, we also write

$$(\varphi) := \{\omega \in \Omega : \varphi(\omega)\} \tag{1.25}$$

$$P(\varphi) := P(\{\omega \in \Omega : \varphi(\omega)\}) \tag{1.26}$$

and when $P(\varphi) = 1$, then we say that φ holds almost surely or, synonymically, that $\varphi(\omega)$ holds for almost all ω. Using these notations, we also define the expectation of a discrete real random variable $Y \geq 0$ as

$$\mathbf{E}Y := \sum_{y:P(Y=y)>0} yP(Y=y) \tag{1.27}$$

Having made these preparations, we can state two important results. First, if a probability distribution $p : \mathbb{X}^* \to \mathbb{R}$ on strings satisfies axioms (1.3)–(1.5), then according to the Kolmogorov process theorem (Theorem 2.7), there exists a unique probability space $(\mathbb{X}^{\mathbb{N}}, \mathcal{X}^{\mathbb{N}}, P)$ such that

$$P(X_1^n = x_1^n) = p(x_1^n) \tag{1.28}$$

Second, for stationary probability distributions, there is some further symmetry. Namely, if a probability distribution $p : \mathbb{X}^* \to \mathbb{R}$ on strings satisfies axioms (1.3)–(1.6), then there exists a unique probability space $(\mathbb{X}^{\mathbb{Z}}, \mathcal{X}^{\mathbb{Z}}, P)$ such that

$$P(X_{j+1}^{j+n} = x_1^n) = p(x_1^n) \tag{1.29}$$

for any offset $j \in \mathbb{Z}$. The intuitive meaning of all these technicalities is simple. A finite sequence of random variables X_1^n corresponds to a random text generated by the probability distribution p.

Some natural questions we may ask are whether we can infer the probability distribution p from a random text X_1^n with a sufficiently large probability, according to the probability measure P. The answer is positive for stationary ergodic probability distributions, which are a subset of stationary probability distributions.

There are several equivalent characterizations of ergodic probability distributions. The simplest is as follows. A stationary probability distribution p is ergodic if and only if we *cannot* decompose it as

$$p(x_1^n) = q_1 r_1(x_1^n) + q_2 r_2(x_1^n) \tag{1.30}$$

where $r_1 \neq r_2$ are two different stationary distributions, and $0 < q_1 = 1 - q_2 < 1$ are some prior probabilities independent of x_1^n. This definition can be difficult to verify for a concrete probability distribution p. But, as shown in Gray (2009, lemma 7.15), see Theorem 4.9, a stationary probability distribution p is ergodic if and only if

$$\lim_{n \to \infty} \frac{1}{n} \sum_{l=0}^{n-1} \sum_{y_1^l} p(x_1^k y_1^l z_1^m) = p(x_1^k)p(z_1^m) \tag{1.31}$$

for all strings x_1^k and z_1^m. In particular, it follows that a stationary kth order Markov process is ergodic if $k = 0$ or if the transition probability (1.14) satisfies inequality $p_k(x_i|x_{i-k}^{i-1}) \geq c$ for a constant $c > 0$. The third equivalent characterization of stationary ergodic distributions is given by the Birkhoff ergodic theorem (Birkhoff, 1932), see Theorem 4.6. According to this theorem, a stationary probability distribution p is ergodic if and only if for any string x_1^k, we have

$$\lim_{n \to \infty} \frac{1}{n} \sum_{i=0}^{n-1} \mathbf{1}\{y_{i+1}^{i+k} = x_1^k\} = p(x_1^k) \text{ for almost all } (y_i)_{i \in \mathbb{Z}} \tag{1.32}$$

In plain words, for an ergodic process, the relative frequency of any string x_1^k in the infinite sequence $(y_i)_{i \in \mathbb{Z}}$ is equal to the probability $p(x_1^k)$ for almost every sequence $(y_i)_{i \in \mathbb{Z}}$.

What happens for stationary distributions which are not ergodic? In fact, there is a version of the Birkhoff ergodic theorem for stationary nonergodic distributions. Namely, in that case, there exists a function $\phi : \mathbb{X}^* \times \mathbb{X}^{\mathbb{Z}} \ni (x_1^k, (y_i)_{i \in \mathbb{Z}}) \mapsto \phi(x_1^k | (y_i)_{i \in \mathbb{Z}}) \in \mathbb{R}$ such that

$$\lim_{n \to \infty} \frac{1}{n} \sum_{i=0}^{n-1} \mathbf{1}\{y_{i+1}^{i+k} = x_1^k\} = \phi(x_1^k | (y_i)_{i \in \mathbb{Z}}) \text{ for almost all } (y_i)_{i \in \mathbb{Z}} \tag{1.33}$$

and $\phi(x_1^k | (y_i)_{i \in \mathbb{Z}})$ as a function of X_1^n is a stationary ergodic probability distribution for almost all sequences $(y_i)_{i \in \mathbb{Z}}$. We can interpret function ϕ as a random stationary ergodic probability distribution on the probability space $(\mathbb{X}^{\mathbb{Z}}, \mathcal{X}^{\mathbb{Z}}, P)$. In particular, when we use a stationary nonergodic distribution to generate an infinitely long random text, then the relative frequency of any string in this text exists almost surely and is described by a certain random stationary ergodic distribution.

1.6 Two Interpretations of Probability

Here is a convenient place to discuss two distinct interpretations of probability and their links with statistical language modeling and computation. According to axioms (1.3)–(1.5), we have a very large freedom of defining probability distributions. Most of such distributions are very irregular and uncomputable, in the sense that there is no finite computer program which computes function $x_1^n \mapsto p(x_1^n)$ with an arbitrary precision. Consequently, do we expect that natural language should be described by a computable probability distribution? To answer this question, let us recall that there are two ways of interpreting probability, frequentist and Bayesian. We will show that they yield different answers.

Seen from a frequentist perspective, probabilities are limiting relative frequencies of events in the infinite empirical data, like in the setting of the Birkhoff ergodic theorem. Let us suppose that the empirical language data were generated by an arbitrary nonergodic stationary probability distribution. Then according to the Birkhoff ergodic theorem (1.33), relative frequencies $\phi(x_1^n | (y_i)_{i \in \mathbb{Z}})$ constitute a random stationary ergodic distribution. It is a mathematical fact that there are uncountably many stationary ergodic distributions, out of which there are only countably many computable distributions, since there are only countably many distinct programs to generate them. In particular, for a vast majority of sequences $(y_i)_{i \in \mathbb{Z}}$ function $x_1^n \mapsto \phi(x_1^n | (y_i)_{i \in \mathbb{Z}})$ is not computable without knowing the entire sequence $(y_i)_{i \in \mathbb{Z}}$. Thus, according to the frequentist interpretation, natural language should be likely described by an ergodic and uncomputable probability distribution.

A different result follows if we adopt the Bayesian interpretation. According to the Bayesian views, probabilities are some subjective expectations of agents that interact with an unknown environment and learn from their past experiences. We may perceive human beings also as Bayesian agents constantly updating their subjective expectations during conversations and other actions. We may also suppose that the subjective expectations of humans are a computable function of their previous experiences, i.e. the conditional probability $x_1^{n+1} \mapsto p(x_{n+1} | x_1^n) := p(x_1^{n+1}) / p(x_1^n)$ should be a computable function. Thus, according to the Bayesian interpretation, natural language should be likely described by a computable probability distribution, but this distribution need not be ergodic, or even stationary.

Assuming stationarity, however, the Birkhoff ergodic theorem prompts a way to reconcile the Bayesian and the frequentist views. These two competing views are just two different perspectives onto a single reality. Simply speaking, there may exist a computable, nonergodic, and maybe also stationary Bayesian probability distribution p that describes the subjective expectations of humans, given their previous experiences. Moreover, if humans act *randomly* as suggested by these expectations, then the resulting outcome of their random utterances $(y_i)_{i \in \mathbb{Z}}$ may give rise to an uncomputable frequentist ergodic distribution $\phi(\cdot | (y_i)_{i \in \mathbb{Z}})$ sampled from the computable Bayesian nonergodic distribution p. In some sense, both the Bayesian distribution p and the frequentist distribution $\phi(\cdot | (y_i)_{i \in \mathbb{Z}})$ express some information about natural language.

Let us also note that the difference between the Bayesian distribution p and the frequentist distribution $\phi(\cdot | (y_i)_{i \in \mathbb{Z}})$ is a probabilistic version of the distinction between the internalized language (I-language) and the externalized language (E-language) made by Chomsky (1986). The I-language is the natural language operationalized by human minds, whereas the E-language is the natural language observable in texts. In particular, the language operationalized by human minds can be a sort of a conditional probability function – or rather, a finite algorithmic description thereof, whereas the language observable in texts can be the empirical distribution of strings in the collection of all texts, which need not have a finite description.

Like Chomsky, we suppose that the preferred goal of scientific research in human language should be inferring the I-language, as represented by the hypothetical computable Bayesian distribution p, rather than inferring the E-language, as represented by the hypothetical uncomputable frequentist distribution $\phi(\cdot|(y_i)_{i\in\mathbb{Z}})$. The reason for choosing such a research goal is that possibly the Bayesian distribution p can have a finite description, whereas the frequentist distribution $\phi(\cdot|(y_i)_{i\in\mathbb{Z}})$ need not have a finite description. The finite description of distribution p remains in a close analogy with Chomsky's idea of an innate universal grammar, whereas distribution $\phi(\cdot|(y_i)_{i\in\mathbb{Z}})$ would be somewhat similar to the universal grammar instantiated with random binary parameters – though the important difference to Chomsky's ideas is that we do not assume a priori that the total number of random binary parameters can be upper bounded.

The challenge for so redefined Chomskyan program is that the only input for our scientific inference are empirical data being just a section of the infinite sequence $(y_i)_{i\in\mathbb{Z}}$, and we would like to make a more intelligent use of them than simply estimating the relative frequencies $\phi(\cdot|(y_i)_{i\in\mathbb{Z}})$. We should admit that it is not completely clear yet what this more intelligent use of data is exactly.

1.7 Insights from Information Theory

Information theory is another useful tool in the analysis of discrete stochastic processes. The founding concept of this theory is the amount of information contained in an object, such as a binary string or a random variable. The intuition is that the amount of information in an object is equal to the degree of unpredictability of this object, given some prior object. It is important to note that when measuring the information content we cannot get rid of this prior object, i.e. the amount of information is always relative to something. Formally, there are two standard ways of measuring the amount of information: one, the Shannon entropy (Shannon, 1948, 1951) – to be treated half-formally in this section, is relative to a probability distribution, whereas another, the Kolmogorov complexity (Solomonoff, 1964; Kolmogorov, 1965) – to be treated half-formally in Section 1.10, is relative to a universal computer. The material from this section will be developed fully formally in Section 3.4 and Chapters 5 and 7. We will now sketch the most important ideas.

A discrete probability distribution $p : \mathbb{X} \to \mathbb{R}$ on a countable set $\mathbb{X}$ is a function such that $p(x) \geq 0$ for each $x \in \mathbb{X}$ and

$$\sum_{x\in\mathbb{X}} p(x) = 1 \tag{1.34}$$

In contrast, a discrete semidistribution (or an incomplete distribution) $q : \mathbb{X} \to \mathbb{R}$ on a countable set $\mathbb{X}$ is a function such that $q(x) \geq 0$ for each $x \in \mathbb{X}$ and

$$\sum_{x\in\mathbb{X}} q(x) \leq 1 \tag{1.35}$$

There are two important functionals of a probability distribution p and a semidistribution q. One is the Shannon entropy

$$H(p) := - \sum_{x:p(x)>0} p(x) \log p(x) \tag{1.36}$$

introduced by Shannon (1948), and another is the Kullback–Leibler (KL) divergence

$$D(p||q) := \sum_{x:p(x)>0} p(x) \log \frac{p(x)}{q(x)} \tag{1.37}$$

introduced by Kullback and Leibler (1951). We have $H(p) \geq 0$ and $D(p||q) \geq 0$.

If we take the binary logarithms in the respective definitions, the Shannon entropy and the KL divergence can be related to coding and lossless data compression. A prefix-free set $A \subseteq \{0, 1\}^*$ is a set of binary strings such that for any $a_1, a_2 \in A$ condition $a_1 \neq a_2$ implies $a_1 \neq a_2 w$ for any string $w \in \{0, 1\}^*$. The intuition behind this concept is that the mapping between a list of strings $(a_1, a_2, \dots, a_n)$, where $a_i \in A$, and the concatenation of strings $a_1 a_2 \dots a_n$ is one-to-one if A is a prefix-free set. Simply speaking, given a concatenation of strings $a_1 a_2 \dots a_n$, we can identify which strings a_i were concatenated if a_i belong to a prefix-free set. We recall that we denote the length of a binary string $|w| := k$ for $w \in \{0, 1\}^k$. An important property of prefix-free sets is the Kraft inequality

$$\sum_{a \in A} 2^{-|a|} \leq 1 \tag{1.38}$$

which holds if set A is prefix-free (McMillan, 1956).

A prefix-free code is a one-to-one mapping $C : \mathbb{X} \to A$, where $A \subseteq \{0, 1\}^*$ is a certain prefix-free set. Plugging the semidistribution $q(x) = 2^{-|C(x)|}$ into inequality $D(p||q) \geq 0$, we obtain for any prefix-free code C that

$$H(p) \leq \sum_{x \in \mathbb{X}} p(x)|C(x)| \tag{1.39}$$

Thus, the expected length of any prefix-free code is greater than the Shannon entropy of the distribution. Conversely, for a given probability distribution p, there exists a code C_p, called the Shannon–Fano code, such that $0 \leq |C_p(x)| + \log p(x) \leq 1$ (Shannon, 1948). In this case, we obtain

$$H(p) \leq \sum_{x \in \mathbb{X}} p(x)|C_p(x)| \leq H(p) + 1 \tag{1.40}$$

Thus, the Shannon entropy is roughly the minimal expected length of a prefix-free code on a countable domain $\mathbb{X}$.

The Shannon entropy can be developed further. For a discrete random variable $X : \Omega \to \mathbb{X}$, we define the Shannon entropy $H(X) := H(p)$ for $p(x) = P(X = x)$. The following functionals of random variables X, Y, and Z are also useful:

- conditional entropy

$$H(X|Z) := H(X,Z) - H(Z) \qquad (1.41)$$

- mutual information

$$I(X;Y) := H(X) + H(Y) - H(X,Y) \qquad (1.42)$$

- conditional mutual information

$$I(X;Y|Z) := H(X|Z) + H(Y|Z) - H(X,Y|Z) \qquad (1.43)$$

The above definitions make sense only if the respective entropies are finite, but they can be generalized for arbitrary random variables (Gelfand et al., 1956; Dobrushin, 1959; Pinsker, 1964; Wyner, 1978), see Section 5.3. In the general case, we have $H(X), H(X|Z), I(X;Y), I(X;Y|Z) \geq 0$.

Subsequently, we will sketch how information theory can be applied to the analysis of stationary stochastic processes, for the formal exposition see Sections 5.2 and 5.4. Suppose that we have a stationary probability distribution p on strings, i.e. one which satisfies axioms (1.3)–(1.6). The Shannon entropy of a block of length n is the function $H(n) := H(X_1^n)$, i.e.

$$H(n) := -\sum_{x_n} p(x_1^n) \log p(x_1^n) \qquad (1.44)$$

It can be easily shown that for a stationary probability distribution, the Shannon block entropy is positive, growing, and concave since

$$H(n) = H(X_1^n) \geq 0 \qquad (1.45)$$

$$\Delta H(n) := H(n) - H(n-1) = H(X_n|X_1^{n-1}) \geq 0 \qquad (1.46)$$

$$\Delta^2 H(n) := H(n) - 2H(n-1) + H(n-1) = -I(X_1;X_n|X_2^{n-1}) \leq 0 \qquad (1.47)$$

In view of the above inequalities, there exists limit

$$h := \lim_{n\to\infty} \frac{H(n)}{n} = \lim_{n\to\infty} \Delta H(n) = H(1) + \sum_{n=2}^{\infty} \Delta^2 H(n) \geq 0 \qquad (1.48)$$

called the entropy rate. The entropy rate is the conditional entropy of a single symbol, given the infinite past, i.e.

$$h = H(X_i|X_{-\infty}^{i-1}) \qquad (1.49)$$

Intuitively, the entropy rate is an amount of limiting unpredictability of the stationary process $(X_i)_{i\in\mathbb{Z}}$ per single symbol.

There is also another asymptotic measure

$$E := \lim_{n\to\infty} [H(n) - hn] = \lim_{n\to\infty} [2H(n) - H(2n)] = -\sum_{n=2}^{\infty} (n-1)\Delta^2 H(n) \geq 0$$

$$(1.50)$$

called the excess entropy (Crutchfield and Feldman, 2003). Let us observe that $2H(n) - H(2n) = I(X_1^n; X_{n+1}^{2n})$. Hence, the excess entropy is the mutual information between the infinite past and the infinite future, i.e.

$$E = I(X_{-\infty}^{i-1}; X_i^{\infty}) \tag{1.51}$$

Intuitively, the excess entropy is an asymptotic amount of memory stored by the stationary process $(X_i)_{i \in \mathbb{Z}}$.

The entropy rate and the excess entropy can be easily evaluated for Markov processes. For example, for the k-th order Markov process (1.13), we have

$$H(n) = (n - k)H(k + 1) - (n - k - 1)H(k) \text{ for } n > k \tag{1.52}$$

$$h = H(k + 1) - H(k) \in [0, \log \#\mathbb{X}] \tag{1.53}$$

$$E = (k + 1)H(k) - kH(k + 1) = H(k) - kh \in [0, k \log \#\mathbb{X}] \tag{1.54}$$

where $\#\mathbb{X}$ is the cardinality of $\mathbb{X}$, i.e. the number of elements in set $\mathbb{X}$. For hidden Markov processes, there are no so neat formulas, but in general, for the hidden Markov distribution (1.18), we have

$$h \in [0, \log \#\mathbb{X}] \tag{1.55}$$

$$E \in [0, \log \#\mathbb{Y}] \tag{1.56}$$

In particular, the excess entropy is finite for finite-state processes, but it can be infinite for a hidden Markov process if the alphabet of hidden states $\mathbb{Y}$ is infinite, see Travers and Crutchfield (2014) and Dębowski (2014). Moreover, as we will argue in Sections 1.9 and 1.11, the natural language, if it can be modeled by a stationary distribution, probably has the infinite excess entropy.

1.8 Estimation of Entropy Rate

It turns out that the entropy rate of a stationary ergodic distribution can be effectively estimated. That is, there is a computable procedure which, given a sufficiently long text generated by any stationary ergodic distribution, returns the estimate of the distribution's entropy rate with an arbitrary precision. This result is a consequence of the Birkhoff ergodic theorem. There are a few particular estimators of the entropy rate (Ziv and Lempel, 1977; Ryabko, 1984; Cleary and Witten, 1984; Kontoyiannis, 1997; Kontoyiannis et al., 1998; Kieffer and Yang, 2000). In this section, we will present a certain procedure of entropy estimation called the PPM algorithm. The acronym PPM stands for *Prediction by Partial Matching*. A rigorous development of this section will be done in Section 6.3.

We remind that we denote strings of symbols $x_j^k := (x_j, x_{j+1}, \dots, x_k)$. Moreover, we will only consider strings over a finite alphabet, say, $x_i \in \mathbb{X} = \{a_1, a_2, \dots, a_D\}$.

In contrast to the stationary frequency $N_s(w_1^k|x_1^n)$ from equation (1.9), which assumed periodic boundary conditions, we define the plain frequency of a substring w_1^k in a string x_1^n where $1 \le k \le n$ as

$$N(w_1^k|x_1^n) := \sum_{i=1}^{n-k+1} \mathbf{1}\{x_i^{i+k-1} = w_1^k\} \tag{1.57}$$

The PPM probabilities were subsequently developed by Krichevsky and Trofimov (1981), Ryabko (1984), and Cleary and Witten (1984) in slightly different versions. Our version is as follows. For $x_1^n \in \mathbb{X}^n$ and $k \in \{0, 1, ...\}$, we put

$$\text{PPM}_k(x_i|x_1^{i-1}) := \begin{cases} \dfrac{1}{D}, & k > i - 2 \\ \dfrac{N(x_{i-k}^i|x_1^{i-1}) + 1}{N(x_{i-k}^{i-1}|x_1^{i-2}) + D}, & \text{else} \end{cases} \tag{1.58}$$

Quantity $\text{PPM}_k(x_i|x_1^{i-1})$ is called the conditional PPM probability of order k of symbol x_i, given string x_1^{i-1}. Next, we put

$$\text{PPM}_k(x_1^n) := \prod_{i=1}^{n} \text{PPM}_k(x_i|x_1^{i-1}) \tag{1.59}$$

Quantity $\text{PPM}_k(x_1^n)$ is called the PPM probability of order k of string X_1^n. Finally, we put

$$\text{PPM}(x_1^n) := \frac{6}{\pi^2} \left[D^{-n} + \sum_{k=0}^{\infty} \frac{\text{PPM}_k(x_1^n)}{(k+2)^2} \right] \tag{1.60}$$

Quantity $\text{PPM}(x_1^n)$ is called the (total) PPM probability of the string X_1^n.

As we can see, quantity $\text{PPM}_k(x_1^n)$ is a sort of an incremental Markov model of string x_1^n of order k, where the imitated string is x_1^n itself. In contrast, the total probability $\text{PPM}(x_1^n)$ is a mixture of such Markov models for all finite orders. In general, quantities $\text{PPM}_k(x_1^n)$ and $\text{PPM}(x_1^n)$ are probability distributions on strings, but they are neither stationary nor ergodic. Let us observe that by definition, $\text{PPM}_k(x_1^n) = D^{-n}$ for $k > n - 2$. Thus, the total PPM probability can be effectively computed, i.e. the summation in definition (1.60) can be rewritten as a finite sum

$$\text{PPM}(x_1^n) = D^{-n} + \frac{6}{\pi^2} \sum_{k=0}^{n-2} \frac{\text{PPM}_k(x_1^n) - D^n}{(k+2)^2} \tag{1.61}$$

For practical use, this sum can be further truncated using the concept of maximal repetition, to be introduced in Section 1.13.

The PPM probability satisfies axioms (1.3)–(1.5). Hence, if X_1^n is a random string drawn from a stochastic process $(X_i)_{i \in \mathbb{N}}$, then by positivity of KL divergence, we obtain

$$\mathbf{E}[-\log \mathrm{PPM}(X_1^n)] \geq H(n) \tag{1.62}$$

where $H(n)$ is the entropy of block X_1^n. This inequality can be further strengthened. If X_1^n is a string drawn from a stationary ergodic distribution p, then statistic $-\log \mathrm{PPM}(X_1^n)$ divided by the string length n consistently estimates the entropy rate of the process $(X_i)_{i\in\mathbb{Z}}$. Precisely, equality

$$\lim_{n\to\infty} \frac{1}{n}[-\log \mathrm{PPM}(X_1^n)] = h \text{ almost surely} \tag{1.63}$$

holds for any stationary ergodic process $(X_i)_{i\in\mathbb{Z}}$ over a finite alphabet. A similar equality is true in expectation, namely, equality

$$\lim_{n\to\infty} \frac{1}{n}\mathbf{E}[-\log \mathrm{PPM}(X_1^n)] = h \tag{1.64}$$

holds for any stationary (also nonergodic) process $(X_i)_{i\in\mathbb{Z}}$ over a finite alphabet, see Section 6.3. Hence, PPM probability can be turned into a universal data compression algorithm when we apply the Shannon–Fano coding, mentioned in Section 1.7.

With its capacity to correctly estimate the entropy rate of any ergodic component of a stationary process, the PPM probability can be treated a baseline statistical model of the subjective I-language. As requested for the statistical I-language model, the PPM probability is computable, and it converges in some precise sense to the frequentist E-language model if we have a sufficiently large amount of data. The problem is that the PPM algorithm is not particularly intelligent, namely, it computes the relative frequencies of symbols in quite simple contexts. Probably, a better statistical model of the subjective I-language can be proposed which in some other precise sense converges to the objective E-language faster than the PPM algorithm while taking into account some specific universal statistical patterns of natural language.

It is interesting to analyze what kind of claims about natural texts may be drawn from running the PPM algorithm on texts in natural language. In fact, in Section 1.12, we will apply the PPM algorithm to partly verify Hilberg's hypothesis about natural language, which will be introduced in Section 1.9.

1.9 Entropy of Natural Language

Long before the discovery of the PPM algorithm, Claude Elwood Shannon, the father of information theory, proposed the first experimental method of estimating conditional entropy $\Delta H(n)$ of natural language (Shannon, 1951). According to this method, we show randomly sampled text fragments of length $n - 1$ characters to human subjects, and we ask them to guess the next character of each text fragment.

For each text fragment, the subjects may make one or more guessing attempts by proposing a single character until they name the correct character. Assuming that texts in natural language can be modeled by a stochastic process, the best strategy for doing such guessing is according to diminishing conditional probabilities. Moreover, if the subjects use this strategy, we can obtain an estimate of the conditional entropy from the empirical distribution of guessing attempts needed by the subjects.

Let us write $r_i(x_1^{n-1}) := x_n$ when character x_n has the ith largest conditional probability $p(x_n|x_1^{n-1}) = p(x_1^n)/p(x_1^{n-1})$, given a fixed text fragment x_1^{n-1}. In the optimal guessing strategy, we first guess the character $r_1(x_1^{n-1})$, then $r_2(x_1^{n-1})$, then $r_3(x_1^{n-1})$, and so on. Let $q_i(n)$ be the probability of guessing the correct character in i attempts, given a text fragment of length $n-1$ in this guessing strategy. Shannon (1951) showed that

$$q_i(n) = \sum_{x_1^n} p(x_1^n)\mathbf{1}\{r_i(x_1^{n-1}) = x_n\} \tag{1.65}$$

and consequently,

$$\sum_{i\geq 1}[q_i(n) - q_{i+1}(n)]i\log i \leq \Delta H(n) \leq -\sum_{i\geq 1} q_i(n)\log q_i(n) \tag{1.66}$$

Let $Q_i(n)$ be the number of times that the actual human subjects needed i attempts to guess the next character, given text fragments of length $n-1$ characters. Assuming that human subjects are good enough in guessing the right characters, we may try to estimate

$$q_i(n) \approx \frac{Q_i(n)}{\sum_{k\geq 1} Q_k(n)} \tag{1.67}$$

and consequently plug this estimate into formula (1.66) to obtain a sandwich bound for the conditional entropy. Shannon (1951) performed such an experiment indeed and published a famous graph presenting some upper and lower estimates of conditional entropy $\Delta H(n)$. The text guessed by the human subjects was *Jefferson the Virginian*, a novel by Dumas Malone. The graph, reproduced here in Figure 1.4, suggests that the entropy rate h of an average text in English equals approximately 1 bit per letter. Many later studies reported similar estimates of entropy rate, also for other human languages, using various methods (Cover and King, 1978; Brown et al., 1992; Kontoyiannis, 1997; Kontoyiannis et al., 1998; Grassberger, 2002; Behr et al., 2003; Takahira et al., 2016; Coupé et al., 2019).

A closer inspection of Figure 1.4 shows that the convergence of conditional entropy $\Delta H(n)$ to the entropy rate h is quite slow. In particular, four decades after Shannon, Wolfgang Hilberg, a German electric engineer, replotted this graph in article (Hilberg, 1990) in a doubly logarithmic scale, as can be seen in the lower plane of Figure 1.4. He observed that both the upper and the lower estimates

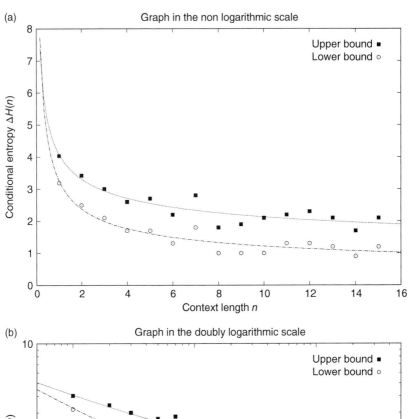

(a) Graph in the non logarithmic scale

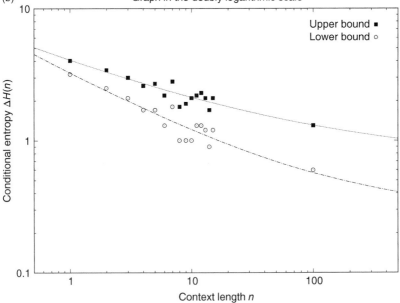

(b) Graph in the doubly logarithmic scale

Figure 1.4 Estimates of conditional entropy for English.

Graphs of the bounds for conditional entropy $\Delta H(n)$ for English in function of the block length n. The lower and the upper bounds are approximated by functions $f_1(n) = 2.9838n^{0.4985-1} + 0.2748$ and $f_2(n) = 3.3776n^{0.6263-1} + 0.7018$ respectively. Source: Author's figure based on the data points taken from Shannon (1951). Functions fitted by the author.

lie on a straight line in a crude approximation. Consequently, Hilberg supposed that conditional entropy $\Delta H(n)$ is inversely proportional to the square root of the context length n,

$$\Delta H(n) \propto n^{-1/2} \tag{1.68}$$

where n can be arbitrarily large. We will call this statement Hilberg's hypothesis. Let us stress that the empirical basis on which this hypothesis is founded is quite weak. Relationship (1.68) is a very rough interpolation of the data points of Shannon (1951), so its extrapolation power should not be believed too much unless we have a good reason to do so. In Figure 1.4, we have fitted two three-parameter functions for the upper and the lower bound, respectively, and we have obtained that

$$2.9838 \; n^{0.4985-1} + 0.2748 \leq \Delta H(n) \leq 3.3776 \; n^{0.6263-1} + 0.7018 \tag{1.69}$$

for block lengths $n \in [1,100]$.

Let us inspect some consequences of Hilberg's hypothesis. First of all, let us notice that Hilberg's hypothesis (1.68) implies a power-law growth of block entropy,

$$H(n) \propto \sum_{k=1}^{n} k^{-1/2} \propto n^{1/2} \tag{1.70}$$

As such, Hilberg's hypothesis (1.68) implies that the entropy rate h equals zero. Is it then a plausible statement? As it can be shown, using tools of measure-theoretic probability, condition $h = 0$ is equivalent to each random letter X_i being almost surely a measurable function of the infinite past $X_{-\infty}^{i-1}$. That is, Hilberg's hypothesis implies asymptotic determinism of human utterances. Such determinism may seem unlikely. For this reason, a few researchers who noticed Hilberg's paper (Ebeling and Nicolis, 1991, 1992; Ebeling and Pöschel, 1994; Bialek et al., 2001a, b; Crutchfield and Feldman, 2003) tended to suppose that the conditional entropy of natural language satisfies rather relationship

$$\Delta H(n) - h \propto n^{-1/2} \tag{1.71}$$

Relationship (1.71) will be called the relaxed Hilberg hypothesis. It is exactly the kind of a functional dependence that we have fitted in (1.69), and it seems to describe Shannon's data slightly better that the original hypothesis by Hilberg.

The relaxed Hilberg hypothesis is also harder to reject on an rational basis. On the contrary, it has some potentially interesting properties. Let us observe first that relationship (1.71) implies

$$H(n) - nh \propto \sum_{k=1}^{n} k^{-1/2} \propto n^{1/2} \tag{1.72}$$

If we could extrapolate relationship (1.72) for an arbitrarily large n, then we would obtain the infinite excess entropy $E = \infty$. Confronting this hypothesis with a statement made in Section 1.7, we may conclude that natural language could not be adequately described by finite-state processes, which have a finite excess entropy. Therefore, the relaxed Hilberg hypothesis, if it is true, strengthens our earlier hypothesis that natural language is non-Markovian and brings us back to the argument of Chomsky (1957) against Skinner (1957), where the validity of finite-state processes as models for natural language has been challenged as well. Recent independent computational experiments also show more directly that the excess entropy of natural language is very large at least, see Hahn and Futrell (2019) and Braverman et al. (2019).

The relaxed Hilberg hypothesis says something more than that natural language is not a finite-state process. An approximate power-law growth of the difference $H(n) - nh$ is the essential feature of the relaxed Hilberg hypothesis. Let us introduce some formal concept which will be used later. To capture the approximate power-law growth of an arbitrary function $\mathfrak{S} : \mathbb{N} \to \mathbb{R}$, we will denote the so called Hilberg exponent

$$\operatorname*{hilb}_{n\to\infty} \mathfrak{S}(n) := \limsup_{n\to\infty} \frac{\log^+ \mathfrak{S}(n)}{\log n} \tag{1.73}$$

where $\log^+ x := \log(x+1)$ for $x \geq 0$ and $\log^+ x := 0$ for $x < 0$, see Dębowski (2018a). We observe that for the exact power-law growth $\mathfrak{S}(n) = n^\beta$ with $\beta \geq 0$, we have $\operatorname*{hilb}_{n\to\infty} \mathfrak{S}(n) = \beta$. More generally, the Hilberg exponent captures an asymptotic power-law growth of the sequence. Let us define $\mathfrak{J}(n) := 2\mathfrak{S}(n) - \mathfrak{S}(2n)$. It can be also shown that if there exists limit $\lim_{n\to\infty} \mathfrak{S}(n)/n = \mathfrak{s}$ then

$$\operatorname*{hilb}_{n\to\infty} (\mathfrak{S}(n) - n\mathfrak{s}) \leq \operatorname*{hilb}_{n\to\infty} \mathfrak{J}(n) \tag{1.74}$$

with the equality if $\mathfrak{J}(n) \geq 0$, see Section 8.1.

A corollary of the above is that for the block entropy $H(X_1^n) = H(n)$, the entropy rate h, and the block mutual information $I(X_1^n; X_{n+1}^{2n}) = 2H(n) - H(2n) \geq 0$, we may equivalently define the Hilberg exponent

$$\beta := \operatorname*{hilb}_{n\to\infty} [H(X_1^n) - hn] = \operatorname*{hilb}_{n\to\infty} I(X_1^n; X_{n+1}^{2n}) \tag{1.75}$$

Thus, the relaxed Hilberg hypothesis for natural language can be stated as

$$\beta \approx 1/2 \tag{1.76}$$

In Section 1.11, we will discuss some linguistically motivated stochastic processes which satisfy this relationship. Prior to this, in Section 1.10, we need to introduce some basic concepts from algorithmic information theory – another approach to defining the amount of information.

1.10 Algorithmic Information Theory

As we have announced in Section 1.7, there are two ways of measuring the amount of information: the Shannon entropy (Shannon, 1948, 1951), treated in Section 1.7, and the Kolmogorov complexity (Solomonoff, 1964; Kolmogorov, 1965), which is relative to a universal computer and which we would like to briefly discuss in this section. The respective material will be developed formally in Chapter 7.

Formally, an arbitrary computer can be imagined as some mapping $S : \{0,1\}^* \to \mathbb{X} \cup \{\perp\}$ with the intended meaning that given a binary program $w \in \{0,1\}^*$ the computer prints out in finite time some object $x \in \mathbb{X}$ if $S(w) = x$ or it does not halt (i.e. it does not terminate computations) if $S(w) = \perp$. The Kolmogorov complexity $\mathbf{H}(x)$ of an object $x \in \mathbb{X}$ is defined as the length of the shortest program that makes the computer print out object x, i.e.

$$\mathbf{H}(x) := \min\{|w| : S(w) = x\} \tag{1.77}$$

By x^*, we denote some fixed program such that $S(x^*) = x$ and $|x^*| = \mathbf{H}(x)$.

A universal computer is a computer that can simulate any other computer, i.e. a computer S is called universal if for each computer V, there exists a string $v \in \{0,1\}^*$ such that $S(vw) = V(w)$ for all strings $w \in \{0,1\}^*$. It turns out that for universal computers functions $x \mapsto \mathbf{H}(x)$ and $x \mapsto x^*$ are not computable. Nevertheless, if we make a restriction that the universal computer S halts only on programs forming a prefix-free set, i.e. if the set

$$\mathbf{A} := \{w \in \{0,1\}^* : S(w) \neq \perp\} \tag{1.78}$$

is prefix-free, then formally, there exists an uncomputable prefix-free code $\mathbf{C} : \mathbb{X} \ni x \mapsto x^* \in \mathbf{A}$ such that $|\mathbf{C}(x)| = \mathbf{H}(x)$. In this case, $\mathbf{H}(x)$ is called the prefix-free Kolmogorov complexity.

A rational function $f : \mathbb{X} \to \mathbb{Q}$ is called recursive if there exists a computer which, given an argument $x \in \mathbb{X}$ computes the value $f(x) \in \mathbb{Q}$ in finite time. Subsequently, a probability distribution p on a countable domain $\mathbb{X}$ will be called recursive if the function $\mathbb{X} \ni x \mapsto p(x) \in \mathbb{Q}$ is recursive. Moreover, if a distribution $p : \mathbb{X} \to \mathbb{R}$ is recursive, then we can write a program that given an algorithmic definition of p and a Shannon–Fano code word $C_p(x)$ computes the string x. The latter implies

$$\mathbf{H}(x) \leq -\log p(x) + \mathbf{H}(p) + c \tag{1.79}$$

for some constant $c > 0$, where $\mathbf{H}(p)$ is the prefix-free Kolmogorov complexity of the distribution p itself, i.e. the length of the algorithmic definition of p, not to be confused with Shannon entropy $H(p)$. In this way, we obtain the inequality

$$H(p) \leq \sum_{x \in \mathbb{X}} p(x)\mathbf{H}(x) \leq H(p) + \mathbf{H}(p) + c \tag{1.80}$$

Thus, the expected prefix-free Kolmogorov complexity is roughly equal to the Shannon entropy, if the algorithmic definition of the distribution is not too long.

Like the Shannon entropy, the Kolmogorov complexity can be developed further. In the following, we will assume that S is a universal computer with $\mathbf{A}$ being a prefix-free set and consequently $\mathbf{H}(x)$ is the prefix-free Kolmogorov complexity. The following functions of finite sequences x, y, and z can be defined:

- conditional Kolmogorov complexity $\mathbf{H}(x|z)$, which satisfies

$$|\mathbf{H}(z) + \mathbf{H}(x|z, \mathbf{H}(z)) - \mathbf{H}(x, z)| \leq c \tag{1.81}$$

- algorithmic mutual information

$$\mathbf{I}(x;y) := \mathbf{H}(x) + \mathbf{H}(y) - \mathbf{H}(x, y) \tag{1.82}$$

- conditional algorithmic mutual information

$$\mathbf{I}(x;y|z) := \mathbf{H}(x|z) + \mathbf{H}(y|z) - \mathbf{H}(x, y|z) \tag{1.83}$$

These definitions make sense only if the sequences are finite, but they can be partly generalized for infinite sequences, see later in this book. In general, we have $\mathbf{H}(x), \mathbf{H}(x|z), \mathbf{I}(x;y), \mathbf{I}(x;y|z) \geq -c$, where $c \geq 0$ is a finite constant (Gács, 1974; Chaitin, 1975a), see Section 7.3.

Subsequently, we would like to show that the prefix-free Kolmogorov complexity sheds some light onto the concept of randomness of a particular infinite sequence of binary digits. We will say that an infinite sequence of binary random variables $(Z_k)_{k\in\mathbb{N}}$ is a fair-coin process if

$$P(Z_1^k = z_1^k) = 2^{-k} \tag{1.84}$$

for all strings $z_1^k \in \{0, 1\}^*$. In other words, $(Z_k)_{k\in\mathbb{N}}$ is a sequence of independent random variables uniformly distributed on set $\{0, 1\}$. In contrast, an infinite sequence of particular bits, i.e. binary digits $(z_k)_{k\in\mathbb{N}}$ is called algorithmically incompressible when there exists a constant $c \geq 0$ such that

$$\mathbf{H}(z_1^k) \geq k - c \tag{1.85}$$

for all $k \in \mathbb{N}$. It turns out that a fair-coin process $(Z_k)_{k\in\mathbb{N}}$ is algorithmically incompressible almost surely so, informally speaking, algorithmically incompressible sequences are typical realizations of a fair-coin process $(Z_k)_{k\in\mathbb{N}}$. Moreover, by the Schnorr theorem, to be discussed in Section 7.5, all individual algorithmically incompressible sequences are indistinguishable from typical fair coin tosses by means of effective statistical tests of randomness. In other words, all algorithmically incompressible sequences are perfectly unpredictable using computable prediction schemes.

It is instructive to exhibit some particular example of an algorithmically incompressible sequence, or, equivalently an algorithmically incompressible number.

We will say that a real number $z \in (0,1)$ is algorithmically incompressible when its binary expansion is algorithmically incompressible, i.e. when for $z = \sum_{k=1}^{\infty} 2^{-k} z_k$ and $z_k \in \{0,1\}$, sequence $(z_k)_{k \in \mathbb{N}}$ is algorithmically incompressible. The halting probability, introduced by Chaitin (1975a) and traditionally denoted as

$$\mathbf{\Omega} := \sum_{w \in A} 2^{-|w|} \in (0,1) \tag{1.86}$$

is an important example of an algorithmically incompressible number. To clearly differentiate number $\mathbf{\Omega}$ from the even more traditional notation Ω for the event space of probability spaces, we will always write the former in boldface. We will also denote the binary expansion of the halting probability as $\mathbf{\Omega} = \sum_{k=1}^{\infty} 2^{-k} \mathbf{\Omega}_k$. Number $\mathbf{\Omega}$ has a kind of magical and paradoxical properties, which will be proved in Section 7.4:

1. First of all, $\mathbf{\Omega}$ is algorithmically incompressible.
2. Yet, given the first k bits $\mathbf{\Omega}_1, \mathbf{\Omega}_2, \dots, \mathbf{\Omega}_k$, we can decide which mathematical statement written down using less than k binary symbols is true or false.
3. Yet, given a formal inference system, i.e. a finite set of axioms and inference rules, we cannot deduce values of bits $\mathbf{\Omega}_k$ for k much larger than the length of the formal inference system, when written down using binary symbols. In other words, if we do formal deduction only, we will never learn distant bits $\mathbf{\Omega}_k$.
4. Yet, there exists a simple recursive function $\omega : \mathbb{N} \times \mathbb{N} \to \{0,1\}$ such that

$$\lim_{n \to \infty} \omega(k,n) = \mathbf{\Omega}_k \tag{1.87}$$

This is possible since convergence of $\omega(k,n)$ to $\mathbf{\Omega}_k$ is very slow.

Thus, number $\mathbf{\Omega}$ is a kind of an elusive oracle or philosophers' stone. As we can see, algorithmically incompressible sequences, though being in some sense random, may represent some highly useful information, and the amount of useful information can be in principle infinite.

1.11 Descriptions of a Random World

This section starts with a philosophical excursion. Its aim is to provide a reason why natural language should have a strictly positive Hilberg exponent $\beta > 0$. As mentioned in Section 1.1, the reason may be the existence of an infinite immutable truth which constitutes a very complex statistic inferrable from both the approximately infinite past and the approximately infinite future of our usage of language. As we will see in Section 1.12, the existence of the infinite immutable truth, as reflected in our utterances, causes a measurable effect. In particular, we may obtain inequality $\beta > 0$ in a natural way. The exact mathematical model will be developed in Sections 8.3 and 8.4.

To begin our considerations, we may suppose that the typical purpose of a text in natural language is to refer to some external reality that is not the text itself. In texts written by mathematicians, this external reality is the Platonic world of mathematical ideas, whereas in texts written by ordinary people, this external reality is the physical world entailing space, time, and matter, or some partly imaginary versions thereof. The crucial idea is that in either case this external reality seems very complex and quite slowly evolving. In some approximation, we may treat it as infinitely complex and immutable. That is, to describe this reality completely, we may need practically an infinite sequence of bits, i.e. digits that take values of 0 or 1 – and this sequence practically does not change in time.

We can ask whether such a necessarily infinite immutable sequence that describes some external reality exists indeed. In the domain of pure mathematics, we can be sure that the answer is positive. It turns out so since what we are asking for is some sort of meaningful infinite randomness, where randomness is understood as algorithmic incompressibility. Thus, the binary expansion of number Ω, introduced in Section 1.10, provides an example of the desired random, but meaningful object. As we have stated in Section 1.10, number Ω stores in the most compact way the information which mathematical statements are true and which are false. Moreover, Ω is incompressible and its binary expansion is infinitely long.

However, we have mentioned that number Ω cannot be fully learned in a deductive way. If we exercise mathematics in a formal way, as it is usually done, then we prove theorems starting from a finite number of axioms and applying a finite number of inference rules. Given such a formal system, i.e. a finite set of axioms and inference rules, we cannot deduce the values of bits Ω_k for k much larger than the length of the formal system when written down using binary symbols. In other words, if we do formal deduction only, we will never learn distant bits Ω_k. Thus, number Ω is some nice guiding model of an infinitely complex meaningful reality, but it is too elusive.

How about the physical world described in nonmathematical texts? Here we do not only do formal deduction, but we also make direct observations or experiments. To some extent, we can simply observe what appears to be the present state of the world and report that state in language utterances. Is there an infinite amount of immutable elementary truths in what we can observe?

To be very concrete, let us imagine, for example, a very long row of chairs randomly painted white or black. The state of this row, if the row is practically infinite, could be described by an algorithmically incompressible collection of bits $(z_k)_{k \in \mathbb{N}}$, indexed by indices $k = 1, 2, 3, ...$, where

$$z_k = \begin{cases} 0 & \text{if } k\text{th chair is white} \\ 1 & \text{if } k\text{th chair is black} \end{cases} \tag{1.88}$$

If these chairs never rot and never can be repainted or rearranged, then their state represents some immutable physical truth. Moreover, if the chairs are painted at random white or black, then this state is infinitely complex, i.e. its description cannot be compressed to a finite sequence of bits. In contrast, if there were no randomness, we could describe the color of some chairs using the knowledge of the color of other chairs. These are our intuitions of links between the physical truth and randomness. In our opinion, the physical truth is also a sort of algorithmic randomness, but it is not necessarily equal to the number Ω.

Can we really paint an infinite row of chairs at random or, rather, observe such an infinite row? Physicists claim that there is some inherent randomness of the physical world at the microscopic level – be it ruled by the classical or quantum mechanics. Still, can an *infinite* amount of algorithmic randomness be extracted from our physical reality? Honestly, we do not know it for sure. However, let us put on the hats of applied mathematicians and let us assume that algorithmic randomness can be extracted and reported in texts.

Formally, we will assume that we have a probability space $(\Omega, \mathcal{J}, P)$ and some random variables. We will suppose that the state of the physical world can be partly described by an algorithmically incompressible collection of bits $(z_k)_{k \in \mathbb{N}}$, indexed by indices k belonging to the infinite set of natural numbers $\mathbb{N}$. These algorithmically incompressible bits z_k will be called facts. Facts should be contrasted with texts. Whereas in our formal model, the collection of facts $(z_k)_{k \in \mathbb{N}}$ represents some truth which is immutable and ingraspable in its infinity, texts are random strings of symbols $X_{j+1}^{j+n} = (X_{j+1}, X_{j+2}, \dots, X_{j+n})$ which refer to some particular facts. Formally, symbols X_i will be assumed to be random variables taking values in some countable set $\mathbb{X}$, i.e. $X_i : \Omega \to \mathbb{X}$. If $\mathbb{X}$ is finite, we can imagine that $\mathbb{X}$ is the set of letters or characters. If $\mathbb{X}$ is infinite, we can imagine that $\mathbb{X}$ is the set of words or propositions. We will also assume that random symbols X_i can be organized as a two-sided infinite sequence $(X_i)_{i \in \mathbb{Z}}$.

What is then a plausible relationship between facts and texts? We suppose that texts allow us to learn some particular facts. That is, there is a fixed method of interpreting finite texts to infer what some particular facts are. Formally, such a method of interpretation can be represented as a recursive function $g : \mathbb{N} \times \mathbb{X}^* \to \{0, 1, 2\}$, where $g(k, X_{j+1}^{j+n})$ would be our guess of fact z_k based on text X_{j+1}^{j+n} with $g(k, X_{j+1}^{j+n}) = 2$ if we do not know the answer. Now, what kind of facts can be called general? General facts seem to sound like immutable facts. Hence, some idea we can initially agree upon is as follows. Given any sufficiently long text X_{j+1}^{j+n} and a fact z_k, the probability of inferring this fact correctly should approach 1 when the text length tends to infinity,

$$\lim_{n \to \infty} P(g(k, X_{j+1}^{j+n}) = z_k) = 1 \tag{1.89}$$

If the collection of facts $(z_k)_{k\in\mathbb{N}}$ represents some infinitely complex truth which is immutable and incrementally learnable, then we should expect that condition (1.89) holds regardless where we start reading, i.e. for any offset $j \in \mathbb{Z}$, and regardless which fact we want to infer, i.e. for any index $k \in \mathbb{N}$.

Let us observe that condition (1.89) is trivially satisfied for any stochastic process $(X_i)_{i\in\mathbb{Z}}$ for recursive function $g(k, x_1^n) = \omega(k, n)$ and the algorithmically incompressible sequence $(z_k)_{k\in\mathbb{N}} = (\mathbf{\Omega}_k)_{k\in\mathbb{N}}$, where $\omega(k, n)$ is the recursive function that satisfies condition (1.87). In spite of this negative result, the power-law growth of the number of inferrable algorithmically incompressible facts corresponds itself to some nontrivial property. For a recursive function $g : \mathbb{N} \times \mathbb{X}^* \rightarrow \{0, 1, 2\}$ and an algorithmically incompressible sequence $(z_k)_{k\in\mathbb{N}} \in \{0, 1\}^{\mathbb{N}}$, the set of initial facts inferrable from a finite text X_1^n will be defined as

$$\mathbf{U}(X_1^n) := \{l \in \mathbb{N} : g(k, X_1^n) = z_k \text{ for all } k \leq l\} \tag{1.90}$$

In other words, we have $\mathbf{U}(X_1^n) = \{1, 2, ..., l\}$, where l is the largest number such that $g(k, X_1^n) = z_k$ for all $k \leq l$. Subsequently, we will call a stationary process $(X_i)_{i\in\mathbb{Z}}$ perigraphic if the Hilberg exponent of the expectation of cardinality of $\mathbf{U}(X_1^n)$ is strictly positive,

$$\operatorname*{hilb}_{n\to\infty} \mathbf{E}\#\mathbf{U}(X_1^n) > 0 \tag{1.91}$$

for a recursive function $g : \mathbb{N} \times \mathbb{X}^* \rightarrow \{0, 1, 2\}$ and an algorithmically incompressible sequence of binary digits $(z_k)_{k\in\mathbb{N}}$.

It can be shown, see Section 8.3, that perigraphic processes are a nontrivial concept since any perigraphic process $(X_i)_{i\in\mathbb{Z}}$ has a nonrecursive distribution. Moreover, we can present a very simple example of a perigraphic process, which comes under an umbrella of so-called Santa Fe processes, discovered by the author of this book during a visit at the Santa Fe Institute in August 2002 – hence comes their name. The first published mention of Santa Fe processes appears in Dębowski (2009, 2011). Let $(K_i)_{i\in\mathbb{Z}}$ be a sequence of probabilistically independent random variables assuming values in natural numbers with a power-law distribution,

$$P(K_i = k) \propto \frac{1}{k^{\alpha}}, \quad \alpha > 1 \tag{1.92}$$

This power law resembles Zipf's law (1.1). Besides process $(K_i)_{i\in\mathbb{Z}}$, we choose an arbitrary algorithmically incompressible sequence of binary digits $(z_k)_{k\in\mathbb{N}}$. To fix our attention, we may choose $(z_k)_{k\in\mathbb{N}} = (\mathbf{\Omega}_k)_{k\in\mathbb{N}}$ but this is not necessary. Subsequently, the individual Zipfian Santa Fe process with an exponent α is a sequence $(X_i)_{i\in\mathbb{Z}}$, where

$$X_i = (K_i, z_{K_i}) \tag{1.93}$$

are pairs of a random number K_i and the fact z_{K_i} with the corresponding index.

It can be shown, see Section 8.3, that the individual Zipfian Santa Fe process is ergodic. Moreover, it is perigraphic since condition

$$\operatorname*{hilb}_{n \to \infty} \mathbf{E}\#\mathbf{U}(X_1^n) = 1/\alpha \in (0, 1) \qquad (1.94)$$

Holds, for example for the recursive guessing function

$$g(k, x_1^n) = \begin{cases} 0 & \text{if } \exists_{1 \le i \le n} x_i = (k, 0) \text{ and } \neg\exists_{1 \le i \le n} x_i = (k, 1) \\ 1 & \text{if } \exists_{1 \le i \le n} x_i = (k, 1) \text{ and } \neg\exists_{1 \le i \le n} x_i = (k, 0) \\ 2 & \text{else} \end{cases} \qquad (1.95)$$

Simply speaking, function $g(k, x_1^n)$ returns 0 or 1 when an unambiguous value of the second constituent can be read off from pairs $x_i = (k, \cdot)$ and returns 2 when there is some ambiguity.

In the individual Zipfian Santa Fe processes, the relationship between a text and the pool of facts is quite simple. Imagine that the units X_i of a text are not atomic, but they have some structure. Namely, they are statements of form "the kth fact equals z" and more specifically, they only make true assertions "the kth fact equals z_k." To write down such a statement formally, we need two pieces of information – the index k and the value z_k. In fact, the relationship between the form and the meaning of a statement in Santa Fe processes is the simplest possible: every statement is nothing but a pair (k, z_k). Subsequently, if we admit some randomness of statements, namely, if statement X_i is a random variable, then index k should be replaced by a random variable K_i. Resuming, our process $(X_i)_{i \in \mathbb{Z}}$ should take form (1.93). Finally, let us observe that descriptions $X_i = (k, z)$ can be interpreted as logically consistent. That is, if two statements $X_i = (k, z)$ and $X_j = (k', z')$ describe bits of the same address $(k = k')$, then they always assert the same value $(z = z')$.

Although the relationship between texts and facts for natural texts is much more complex, the individual Zipfian Santa Fe process seems to capture some intuition about the natural language. When we write a text, we have a strong feeling that we articulate a sequence of propositions that refer to some external reality. The construction of an individual Zipfian Santa Fe process suggests that we may split our propositions $(X_i)_{i \in \mathbb{Z}}$ into two parts: a static infinite description of some fixed external reality $(z_k)_{k \in \mathbb{N}}$ and a narration process $(K_i)_{i \in \mathbb{Z}}$ that describes the flow of our attention. We cannot exclude the possibility that the narration process $(K_i)_{i \in \mathbb{Z}}$ in natural language is quite a complex process, also depending on the state of the external reality $(z_k)_{k \in \mathbb{N}}$.

As we have mentioned, any perigraphic process has a nonrecursive distribution p. If we want to model the subjective I-language rather than the objective E-language, as discussed in the conclusion of Section 1.5, we should likely consider recursive distributions only. Thus, in the first move, we should replace the sequence of algorithmically incompressible facts $(z_k)_{k\in\mathbb{N}}$ by something computable. A natural probabilistic model of an algorithmically incompressible sequence $(z_k)_{k\in\mathbb{N}}$ is a fair-coin process $(Z_k)_{k\in\mathbb{N}}$, i.e. a sequence of independent uniformly distributed binary random variables, formally, satisfying condition (1.84). Let us analyze what happens to our theory when we substitute a fair-coin process $(Z_k)_{k\in\mathbb{N}}$ for the algorithmically incompressible sequence $(z_k)_{k\in\mathbb{N}}$. A Bayesian statistician would say that we are going to put a prior probability distribution onto the sequence $(z_k)_{k\in\mathbb{N}}$.

It can be shown, see Section 8.3, that a stationary process $(X_i)_{i\in\mathbb{Z}}$ is not ergodic if and only if there exists a single random variable Z_k which is not constant and satisfies condition

$$\lim_{n\to\infty} P(g(k, X_{j+1}^{j+n}) = Z_k) = 1 \tag{1.96}$$

for any offset $j \in \mathbb{Z}$. For this reason, a stationary process $(X_i)_{i\in\mathbb{Z}}$ which satisfies condition (1.96) for a fair-coin process $(Z_k)_{k\in\mathbb{N}}$ for any offset $j \in \mathbb{Z}$ and any index $k \in \mathbb{N}$ will be called strongly nonergodic. There is also a more abstract equivalent definition of strongly nonergodic processes, which will be introduced in Section 5.6.

A simple example of a strongly nonergodic process can be formed by modifying the individual Zipfian Santa Fe process. Let process $(K_i)_{i\in\mathbb{Z}}$ be as before and let process $(Z_k)_{k\in\mathbb{N}}$ be probabilistically independent from process $(K_i)_{i\in\mathbb{Z}}$. Then the process $(X_i)_{i\in\mathbb{Z}}$ defined as

$$X_i = (K_i, Z_{K_i}) \tag{1.97}$$

is strongly nonergodic for the same guessing function (1.95). We will call this process the mixture Zipfian Santa Fe process. In contrast to the individual Zipfian Santa Fe processes, the probability distribution of the mixture Zipfian Santa Fe process is recursive. Moreover, if we define the set of initial independent probabilistic facts inferrable from a finite text X_1^n as

$$U(X_1^n) := \{l \in \mathbb{N} : g(k, X_1^n) = Z_k \text{ for all } k \leq l\} \tag{1.98}$$

then for the mixture Zipfian Santa Fe, we obtain

$$\operatorname*{hilb}_{n\to\infty} \mathbf{E}\#U(X_1^n) = 1/\alpha \in (0,1) \tag{1.99}$$

see Section 8.3.

For the mixture Zipfian Santa Fe process, there is an alternative interpretation of process $(Z_k)_{k\in\mathbb{N}}$. As stated in Section 1.5, a stationary process is ergodic if the relative frequencies of all finite substrings in the infinite text generated by the process converge in the long run with probability one to some constants – the probabilities of the respective substrings. Now, some basic linguistic intuition suggests that this convergence can be very slow for natural language. As Yaglom and Yaglom

(1983) have observed, there is hardly any book that contains a large number of occurrences of both word *lemma* and word *love*. In particular, counting relative frequencies of key words such as *lemma* or *love*, we can recognize the topic of a book, be it a mathematical monograph or a love story. Said more abstractly, there is a variation of topics of finite texts in natural language, and these topics can be effectively distinguished by counting relative frequencies of certain words or using some other effective computational methods.

In particular, function $g(k, x_1^n)$ can be seen as an effective procedure of recognizing the kth bit of some abstract description of a particular topic of an arbitrary text X_1^n. Consequently, we can imagine the process $(Z_k)_{k \in \mathbb{N}}$ in the mixture Zipfian Santa Fe process as an abstract description of one of a continuum of possible persistent topics of an infinitely long random text $(X_i)_{i \in \mathbb{Z}}$. The random topic interpretation of process $(Z_k)_{k \in \mathbb{N}}$ in the mixture Zipfian Santa Fe process may be more appealing to some readers than immutable algorithmically incompressible truth $(z_k)_{k \in \mathbb{N}}$ in the individual Zipfian Santa Fe process. But, in fact, the practical difference between the continuum of persistent random topics and the single immutable algorithmically incompressible truth is not so large. Once we are given a particular realization of an infinitely long text $(X_i)_{i \in \mathbb{Z}}$, we cannot observe any other realization of processes $(X_i)_{i \in \mathbb{Z}}$ and $(Z_k)_{k \in \mathbb{N}}$ and the realization of process $(Z_k)_{k \in \mathbb{N}}$ is algorithmically incompressible almost surely.

Resuming this section, our investigations of the meaning of texts and its relationship to the concept of immutable infinitely complex physical truth turn out to be closely linked to the distinction between recursive and nonrecursive processes, as well as, ergodic and nonergodic processes. However, our perspective is a complete opposite of the usual mathematical point of view. Whereas mathematicians are mostly interested in recursive and ergodic probability measures, being simpler and more regular cases, nonrecursive and nonergodic processes seem far more natural for investigation of human language.

1.12   Facts and Words Related

It may be baffling to some readers that the existence of immutable infinitely complex reality that is repeatedly referred in a text causes a measurable effect. In this section, we will present some results about stationary processes, which we call the theorems about facts and words. These propositions state that the expected number of independent probabilistic or algorithmic facts inferrable from the text drawn from a stationary process is roughly less than the expected number of distinct word-like substrings detectable in the text via the PPM algorithm introduced in Section 1.8. In particular, an asymptotic power law growth of the number of inferrable probabilistic or algorithmic facts implies an asymptotic power law growth of the number of word-like substrings, which is a statistically measurable effect, closely related to Herdan's law introduced in Section 1.3 and Hilberg's hypothesis introduced in Section 1.9. The rigorous development of the results of this section will be done in Sections 6.5 and 8.4.

Continuing the notations from Section 1.8, we define the largest of probabilities $\text{PPM}_k(x_1^n)$ for a fixed string X_1^n as

$$\text{PPM}_G(x_1^n) := \max\{\text{PPM}_k(x_1^n) : k \geq -1\} \tag{1.100}$$

where we put $\text{PPM}_{-1}(x_1^n) := D^{-n}$, and we define the PPM order of a string x_1^n as

$$G(x_1^n) := \min\{k \geq -1 : \text{PPM}_G(x_1^n) = \text{PPM}_k(x_1^n)\} \tag{1.101}$$

Since $\text{PPM}_k(x_1^n) = D^{-n} = \text{PPM}_{-1}(x_1^n)$ for $k > n - 2$, the PPM order is less than the string length, i.e.

$$G(x_1^n) \leq n - 2 < n \tag{1.102}$$

Hence, we obtain the inequality

$$0 \leq -\log \text{PPM}(x_1^n) + \log \text{PPM}_G(x_1^n) < 2\log(n + 2) + \log \frac{\pi^2}{6} \tag{1.103}$$

Thus, $\text{PPM}_G(x_1^n)$ is roughly equal to the probability $\text{PPM}(x_1^n)$. We can effectively think of the PPM probability as the maximal probability $\text{PPM}_G(x_1^n)$.

In fact, the PPM order is a semiconsistent estimator of the Markov order of the process. That is, for any ergodic kth order Markov process, the PPM order $G(X_1^n)$ is almost surely greater than k for block length n tending to infinity, whereas for non-Markovian processes, the PPM order $G(X_1^n)$ diverges to infinity.

For $k \geq 0$, the set of distinct substrings of length k in string $w \in \mathbb{X}^*$ is

$$V_k(w) := \{u \in \mathbb{X}^k : w = vuz \text{ for some } v, z \in \mathbb{X}^*\} \tag{1.104}$$

The cardinality of set $V_k(w)$ as a function of substring length k is called the subword complexity of string w (de Luca, 1999). The readers working in quantitative linguistics will notice that the quotient $\#V_k(w)/(n - |w| + 1) \in (0, 1]$ is the type-token ratio for substrings of length k (Baayen, 2001). In the following, let us apply the concept of the PPM order to define some special set of substrings of an arbitrary string x_1^n. The set of distinct PPM words detected in x_1^n will be defined as the set $V_k(x_1^n)$ for the substring length $k = G(x_1^n)$, i.e.

$$V_G(x_1^n) := V_{G(x_1^n)}(x_1^n) \tag{1.105}$$

Now we may write down the theorems about facts and words. These theorems come in several versions and date back to Dębowski (2006, 2011). Here we present two relatively strong versions, one for probabilistic facts and one for algorithmic facts, proved by Dębowski (2018a). Let $(X_i)_{i \in \mathbb{Z}}$ be a stationary strongly nonergodic process over a finite alphabet. The probabilistic theorem about facts and words reads

$$\operatorname*{hilb}_{n \to \infty} \mathbf{E}\# U(X_1^n) \leq \operatorname*{hilb}_{n \to \infty} I(X_1^n; X_{n+1}^{2n})$$

$$\leq \operatorname*{hilb}_{n \to \infty} \mathbf{E}[G(X_1^n) + \# V_G(X_1^n)] \qquad (1.106)$$

In contrast, the algorithmic theorem about facts and words is a bit more general. Let $(X_i)_{i \in \mathbb{Z}}$ be an arbitrary stationary process over a finite alphabet. We have inequalities

$$\operatorname*{hilb}_{n \to \infty} \mathbf{E}\# \mathbf{U}(X_1^n) \leq \operatorname*{hilb}_{n \to \infty} \mathbf{EI}(X_1^n; X_{n+1}^{2n})$$

$$\leq \operatorname*{hilb}_{n \to \infty} \mathbf{E}[G(X_1^n) + \# V_G(X_1^n)] \qquad (1.107)$$

Resuming, the theorems about facts and words state that the expected number of independent inferrable facts is roughly less than the block Shannon or algorithmic mutual information, and this is roughly less than the expected number of distinct PPM words increased by the PPM order. The precise statements are expressed using the Hilberg exponents for the respective quantities.

From what we have written so far, it follows that if relationship

$$\operatorname*{hilb}_{n \to \infty} \mathbf{E}\# U(X_1^n) > 0 \qquad (1.108)$$

holds true, then the stochastic process is not only strongly nonergodic, but it also has the infinite excess entropy, and consequently, it cannot be a finite-state process. As we will see in Section 11.2, there are some counterexamples of processes which do not satisfy condition (1.108) but satisfy a power-law growth of the block mutual information and hence satisfy condition

$$\operatorname*{hilb}_{n \to \infty} \mathbf{E}[G(X_1^n) + \# V_G(X_1^n)] > 0 \qquad (1.109)$$

Such counterexamples can be obtained by modifying the construction of Santa Fe processes by allowing the described reality to slowly evolve. Hence, whereas we can deduce condition (1.109) from condition (1.108), the implication in the opposite direction does hold. However, we may ask an open question whether the Hilberg exponent for the block algorithmic mutual information is in general close to the Hilberg exponent for the number of PPM words. We doubt that this is true, but understanding why this may fail could give some insight into what these two statistics do measure.

Thus, the theorems about facts and words provide a convincing but only rational explanation of the relaxed Hilberg hypothesis. Namely, if the number of independent inferrable facts for natural language grows like a power of the text length, then we have an asymptotic power law growth of the block mutual information and also an asymptotic power law growth of the number of distinct PPM words. In other words, if natural language is a perigraphic process and the number of PPM words is close to the number of orthographic words, then the theorems about facts and words provide an explanation of both Hilberg's relaxed hypothesis discussed in Section 1.9 and Herdan's law discussed in Section 1.3.

In Figures 1.5 and 1.6, we have presented the data for Shakespeare's First Folio/ 35 Plays, our diagnostic text in natural language, and a random permutation of its characters. For the text in natural language, we seem to have indeed a stepwise power law growth of the number of distinct PPM words with an exponent close to 0.8, i.e.

$$\text{hilb}_{n\to\infty} \#V_G(x_1^n) \approx 0.8 \qquad (1.110)$$

In view of the results of Takahira et al. (2016), who used the PPM algorithm to compress very large corpora of texts in natural language ($n \leq 10^{10}$), we may suppose that relationship (1.110) holds for human languages more generally. In contrast, for the random permutation of characters, we obtain

$$\text{hilb}_{n\to\infty} \#V_G(x_1^n) = 0 \qquad (1.111)$$

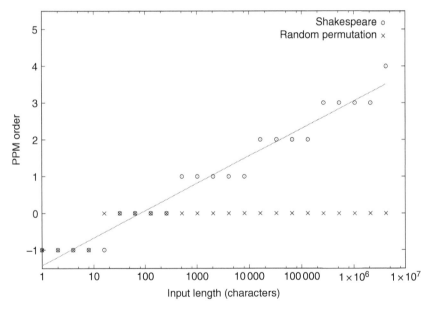

Figure 1.5 The PPM order.

The PPM order $G(x_1^n)$ versus the input length n for Shakespeare's First Folio/35 Plays and a random permutation of the text's characters. Considered is the alphabet size $D = 27$ (26 letters and the space). The straight line is the regression $G(x_1^n) \approx -1.423 + 0.323 \ln n$. Source: Author's figure based on the text downloaded from Project Gutenberg (https://www.gutenberg.org).

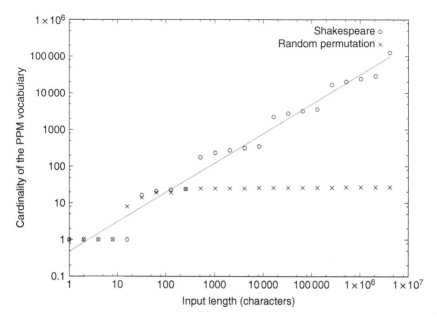

Figure 1.6 The PPM vocabulary.

The cardinality of the PPM vocabulary $\#V_G(x_1^n)$ versus the input length n for Shakespeare's First Folio/35 Plays and a random permutation of the text's characters. Considered is the alphabet size $D = 27$ (26 letters and the space). The straight line is the regression $\ln \#V_G(x_1^n) \approx -0.737 + 0.801 \ln n$. Source: Author's figure based on the text downloaded from Project Gutenberg (https://www.gutenberg.org).

Seeing the striking difference between observations (1.111) and (1.110), we may be tempted to suppose that the number of inferrable probabilistic or algorithmic facts for texts in natural language also obeys a power-law growth. Formally speaking, however, we cannot deduce condition (1.108) from observing mere relationship (1.110). The implication goes in the reverse direction.

1.13 Repetitions and Entropies

As we have tried to argue so far, some important statistical regularities of natural texts, such as Herdan's and Zipf' laws as well as the relaxed Hilberg hypothesis, may be a direct consequence of relatively abstract semantic constraints, simply speaking, that natural texts convey some meaning. Basically, we can identify a few distinct modes of a text being meaningful, such as

1. Meaningfulness by a consistent description of an external or imagined reality.
2. Meaningfulness by an internal cohesion of the narration or the discourse.

3. Meaningfulness by an effective control of an external reality (in particular, reaching or striving toward some goal).

The first mode of meaningfulness can be roughly captured by the notion of a perigraphic process and power laws for facts, mutual information, and words, as developed in Sections 1.11 and 1.12. The second and the third modes of meaningfulness matter as well, but they are harder to quantify and measure, especially if we are given only texts as an object of our investigation.

In fact, we may suppose that the third mode of meaningfulness is a primary one in natural language, see Rączaszek-Leonardi (2012). In particular, our need to describe any imagined reality or to conduct a cohesive narration or discourse is a mere derivative of our need to control the external reality. Here, let us also observe that, to effectively control the external reality, we do not need too accurate descriptions thereof, since we can force some facts about the external reality to be true by the power of our will. In any case, understanding of the third mode of meaningfulness posits a distinction between the external and the imagined realities and demands studying complicated interactions between them, see Seth (2019). This setting seems much harder to model, especially if we assume that the external and the imagined realities are far from being random, so let us intentionally not consider this setting here.

Focusing back on the second mode of meaningfulness, we may wonder whether this mode implies the first mode in some sense. We have an intuition that mere internal cohesion of a narration or a discourse might directly imply that there is a sort of an imagined reality, unbounded or potentially infinite, that is described by this internally cohesive discourse. From a mathematical point of view, this problem becomes quite interesting since we could measure the internal cohesion of a text using different statistics than for the consistent description of external or imagined realities, and we may ask whether we can prove that these two distinct statistics are related.

Some intuitive measure of text cohesion can be various statistics pertaining to string repetitions. In fact, concerning string repetitions, there seems to be yet another quantitative law of language which is strongly specific for natural texts as contrasted with unigram texts. This law concerns the maximal repetition. For a given string, the maximal repetition is the maximal length of a substring that is repeated in the string on possibly overlapping positions. For example, the maximal repetition of string "How much wood would a woodchuck chuck if a woodchuck could chuck wood?" is the length of the repeated string " a woodchuck c," which is 14.

Put formally, the maximal repetition of a string $x_1^n \in \mathbb{X}^*$ is defined as

$$L(x_1^n) := \max\{k : x_{i+1}^{i+k} = x_{j+1}^{j+k} \text{ for some } 0 \leq i < j \leq n - k\} \qquad (1.112)$$

Given a string x_1^n, maximal repetition can be computed in time which grows linearly with the length n (Kolpakov and Kucherov, 1999a, b). Thus, we can easily compute the maximal repetition for empirical data.

For empirical data, being some fixed text y_1^m, an interesting Figure, we can make is a plot of the maximal repetition $L(x_1^n)$ with respect to the substring length n for randomly selected substrings X_1^n of the text y_1^m. In particular, we can make such plots for both texts in natural language and some unigram texts, see Figure 1.7. In the plots, we can observe the approximate empirical law

$$L(x_1^n) \propto (\log n)^\alpha \tag{1.113}$$

where $\alpha \approx 1$ for unigram texts and $\alpha \approx 3$ for natural texts. In other words, the growth of maximal repetition is logarithmic for unigram texts and power-law logarithmic for natural texts. In fact, the power-law logarithmic growth of maximal

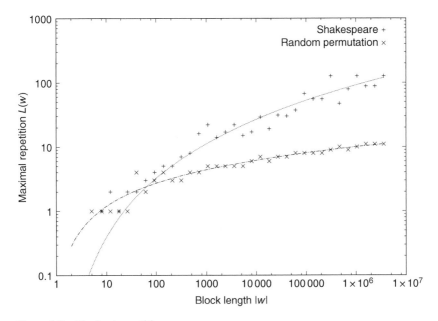

Figure 1.7 Maximal repetition.

Maximal repetition for Shakespeare's First Folio/35 Plays and a random permutation of characters. To smooth the results, we sampled blocks $w = y_{c_n+1}^{c_n+n}$ of length $|w| = n$ from both sources at random assuming a uniform probability distribution on $c_n = 0, 1, \dots,$ $m - n$, where y_1^m is the investigated text. For each length $|w| = n$, only one block $w = y_{c_n+1}^{c_n+n}$ was sampled. The fitted model is $L(y_{c_n+1}^{c_n+n}) \approx 0.03148(\ln n)^{3.036}$ for Shakespeare and $L(y_{c_n+1}^{c_n+n}) \approx 0.4415(\ln n)^{1.187}$ for the random permutation. Source: Author's figure based on the text downloaded from Project Gutenberg (https://www.gutenberg.org).

repetition for natural texts seems to be text-independent and is almost the same for three languages: English, German, and French (Dębowski, 2012b, 2015b). Hence, the rate of growth of maximal repetition may be a language universal, and it may distinguish natural texts from unigram texts.

Consequently, we may ask what the empirical law (1.113) can tell us about the stochastic mechanism of natural language generation. We consider it an important open problem whether the power-law logarithmic growth of maximal repetition (1.113) with $\alpha > 1$ implies in a sufficiently general case that the generating stochastic process with a strictly positive Shannon entropy rate has an infinite excess entropy and obeys a sort of the relaxed Hilberg condition (1.71). We suppose that the power-law logarithmic growth of maximal repetition is a more fundamental property of natural language than the power-law growth of the number of words, i.e. Herdan's law, or the power-law growth of the number of described facts, i.e. natural language being a perigraphic process. In particular, the infinitely complex world repeatedly described in natural texts may emerge in the limit as a mere result of human tendency to imitate previously encountered chunks of text. We cannot exclude the possibility that the world described by texts is by large dynamically created by texts themselves.

Our rigorous results concerning the above hypothesis are modest so far. As we will show now, the power-law logarithmic growth of maximal repetition can be linked to a power-law growth of some generalized block entropies and vanishing of some generalized entropy rates. Let us observe that scaling law (1.113) can be written in a somewhat fancy form as

$$\lim_{n \to \infty} \text{hilb } \mathcal{P}_{\bullet}(L(x_1^n)) \approx \text{const} \tag{1.114}$$

for some generalized block entropy function $H_{\bullet}(k) = k^{1/\alpha}$ and its related entropy power $\mathcal{P}_{\bullet}(k) := 2^{H_{\bullet}(k)}$. In other words, if the Hilberg exponent of function $\mathcal{P}_{\bullet}(L(x_1^n))$ for some generalized entropy $H_{\bullet}(k)$ is bounded away from zero and infinity, then the power-law logarithmic growth of maximal repetition is equivalent to the power-law growth of the generalized entropy function $H_{\bullet}(k)$ and implies in particular that the generalized entropy rate $\lim_{n \to \infty} H_{\bullet}(k)/k$ is zero.

This idea is fruitful to some extent since we can consider different generalized entropies and prove some bounds for the Hilberg exponent of random variables $\mathcal{P}_{\bullet}(L(x_1^n))$ in case when $(X_i)_{i \in \mathbb{Z}}$ is a stationary process. In particular, if $H_{\bullet}(k)$ is a growing function of block length k, then the Hilberg exponent of random variable $\mathcal{P}_{\bullet}(L(x_1^n))$ is almost surely constant for a stationary ergodic process. Moreover, in the case of a stationary ergodic process, we obtain two important bounds, a lower and an upper one,

$$\lim_{n \to \infty} \text{hilb } \mathcal{P}(L(X_1^n)) \geq 1 \text{ almost surely} \tag{1.115}$$

$$\lim_{n \to \infty} \text{hilb } \mathcal{P}_{\infty}^{\text{cond}}(L(X_1^n)) \leq 2 \text{ almost surely} \tag{1.116}$$

where $\mathcal{P}(k) := 2^{H(k)}$ is the power of the Shannon block entropy $H(k) := H(X_1^k)$ and $\mathcal{P}_\infty^{\text{cond}}(k) := 2^{H_\infty^{\text{cond}}(k)}$ is the power of the conditional block min-entropy

$$H_\infty^{\text{cond}}(k) := -\log \mathbf{E} \max_{x_1^k} P(X_1^k = x_1^k | X_{k+1}^\infty) \tag{1.117}$$

Since the difference between entropies $H(k)$ and $H_\infty^{\text{cond}}(k)$ can be arbitrarily large, we can ask if the gap between the lower and the upper bound for the maximal repetition can be narrowed down. The natural next step is to consider other generalized entropies, such as the Rényi entropies (Rényi, 1961). As we will see later in Chapter 9, this idea can be partly carried out.

In this way, using the scaling of maximal repetition, we can show that natural language is not a typical finite-state process. The reasoning goes as follows. Processes that satisfy condition

$$H_\infty^{\text{cond}}(k) \geq Bk \tag{1.118}$$

for a $B > 0$ are called finite energy. In fact, finite energy processes include typical hidden Markov processes. For clarity, consider a hidden Markov probability distribution (1.18). We will prove in Section 11.1 that this distribution is finite energy if the emission probabilities satisfy

$$\sup_{x_i, y_i} e(x_i | y_i) < 1 \tag{1.119}$$

Thus, we can formally demonstrate that natural language is not a hidden Markov process which satisfies condition (1.119). Simply, as we have stated in this paragraph, these hidden Markov processes are finite energy, whereas by bound (1.116), the power-law logarithmic growth of maximal repetition (1.113) with an exponent $\alpha > 1$ excludes the class of finite energy processes.

As we will show in Section 11.1, also the Santa Fe processes, introduced in Section 1.11, are finite energy and at least for that reason, they cannot be adequate models of natural language. In Section 11.4, we will exhibit another class of processes, called RHA processes, which satisfy the power-law logarithmic growth of maximal repetition (1.113) with an exponent $\alpha > 1$. Unfortunately, in the considered particular examples, these processes have zero Shannon entropy rate, $h = 0$, so they cannot be adequate models of natural language, either. It is an open problem how to construct a more linguistically motivated class of stochastic processes.

However, before we reject the particular examples of RHA processes discussed in Section 11.4, we should note that these processes satisfy not only the power-law logarithmic growth of maximal repetition (1.113) but also the power-law growth of mutual information, as stated in the relaxed Hilberg hypothesis (1.71). Moreover, these particular processes are nonergodic, probably they are strongly nonergodic, and we suppose that their ergodic components are also perigraphic almost surely. Thus, the general model of RHA processes can be possibly used further to mathematically study the relationship between the meaningfulness by a consistent description and the meaningfulness by an internal cohesion, as sketched in the beginning of this section. Such a study is also left as an open field of future research.

1.14 Decay of Correlations

The relaxed Hilberg hypothesis (1.71) and the power-law logarithmic growth of maximal repetition (1.113) with an exponent $\alpha > 1$ suggest that natural texts should be described by a stochastic process with a sort of a long memory. Thus, conditions (1.71) and (1.113) should be compared with some more standard ways of defining long memory in stochastic processes, such as those discussed in time series analysis. In time series analysis, which mostly deals with real-valued stochastic processes, a pretty standard way of defining stationary processes with a long-range dependence is through their autocorrelation function. The autocorrelation function of a real-valued stationary process $(Y_i)_{i \in \mathbb{Z}}$ is defined as

$$\rho(n) := \mathrm{Corr}(Y_0, Y_n) \tag{1.120}$$

where we use the correlation coefficient

$$\mathrm{Corr}(X, Y) := \frac{EXY - EXEY}{\sqrt{EX^2 - (EX)^2}\sqrt{EY^2 - (EY)^2}} \tag{1.121}$$

The process is usually called to exhibit the long-range dependence if the series of autocorrelations does not converge, $\sum_{n=-\infty}^{\infty} \rho(n) = \infty$, or simplifying somewhat the phenomenology, if $\rho(n) \propto n^{-\gamma}$ for an exponent $\gamma \leq 1$ (Beran, 1994). This property has a relatively simple expression in the information-theoretic terms if the process is Gaussian. Namely, for complex-valued Gaussian processes, the Shannon mutual information between two random variables separated by n positions can be expressed as

$$I(Y_0; Y_n) = -\log(1 - |\rho(n)|^2) \tag{1.122}$$

In other words, we have a long-range dependence if there holds $I(Y_0; Y_n) \propto n^{-2\gamma}$ for a parameter $\gamma \leq 1$.

In contrast, we can observe by (1.47) and (1.50) that process $(Y_i)_{i \in \mathbb{Z}}$ has an infinite excess entropy if $\sum_{n=1}^{\infty} nI(Y_0; Y_n | Y_1^{n-1}) = \infty$, or simplifying somewhat the phenomenology, if $I(Y_0; Y_n | Y_1^{n-1}) \propto n^{-2\delta}$ for a parameter $\delta \leq 1$. We may further observe that for a complex-valued Gaussian process $(Y_i)_{i \in \mathbb{Z}}$, we have

$$I(Y_0; Y_n | Y_1^{n-1}) = -\log(1 - |\alpha(n)|^2) \tag{1.123}$$

where use the partial autocorrelation function

$$\alpha(n) := \mathrm{Corr}(Y_0 - E[Y_0 | Y_1^{n-1}], Y_n - E[Y_n | Y_1^{n-1}]) \tag{1.124}$$

with the formula applying the concept of conditional expectation, to be explained in Section 3.5. In other words, process $(Y_i)_{i\in\mathbb{Z}}$ has the infinite excess entropy if $\sum_{n=1}^{\infty} n|\alpha(n)|^2 = \infty$, or roughly if $\alpha(n) \propto n^{-\delta}$ for an exponent $\delta \leq 1$.

It is also a fortunate case of the Gaussian processes that the autocorrelation functions $\rho(n)$ and $\alpha(n)$ remain in a one-to-one correspondence given by the Durbin–Levinson algorithm. Using this relationship, it can be shown in particular that we have equality

$$\sum_{n=-\infty}^{\infty} (\pm 1)^n \rho(n) = \prod_{n=1}^{\infty} \frac{1 + (\pm 1)^n \alpha(n)}{1 - (\pm 1)^n \alpha(n)} \qquad (1.125)$$

if the series of absolute partial autocorrelations converges, $\sum_{n=1}^{\infty} |\alpha(n)| < \infty$ (Dębowski, 2007a). Hence, in the latter case, the series of plain correlations also converges, $\sum_{n=-\infty}^{\infty} \rho(n) < \infty$. In other words, simplifying somewhat phenomenology, if a stationary Gaussian process $(Y_i)_{i\in\mathbb{Z}}$ exhibits the long-range dependence, then usually it has the infinite excess entropy. The converse need not be true.

We can ask whether these results can be generalized to a discrete-valued stationary process $(X_i)_{i\in\mathbb{Z}}$. Unfortunately, in this case, we do not have an analogue of the Durbin–Levinson algorithm. Consequently, the mutual information $I(X_0; X_n)$ and the conditional mutual information $I(X_0; X_n | X_1^{n-1})$ cannot be simply related. But by the Jensen inequality (Theorem 3.14) and inequality $\log x \leq (x-1)\log e$, we may bound both by

$$I(X_0; X_n) \leq r(n)\log e, \quad r(n) := \mathbf{E}\left[\frac{P(X_0, X_n)}{P(X_0)P(X_n)} - 1\right] \qquad (1.126)$$

$$I(X_0; X_n | X_1^{n-1}) \leq a(n)\log e, \quad a(n) := \mathbf{E}\left[\frac{P(X_0, X_n | X_1^{n-1})}{P(X_0 | X_1^{n-1})P(X_n | X_1^{n-1})} - 1\right]$$
$$(1.127)$$

Assume that process $(X_i)_{i\in\mathbb{Z}}$ is an irreducible and aperiodic Markov process of order 1 over a finite alphabet. As shown, for example by Lin and Tegmark (2017), in this case, there exist constants $0 \leq \lambda_r < 1$ and $C_r > 0$ such that

$$r(n) \leq C_r \lambda_r^n \qquad (1.128)$$

Hence, mutual information $I(X_0; X_n)$ for these processes decays exponentially with separation n. The same holds also for finite-state processes with an irreducible and aperiodic underlying Markov process, by the data-processing inequality (i.e. claim 6 of Theorem 5.8). Moreover, for any Markov process of order 1, we have $P(X_0, X_n | X_1^{n-1}) = P(X_0 | X_1^{n-1})P(X_n | X_1^{n-1})$ and thus $a(n) = 0$. Hence, the conditional mutual information satisfies $I(X_0; X_n | X_1^{n-1}) = 0$ for these processes.

In contrast, Lin and Tegmark (2017) observed empirically that for samples of natural language texts, we have probably a power law decay of the unconditional mutual information between two symbols,

$$I(X_0; X_n) \propto n^{-1/2} \tag{1.129}$$

Moreover, if the relaxed Hilberg hypothesis (1.71) is true, then the conditional mutual information between two symbols satisfies

$$I(X_0; X_n | X_1^{n-1}) = -\Delta^2 H(n) \propto n^{-3/2} \tag{1.130}$$

It would be interesting to learn what kind of stochastic processes can satisfy condition (1.129). As argued in the previous paragraph, such processes cannot be typical finite-state processes, which provides another argument that natural language cannot be modeled properly by a finite-state process. Searching for a simple example of a process that satisfies condition (1.129), Lin and Tegmark (2017) proposed a half-formal construction which resembles somewhat the RHA processes, which we have mentioned in the previous section and which will be discussed in detail in Section 11.4. Showing that RHA processes actually satisfy condition (1.129) is left as an open problem.

1.15 Recapitulation

In the preceding sections, we have seen that natural texts, i.e. texts written in natural language, exhibit certain statistical regularities. Some of these regularities are similar to statistical regularities of unigram texts, i.e. random permutations of characters, whereas some are strikingly different. Since the question of statistical language modeling is of a general importance for artificial intelligence, as mathematicians, we may ask whether we can explicitly construct stochastic processes which obey some of these language-specific statistical regularities. We should ask, as well, whether our constructions may be useful to engineers who seek for better statistical models of language to implement them in computers. With these questions in mind, in this book, we will develop some theory and construct toy examples of stochastic processes. We hope to draw interest of professional mathematicians as primary intended readers of this book.

In our search for statistical models of natural texts, in this book, we will confine ourselves to processes that satisfy the Birkhoff ergodic theorem, i.e. we will consider only stationary and asymptotically mean stationary processes. Honestly, we do not know whether the process of human culture or the stream of consciousness can be reasonably modeled by such processes, and we cannot exclude that the real probability model of language may lie out of these classes. Nevertheless, we think that stationary and asymptotically mean stationary processes can capture a few linguistically relevant phenomena, and they offer a good starting point to investigate what is mathematically possible. The edifice of science is extended by considering simpler cases first.

Another question is that we will be considering mostly computable processes. As we have said, there is a tension between the frequentist and the Bayesian interpretations of probability. According to the frequentist view, language should be likely modeled by an ergodic and uncomputable process, whereas according to the Bayesian view, language should be likely modeled by a recursive but nonergodic process. We have seen that the distinction between the frequentist and the Bayesian interpretation corresponds approximately to the distinction between the E-language and the I-language made by Noam Chomsky. These two interpretations of probability are just two different perspectives onto the single phenomenon of language, as formally reconciled by the Birkhoff ergodic theorem. While doing the scientific research, we should primarily focus on the I-language and look for a finite computable description of this phenomenon. As Wilhelm von Humboldt, a German linguist, once put it, human language should make *infinite use of finite means* (*unendlicher Gebrauch von endlichen Mitteln* in German) (von Humboldt, 1836).

A book about statistical language modeling could be written in a few different styles, touching different topics and particular problems. This book tries to match pure mathematics and quantitative linguistics since we suppose that a good statistical theory of language should be capable of explaining some formally defined quantitative laws obeyed by natural texts. In this chapter, we have identified four such quantitative phenomena which are likely satisfied by texts in natural language:

- The strictly positive Shannon entropy rate,

$$h = \lim_{n \to \infty} \frac{H(X_1^n)}{n} > 0 \qquad (1.131)$$

- The power-law growth of block mutual information,

$$I(X_1^n; X_{n+1}^{2n}) \propto n^\beta, \quad 0 < \beta < 1 \qquad (1.132)$$

- The power-law logarithmic growth of maximal repetition,

$$L(X_1^n) \propto (\log n)^\alpha, \quad \alpha > 1 \qquad (1.133)$$

- The power-law decay of mutual information between individual symbols,

$$I(X_0; X_n) \propto n^{-2\gamma}, \quad \gamma > 0 \qquad (1.134)$$

In this book, we will mostly analyze condition (1.132) – via the theorems about facts and words – and, to less extent, condition (1.133) – via Rényi entropies. We will not try to analyze condition (1.134), although it may be somehow connected to the previous two. We suppose that provided some additional assumptions such as ergodicity and condition (1.131) hold, conditions related to (1.133) and (1.134) may imply a condition related to (1.132).

We consider proving these implications an important open problem. We suppose that the power-law logarithmic growth of maximal repetition (1.133) and the power-law decay of mutual information between individual symbols (1.134) are more fundamental properties of natural language than the power-law growth of block mutual information (1.132). These more fundamental laws may also possibly imply the power-law growth of the number of distinct words, i.e. Herdan's law, and the power-law growth of the number of described facts, i.e. the natural language being a perigraphic process. In other words, the unbounded or potentially infinite imagined reality repeatedly described in texts created by humans may emerge as a sole result of our tendency to repeat previously generated chunks of text. We cannot either exclude that the reality described in texts is actively created by texts themselves.

Of course, it is desirable to exhibit some linguistically motivated examples of processes that satisfy conditions (1.131), (1.132), (1.133), and (1.134). Our two toy examples of processes, to be analyzed in the final Chapter 11, are Santa Fe processes, already introduced in Section 1.11, and RHA processes, mentioned in Section 1.13. The Santa Fe processes obey conditions (1.131) and (1.132) but do not obey (1.133). In contrast, the RHA processes satisfy both conditions (1.132) and (1.133) but do not satisfy (1.131). Moreover, as sketched by Lin and Tegmark (2017), condition (1.134) may be satisfied by processes resembling the RHA processes. Thus, we still do not have an example of a process that satisfies all four conditions (1.131)–(1.134).

There are a few other open problems that we came across when writing this book. The list of all open problems that we have formulated has been presented at the end of the book. Many of these problems can be stated in a relatively simple way, which indicates that there is still much to do at the interface of natural language and stochastic processes. We hope that this domain will attract more researchers among mathematicians. This book should not be treated as a final report but rather as an inspiration for further work. Moreover, although the book is relatively self-contained, we are quite sure that we have not gathered here all relevant background material for future research.

2

Probabilistic Preliminaries

From this chapter on, we will start building rigorous mathematical constructions, commencing with more abstract notions (Chapters 2–7) and subsequently heading toward idealized models of natural language (Chapters 8–11). In principle, we will not formally refer to the notations introduced in the half-formal Chapter 1. However, we will suppose that the reader has read that chapter, is acquainted with the general ideas, and is interested in the formal exposition. From time to time, we will recall some motivations from Chapter 1, and we will connect them to formal concepts.

In this chapter, we will make the first steps toward measure-theoretic probability, and we will construct some basic stochastic processes of theoretical importance. We assume that the reader is somewhat familiar with elementary probability calculus. To remind, the elementary probability calculus deals with probability on countable spaces of possible outcomes. If we have a collection of certain elementary events $x \in \mathbb{X}$, where set $\mathbb{X}$ is countable, we can define a discrete probability distribution $p(x)$ as an arbitrary function such that $p(x) \geq 0$ for each $x \in \mathbb{X}$ and $\sum_{x \in \mathbb{X}} p(x) = 1$. The underlying idea is that we find the finest partition of the space of interest, and we define the probability distribution for the smallest elements of this space.

The above approach leads to many important intuitions about probability and works well for all countable sets, both finite and infinite, such as the set of all finite sequences $\mathbb{X}^*$, but it cannot be carried out for uncountably infinite sets, such as the sets of infinite sequences $\mathbb{X}^{\mathbb{T}}$, where $\mathbb{T} = \mathbb{N}$ or $\mathbb{T} = \mathbb{Z}$. Such sets of infinite sequences arise naturally, however, when we want to investigate statistical properties of texts of an unbounded length and study possible mechanisms of generating them.

In the latter case, probabilities of elementary events $(x_i)_{i \in \mathbb{T}} \in \mathbb{X}^{\mathbb{T}}$ are equal to zero in many interesting cases, and we cannot reconstruct probabilities of arbitrary unions of elementary events by simply adding their probabilities. A simple solution to this problem is to define probabilities for subsets of infinite sequences

Information Theory Meets Power Laws: Stochastic Processes and Language Models, First Edition.
Łukasz Dębowski.
© 2021 John Wiley & Sons, Inc. Published 2021 by John Wiley & Sons, Inc.

$\mathbb{X}^{\mathbb{T}}$ and to require that so defined function be additive, or even countably additive – to ensure continuity. However, in this setting, there appears another problem, that we cannot reasonably define probabilities for certain sets, see Problem 2.1.

Thus, we need a more refined and quite rigorous foundation for probability calculus. The appropriate construction dates back to the first systematic exposition by Kolmogoroff (1933) and is known as probability measure. There are many excellent in-depth introductions to probability measures, e.g. by Billingsley (1979), Breiman (1992), and Kallenberg (1997). Whereas in this book, we will prove only selected facts concerning the foundations of measure-theoretic probability, the interested reader may find the remaining proofs in the mentioned textbooks. We also abstain from making systematic historical remarks regarding fundamental concepts of measure-theoretic probability. The reader interested in history is referred to von Plato (1994), Kallenberg (1997), and Shafer and Vovk (2006).

More detailed contents of this chapter are as follows: In Section 2.1, we will be preoccupied with a general concept of a probability measure. Modern probability theory is a theory of probability measures, and it is important to state that from a mathematical point of view, there are many alternative probability measures. This stays in contrast with some preformal attitudes, where we are inclined to ask what the objective probability is of this or that particular event in a physical world. Contrary to this intuition, it makes sense to ask what other imaginable probabilities are. This can shed some light onto how we would like to understand not only objective probabilities in the physical world but also subjective probabilities in our minds.

Another important difference between the preformal intuitions and the formal approach to probability is the space of events on which the probabilities live. The physical world or the psychological realm is a sort of a magma, subject to further investigations of an independent interest, whereas mathematical models of probability live on some well-grounded spaces of elementary events, such as infinite sequences of symbols from a countable or a finite alphabet. Formally, we can conceptualize such spaces as products of smaller spaces. This will be the topic of Section 2.2.

Entire elementary events, such as infinite sequences of symbols, may be inaccessible to our direct precise measurements. Hence, there is a need to define some sampling procedures which return graspable outcomes with well-defined probabilities. In modern probability theory, random variables are some standard models of such procedures. Random variables can be concatenated back into infinite sequences, which are known as stochastic processes. General random variables and stochastic processes will be made precise in Section 2.3.

The world of stochastic processes is extremely rich, much richer than the contents of this book. As a primer, in Section 2.4, we will discuss some simple examples of stochastic processes such as independent identically distributed (IID), Markov, and hidden Markov processes. For various reasons, different for each class – such as too much independence or too little memory, these processes cannot be realistic models of natural language. They are, however, important primary models to consider so as to learn to imagine what else could be mathematically possible. We should treat them as useful building blocks or first steps in the ladder of more complex and more complicated phenomena.

2.1 Probability Measures

In this section, we will define probability measures and some related concepts. Let us fix a certain set Ω, which will be called the event space. Elements $\omega \in \Omega$ will be called points or elementary events. Usually, set Ω will consist of uncountably many points, and for many practical applications of probability, it does not matter what these points are exactly. The general intuition, however, is that individual points $\omega \in \Omega$ carry the maximal possible information about an outcome of a random experiment under consideration. Thus, in the context of applications of probability, we may imagine that points $\omega \in \Omega$ are different states of the entire physical world, whereas the event space Ω is the set of possible worlds. For effective mathematical probability models, however, we usually imagine the event space Ω as something better formally defined, such as $\Omega = \mathbb{X}^{\mathbb{Z}}$ being the space of two-sided infinite sequences. We have encountered this space in the introductory Section 1.5.

In any case, subsets $A \subseteq \Omega$ will be called events. Set $2^{\Omega} := \{A : A \subseteq \Omega\}$ is called the power set of set Ω. Sets consisting of events, such as the power set 2^{Ω}, will be called classes (of events or of sets). Quite often, class 2^{Ω} is too large to define an interesting probability measure. We have to consider classes of events which are strictly smaller. There are several interesting kinds of smaller classes.

Definition 2.1 *(π-system, field, σ-field)*: For a class of events $\mathcal{J} \subseteq 2^{\Omega}$ consider conditions:

1. $\Omega \in \mathcal{J}$,
2. $A, B \in \mathcal{J}$ implies $A \cap B \in \mathcal{J}$,
3. $A \in \mathcal{J}$ implies $A^c \in \mathcal{J}$, where $A^c := \Omega \backslash A$,
4. $A_1, A_2, \ldots \in \mathcal{J}$ implies $\bigcap_{n \in \mathbb{N}} A_n \in \mathcal{J}$.

Class $\mathcal{J}$ is called a π-system if it satisfies conditions (1)–(2), a field–if it satisfies conditions (1)–(3), and a σ-field–if it satisfies conditions (1)–(4).

For a σ-field $\mathcal{J}$, pair $(\Omega, \mathcal{J})$ is called a measurable space. Class $\{\emptyset, \Omega\}$ is the smallest σ-field, whereas the power set 2^Ω is the largest σ-field. We note that singleton sets $\{\omega\}$ for points $\omega \in \Omega$ need not belong to a particular σ-field.

Sets A_i for certain $i \in I$ are called disjoint if their intersections are empty, i.e. $A_i \cap A_j = \emptyset$ for $i, j \in I$ and $i \neq j$. A finite measure is a real function of sets which is countably additive for unions of disjoint sets. According to the measure-theoretic approach to probability calculus, probability is a special case of a measure which is normed to 1. The formal definition is as follows:

Definition 2.2 *(Finite and probability measures)*: For a field $\mathcal{J}$ and a function $\mu : \mathcal{J} \to \mathbb{R}$ consider conditions:

1. $\mu(A \cup B) = \mu(A) + \mu(B)$ for disjoint $A, B \in \mathcal{J}$ such that $A \cap B = \emptyset$;
2. $\mu\left(\bigcup_{n \in \mathbb{N}} A_n\right) = \sum_{n \in \mathbb{N}} \mu(A_n)$ for disjoint $A_i \in \mathcal{J}$ and $\bigcup_{n \in \mathbb{N}} A_n \in \mathcal{J}$;
3. $\mu(A) \geq 0$ for $A \in \mathcal{J}$;
4. $\mu(\Omega) = 1$.

Condition (1) is called finite additivity, whereas condition (2) is called countable additivity. Function μ is called a finite measure if it satisfies conditions (1)–(3), whereas it is called a probability measure if it satisfies conditions (1)–(4).

It turns out that countable additivity does not follow from finite additivity and must be assumed independently. Let us stress that in the above definitions, σ-fields $\mathcal{J} \neq 2^\Omega$ are considered on purpose since there exist interesting probability measures which are defined on some σ-field $\mathcal{J} \neq 2^\Omega$ and cannot be extended meaningfully or uniquely beyond this domain. As shown in Billingsley (1979, section 3), an example of such a measure is the Lebesgue measure, to be discussed in Section 3.1.

Thus, probability theory is a special case of measure theory, with its own particular terminology and special conventions. Probability measures usually will be written as P, with some decorations if needed. Usually, we will assume that the measurable space $(\Omega, \mathcal{J})$ is large enough to account for all random phenomena that we are interested in. Triple $(\Omega, \mathcal{J}, P)$, where $(\Omega, \mathcal{J})$ is a measurable space, and P is a probability measure is called a probability space. Similarly, a triple $(\Omega, \mathcal{J}, \mu)$, where $(\Omega, \mathcal{J})$ is a measurable space and μ is a finite measure is called a finite measure space.

Usually, we will work with fields and σ-fields that are generated by some particular classes of sets.

Definition 2.3 *(Generated π-system, field, σ-field)*: For an arbitrary class $\mathcal{A}$, symbol $\pi(\mathcal{A}) \subseteq 2^\Omega$ denotes the π-system containing the sets obtained from elements of $\mathcal{A} \cup \{\Omega\}$ by taking a finite number of intersections. Symbol $\sigma_0(\mathcal{A}) \subseteq 2^\Omega$

denotes the field containing the sets obtained from elements of $\mathcal{A}$ by taking a finite number of intersections and complements. In contrast, symbol $\sigma(\mathcal{A})$ denotes the intersection of all σ-fields $\mathcal{J} \subseteq 2^{\Omega}$ such that $\mathcal{A} \subseteq \mathcal{J}$. Classes $\pi(\mathcal{A})$, $\sigma_0(\mathcal{A})$, and $\sigma(\mathcal{A})$ will be called the π-system, the field, and the σ-field generated by class $\mathcal{A}$, respectively.

We have $\sigma_0(\pi(\mathcal{A})) = \sigma_0(\mathcal{A})$ and $\sigma(\sigma_0(\mathcal{A})) = \sigma(\mathcal{A})$. If class $\mathcal{A}$ is finite, then field $\sigma(\mathcal{A}) = \sigma_0(\mathcal{A})$ is finite as well. The next natural step is to consider a countably infinite generating class.

Definition 2.4 *(Countably generated field and σ-field)*: We say that a field $\mathcal{G}_0$ is countably generated if there is a countable class $\mathcal{A} = \{A_1, A_2, ...\}$ such that $\mathcal{G}_0 = \sigma_0(\mathcal{A})$. Similarly, we say that a σ-field $\mathcal{G}$ is countably generated if there is a countable class $\mathcal{A} = \{A_1, A_2, ...\}$ such that $\mathcal{G} = \sigma(\mathcal{A})$.

Similarly, a probability space $(\Omega, \mathcal{J}, P)$, where $\mathcal{J}$ is countably generated, is called countably generated. Countably generated probability spaces enjoy certain nice properties, such as existence of conditional probability measures and approximability of arbitrary sub-σ-fields by countably generated σ-fields, to be discussed in Sections 3.5 and 3.7. Consequently, these properties have some further implications for ergodic decomposition and Shannon information measures, to be discussed in Sections 4.4 and 5.4. In contrast to the case of a finite class $\mathcal{A}$, if class $\mathcal{A}$ is countably infinite, then classes $\sigma_0(\mathcal{A})$ and $\sigma(\mathcal{A})$ need not be equal. Whereas the generated field $\sigma_0(\mathcal{A})$ is countable in this case, the generated σ-field $\sigma(\mathcal{A})$ need not be so.

The elements of the generated field $\sigma_0(\mathcal{A})$ can be homeomorphically mapped into the set of well-formed formulas in the propositional calculus – for the logical variables corresponding to the individual elements of class $\mathcal{A}$. Thus, the following fact known from logic characterizes the elements of the generated field.

Theorem 2.1 *(Disjunctive normal form)*: *Any set $B \in \sigma_0(\mathcal{A})$ can be written as a finite union $B = \bigcup_{i=1}^{n} B_i$, where sets B_i are disjoint and $B_i = C_{i1} \cap \cdots \cap C_{ik}$, where $C_{ij} = D_j$ or $C_{ij} = D_j^c$ and $D_j \in \mathcal{A}$.*

Proof: See Lemma 3.4.1 of Cohn (2003). □

The above characterization is not usually discussed in expositions of measure-theoretic probability. To satisfy the curiosity of an intrigued more informed reader, the above fact will be used in the proofs of some theorems in Section 4.3, concerning effective criteria for ergodic and mixing probability measures. The standard proofs of these facts apply different tools.

Another way of looking onto field $\sigma_0(\mathcal{A})$ when $\mathcal{A}$ is countable is as follows:

Theorem 2.2: *For a countable class $\mathcal{A} = \{A_1, A_2, ...\}$, we have*

$$\sigma_0(\mathcal{A}) = \bigcup_{n=1}^{\infty} \sigma(\{A_1, ..., A_n\}) \tag{2.1}$$

Proof: Clearly, $B \in \sigma_0(\mathcal{A})$ if and only if $B \in \sigma(\{A_1, ..., A_n\})$ for some n. □

In contrast, there are no simple characterizations of the generated σ-field. Whereas the generated field $\sigma_0(\mathcal{A})$ corresponds to the set of well-formed (i.e. finite in particular) formulas in the propositional calculus, the generated σ-field $\sigma(\mathcal{A})$ for an infinite $\mathcal{A}$ seems to correspond to a set of infinite formulas. This feature is not a drawback, but an advantage, since in this way, we may discuss probabilities of some limiting events – once we have defined a suitable probability measure.

Now, we may ask how we can effectively define a probability measure on an infinite σ-field. In general, to uniquely define a probability measure on a σ-field $\mathcal{J}$, it suffices to define its values on any π-system $\mathcal{A}$ such that $\sigma(\mathcal{A}) = \mathcal{J}$. Namely, we have this general fact.

Theorem 2.3 *(π-λ theorem):* *Let P_1 and P_2 be two probability measures on a measurable space $(\Omega, \mathcal{J})$ such that $\sigma(\mathcal{A}) = \mathcal{J}$ for a certain π-system $\mathcal{A}$. If $P_1(A) = P_2(A)$ for all $A \in \mathcal{A}$ then $P_1 = P_2$.*

Proof: See Billingsley (1979, theorem 3.3). □

The π–λ theorem does not answer, however, whether probability measures on a given measurable space exist at all. It only asserts that if we already have a probability measure on a σ-field $\mathcal{J}$ then it is uniquely determined by its values on the generating π-system. In contrast, if we have a probability measure on a field $\mathcal{K}$ such that $\sigma(\mathcal{K}) = \mathcal{J}$, then according to the next result, called the extension theorem, there exists a unique extension of this probability measure to σ-field $\mathcal{J}$.

Theorem 2.4 *(Extension theorem):* *Let P be a probability measure on a field $\mathcal{K}$, where $\sigma(\mathcal{K}) = \mathcal{J}$. Then there exists a unique measure $\tilde{P}$ on the measurable space $(\Omega, \mathcal{J})$ such that $\tilde{P}(A) = P(A)$ for all $A \in \mathcal{K}$.*

Proof: See Billingsley (1979, theorem 3.1). □

Theorems 2.3 and 2.4 do not prescribe yet how to define a probability measure on a π-system or a field that generate the σ-field $\mathcal{J}$. In Section 2.2, we will see how this can be done effectively by decomposing the probability space into smaller pieces.

2.2 Product Measurable Spaces

Some effective approach to constructing probability measures consists in build-ing infinite probability spaces from finite probability spaces. Suppose that ele-mentary events $\omega \in \Omega$ are collections of some values, $\omega = (x_i)_{i \in \mathbb{T}}$, where $\mathbb{T}$ is an arbitrary set of indices and $x_i \in \mathbb{X}_i$ for some arbitrary sets $\mathbb{X}_i$. Suppose that we have measurable spaces $(\mathbb{X}_i, \mathcal{X}_i)$. We will construct the product measurable space $(\Omega, \mathcal{J}) = (\prod_{i \in \mathbb{T}} \mathbb{X}_i, \bigotimes_{i \in \mathbb{T}} \mathcal{X}_i)$, where the operations are defined as follows:

Definition 2.5 *(Product measurable space)*: We define Cartesian products

$$\prod_{i \in \mathbb{T}} \mathbb{X}_i := \{(x_i)_{i \in \mathbb{T}} : x_j \in \mathbb{X}_j, j \in \mathbb{T}\} \tag{2.2}$$

$$\prod_{i \in \mathbb{T}} \mathcal{X}_i := \{\{(x_i)_{i \in \mathbb{T}} : x_j \in A_j, j \in \mathbb{T}\} : A_k \in \mathcal{X}_k, k \in \mathbb{T}\} \tag{2.3}$$

Moreover, we define cylinder sets

$$[A]_k^{\mathbb{T}} := \left\{(x_i)_{i \in \mathbb{T}} \in \prod_{i \in \mathbb{T}} \mathbb{X}_i : x_k \in A\right\} \tag{2.4}$$

$$[x]_k^{\mathbb{T}} := \left\{(x_i)_{i \in \mathbb{T}} \in \prod_{i \in \mathbb{T}} \mathbb{X}_i : x_k = x\right\} \tag{2.5}$$

Having these, we define the product σ-field

$$\bigotimes_{i \in \mathbb{T}} \mathcal{X}_i := \sigma(\{[A]_k^{\mathbb{T}} : A \in \mathcal{X}_k, k \in \mathbb{T}\}) \tag{2.6}$$

For $\mathbb{T} = \{1, \dots, k\}$, we write

$$\mathbb{X}_1 \times \dots \times \mathbb{X}_k := \prod_{i \in \mathbb{T}} \mathbb{X}_i \tag{2.7}$$

$$\mathcal{X}_1 \times \dots \times \mathcal{X}_k := \prod_{i \in \mathbb{T}} \mathcal{X}_i \tag{2.8}$$

$$\mathcal{X}_1 \otimes \dots \otimes \mathcal{X}_k := \bigotimes_{i \in \mathbb{T}} \mathcal{X}_i \tag{2.9}$$

whereas for $\mathbb{X}_i = \mathbb{X}$ and $\mathcal{X}_i = \mathcal{X}$, we write

$$\mathbb{X}^{\mathbb{T}} := \prod_{i \in \mathbb{T}} \mathbb{X} \tag{2.10}$$

$$\mathcal{X}^{\mathbb{T}} := \bigotimes_{i \in \mathbb{T}} \mathcal{X} \tag{2.11}$$

In this book, we will work only with countably generated product spaces. Let us note a simple fact which asserts when product spaces are countably generated.

Theorem 2.5: *For a countable set* $\mathbb{T}$, *countable sets* $\mathbb{X}_i$, *and* σ-*fields* $\mathcal{X}_i = \sigma(\{\{x\}:$ $x \in \mathbb{X}_i\})$, *measurable space* $(\prod_{i \in \mathbb{T}} \mathbb{X}_i, \bigotimes_{i \in \mathbb{T}} \mathcal{X}_i)$ *is countably generated.*

Proof: The class of cylinders $\mathcal{A} = \{[x]_k^{\mathbb{T}} : x \in \mathbb{X}_k, k \in \mathbb{T}\}$ is countable, and we have $\bigotimes_{i \in \mathbb{T}} \mathcal{X} = \sigma(\mathcal{A})$, so the claim follows by definition. $\qquad \square$

There are two main theorems which answer how to define a probability measure on a product measurable space. The first one is the product measure theorem, which states that we can always assume probabilistic independence for an arbitrary number of combined spaces. To recall from elementary probability calculus, events A and B are called probabilistically independent when

$$P(A \cap B) = P(A)P(B) \tag{2.12}$$

We will generalize this relation to an arbitrary number of events and σ-fields.

Theorem 2.6 *(Product measure):* *Suppose that we have probability spaces* $(\mathbb{X}_i, \mathcal{X}_i, P_i)$, *where* $i \in \mathbb{T}$ *and* $\mathbb{T}$ *is an arbitrary set. Then, there exists a unique probability measure* P *on the measurable space* $(\prod_{i \in \mathbb{T}} \mathbb{X}_i, \bigotimes_{i \in \mathbb{T}} \mathcal{X}_i)$ *such that*

$$P([A_1]_{j_1}^{\mathbb{T}} \cap \cdots \cap [A_k]_{j_k}^{\mathbb{T}}) = P_{j_1}(A_1) \times \cdots \times P_{j_k}(A_k) \tag{2.13}$$

for all cylinder sets $[A_i]_{j_i}^{\mathbb{T}}$, $A_i \in \mathcal{X}_{j_i}$, $j_i \in \mathbb{T}$.

Proof: See Rudin (1974, section 7.7) or Billingsley (1979, theorem 18.2) for finite products and Pap (2002, chapter 2 and section 5.3) for infinite products. $\qquad \square$

We will denote $\prod_{i \in \mathbb{T}} P_i := P$ for the measure P from the above theorem, and will call it the product measure. Space $(\prod_{i \in \mathbb{T}} \mathbb{X}_i, \bigotimes_{i \in \mathbb{T}} \mathcal{X}_i, \prod_{i \in \mathbb{T}} P_i)$ will be called the product probability space.

The existence of product measure is a rationale for a certain thought shortcut. Namely, if we have certain independent probability spaces $(\mathbb{X}_i, \mathcal{X}_i, P_i)$, then we may treat them mentally as some subspaces of a single product probability space $(\prod_{i \in \mathbb{T}} \mathbb{X}_i, \bigotimes_{i \in \mathbb{T}} \mathcal{X}_i, P)$, where $P = \prod_{i \in \mathbb{T}} P_i$. This thought shortcut is applied in Chapters 10 and 11, where we assume that we can always place an asymptotically mean stationary process and its stationary mean as independent processes on the same probability space. We hope that the reader will forgive us this practice, but sometimes it is quicker to grasp two different stochastic processes living on the same probability space than a single stochastic process with two different probability measures.

In the second theorem, called the Kolmogorov process theorem, we will no longer assume probabilistic independence, but we will restrict ourselves to spaces $(\prod_{i \in \mathbb{T}} \mathbb{X}_i, \bigotimes_{i \in \mathbb{T}} \mathcal{X}_i)$ where $\mathbb{X}_i$ are countable sets. A less general version of the

Kolmogorov process theorem was mentioned in Section 1.5 – in connection to stationary processes, which will be discussed in Section 4.2.

Theorem 2.7 *(Kolmogorov process theorem)*: *For an arbitrary set* $\mathbb{T}$ *and countable sets* $\mathbb{X}_i$, *where* $i \in \mathbb{T}$, *let* $p_{j_1 \dots j_k} : \mathbb{X}_{j_1} \times \cdots \times \mathbb{X}_{j_k} \to \mathbb{R}$, *where* $j_i \in \mathbb{T}$, *be functions such that*

1. $p_{j_1 \dots j_k}(x_1 \dots x_k) \geq 0$ *and* $\sum_{x_1 \in \mathbb{X}_{j_1}, \dots, x_k \in \mathbb{X}_{j_k}} p_{j_1 \dots j_k}(x_1 \dots x_k) = 1$;
2. *for any permutation* $\phi : \{1, \dots, k\} \to \{1, \dots, k\}$ *we have* $p_{j_{\phi(1)} \dots j_{\phi(k)}}(x_{\phi(1)} \dots x_{\phi(k)}) = p_{j_1 \dots j_k}(x_1 \dots x_k)$;
3. $\sum_{x_{k+1} \in \mathbb{X}_{j_{k+1}}} p_{j_1 \dots j_k, j_{k+1}}(x_1 \dots x_k x_{k+1}) = p_{j_1 \dots j_k}(x_1 \dots x_k)$.

Then there exists a unique probability measure P on the product measurable space $(\prod_{i \in \mathbb{T}} \mathbb{X}_i, \bigotimes_{i \in \mathbb{T}} \mathcal{X}_i)$, *where* $\mathcal{X}_i = \sigma(\{\{x_i\} : x_i \in \mathbb{X}_i\})$, *such that*

$$P([x_1]_{j_1}^{\mathbb{T}} \cap \cdots \cap [x_k]_{j_k}^{\mathbb{T}}) = p_{j_1 \dots j_k}(x_1 \dots x_k) \tag{2.14}$$

Conversely, if there exists a probability measure P on the measurable space $(\prod_{i \in \mathbb{T}} \mathbb{X}_i, \bigotimes_{i \in \mathbb{T}} \mathcal{X}_i)$, *then functions* $p_{j_1 \dots j_k}(x_1 \dots x_k)$ *defined by equality (2.14) satisfy conditions (1)–(3).*

Proof: See Billingsley (1979, theorems 36.1–36.2). $\qquad\square$

The Kolmogorov process theorem links elementary probability calculus with probability measures. Namely, to construct a probability measure from elementary discrete probability distributions $p_{j_1 \dots j_k}(x_1 \dots x_k)$, it is necessary and it suffices that these distributions satisfy some intuitive consistency conditions. In fact, uniqueness of the probability measure given by Eq. (2.14) follows by the π–λ theorem (Theorem 2.3) since the class of all cylinder sets is a π-system. The nontrivial fact asserted by the Kolmogorov process theorem is that we can actually extend the probability measure beyond the class of cylinder sets.

2.3 Discrete Random Variables

In Section 2.2, we have shown how to construct larger probability spaces from smaller probability spaces. In this section, we will do the converse, namely, we will sample smaller probability spaces from larger probability spaces. The respective sampling operations are called measurable functions or random variables. For example, if elementary events are infinite sequences, some random variable is a function that given an infinite sequence returns the symbol on a fixed position of the sequence. Intuitively, we may say that random variables serve for extracting useful information from elementary events. In a way, random variables are not random – they are deterministic functions of elementary events and they only get interpreted as random through the probability measure defined on certain subsets of elementary events. For this reason, we must be a bit careful about measurability of preimages of random variables.

Let $f : \mathbb{X} \to \mathbb{Y}$ be a function. If there exists a function $f^{-1} : \mathbb{Y} \to \mathbb{X}$ such that $f(x) = y$ if and only if $f^{-1}(y) = x$, function f^{-1} is called the inverse of f. Overloading the notation, for an arbitrary function $f : \mathbb{X} \to \mathbb{Y}$, the preimage of a set $A \subseteq \mathbb{Y}$ under function f is defined as $f^{-1}(A) := \{x \in \mathbb{X} : f(x) \in A\}$.

Definition 2.6 (*Measurable function*): Function f is called measurable from a measurable space $(\mathbb{X}, \mathcal{X})$ to measurable space $(\mathbb{Y}, \mathcal{Y})$ if $f : \mathbb{X} \to \mathbb{Y}$ and $f^{-1}(A) \in \mathcal{X}$ for all $A \in \mathcal{Y}$. Function f is called a measurable bijection from $(\mathbb{X}, \mathcal{X})$ to $(\mathbb{Y}, \mathcal{Y})$ if f is measurable from $(\mathbb{X}, \mathcal{X})$ to $(\mathbb{Y}, \mathcal{Y})$ and the inverse function $f^{-1} : \mathbb{Y} \to \mathbb{X}$ exists and is measurable from $(\mathbb{Y}, \mathcal{Y})$ to $(\mathbb{X}, \mathcal{X})$.

The above definition captures the simple requirement that for a measure space $(\mathbb{X}, \mathcal{X}, \mu)$ and each subset $A \in \mathcal{Y}$ the value of measure $\mu(f \in A) := \mu(f^{-1}(A))$ must be well defined.

Now, we will define random variables. Here, we would also like to preserve measurability of events; hence, we will restrict ourselves to measurable functions.

Definition 2.7 (*Random variable*): Random variable X with an image space $(\mathbb{X}, \mathcal{X})$ on a measure space $(\Omega, \mathcal{J}, \mu)$ is an arbitrary function which is measurable from $(\Omega, \mathcal{J})$ to $(\mathbb{X}, \mathcal{X})$. Set $\mathbb{X}$ is called the image or the alphabet of X. The preimage space of X is the measurable space $(\Omega, X^{-1}(\mathcal{X}))$. In case of a probability space $(\Omega, \mathcal{J}, P)$, probability measure $P_X : \mathcal{X} \ni A \mapsto P(X \in A)$ will be called the distribution of random variable X.

Random variables enjoy a special notation which can be concisely defined as follows: For any random variable being a random proposition $\varphi : \Omega \to \{\text{true, false}\}$, we will use the shorthand notations

$$(\varphi) := \{\omega \in \Omega : \varphi(\omega)\} \tag{2.15}$$

$$\mu(\varphi) := \mu(\{\omega \in \Omega : \varphi(\omega)\}) \tag{2.16}$$

We note that notation $\mu(f \in A) = \mu(f^{-1}(A))$ introduced in the previous paragraph is a special case of this notation.

When we consider some σ-fields $\mathcal{G} \subseteq \mathcal{J}$, it is often a technical requirement that a given random variable X be measurable from $(\Omega, \mathcal{G})$ to $(\mathbb{X}, \mathcal{X})$. This case happens sufficiently often to give it a name.

Definition 2.8 *(𝒢-measurable random variable)*: For brevity, a random variable X with an image space $(\mathbb{X}, \mathcal{X})$ will be called 𝒢-measurable for a σ-field $\mathcal{G} \subseteq \mathcal{J}$ if it is measurable from $(\Omega, \mathcal{G})$ to $(\mathbb{X}, \mathcal{X})$.

As the reader may verify, class $\mathcal{G} = X^{-1}(\mathcal{X})$ is the smallest σ-field such that random variable X is 𝒢-measurable.

There are a few particular standard types of random variables with certain default image spaces. This idea resembles conventions of some programming languages with predefined types of variables, a remark for those readers who do code. We will begin with two types of random variables: discrete random variables and discrete stochastic processes. Real random variables and real random processes, such as random measures, an important concept for our considerations, will be discussed in Sections 3.1 and 3.5.

Discrete random variables are defined as follows:

Definition 2.9 *(Discrete random variable)*: A discrete random variable X is a random variable with an image space $(\mathbb{X}, \mathcal{X})$, where set $\mathbb{X}$ is countable and $\mathcal{X} = \sigma(\{\{x\} : x \in \mathbb{X}\})$. A discrete random variable with a finite image $\mathbb{X}$ is called a simple random variable.

The concept of a discrete random variable is related to the concept of a partition.

Definition 2.10 *(Partition)*: A partition on a measurable space $(\Omega, \mathcal{J})$ is a countable class $\alpha = \{A_1, A_2, ...\}$, where events $A_i \in \mathcal{J}$ are disjoint and $\bigcup_{n \in \mathbb{N}} A_n = \Omega$. The partition is called finite if there are only finitely many events $A_i \neq \emptyset$.

Let us note that discrete random variables $X_i : \Omega \to \mathbb{X}_i$ generate unique partitions $\alpha = \{(X = x) : x \in \mathbb{X}\}$. Moreover, each finite σ-field $\mathcal{A}$ is generated by a unique finite partition α, i.e. $\mathcal{A} = \sigma(\alpha)$. Some natural functions of a partition are Shannon entropy and mutual information, to be discussed in Section 5.1.

To proceed further, according to the next definition, discrete stochastic process are collections of discrete random variables – of the same type.

Definition 2.11 *(Discrete stochastic process)*: A discrete stochastic process $(X_t)_{t \in \mathbb{T}}$ is a collection of discrete random variables X_t indexed by certain indices $t \in \mathbb{T}$ sharing the same image space $(\mathbb{X}, \mathcal{X})$. The image space of process $(X_t)_{t \in \mathbb{T}}$ is pair $(\mathbb{X}^{\mathbb{T}}, \mathcal{X}^{\mathbb{T}})$, where $\mathcal{X}^{\mathbb{T}}$ is the σ-field generated by cylinder sets.

We can see that if X_t are random variables, i.e. they satisfy $X_t^{-1}(\mathcal{X}) \in \mathcal{J}$, then process $(X_t)_{t \in \mathbb{T}}$ is also a random variable, since $(X_t)_{t \in \mathbb{T}}^{-1}(\mathcal{X}^{\mathbb{T}}) \in \mathcal{J}$.

We can realize as well that the Kolmogorov process theorem (Theorem 2.7) is in fact a proposition about stochastic processes. Namely, if we have a probability space $(\mathbb{X}^{\mathbb{T}}, \mathcal{X}^{\mathbb{T}}, P)$ and define random variables

$$X_t : \mathbb{X}^{\mathbb{T}} \ni (x_i)_{i \in \mathbb{T}} \mapsto x_t \in \mathbb{X} \qquad (2.17)$$

then probability measure P can be uniquely defined via joint finitely dimensional distributions $p_{j_1 \dots j_k}(x_1 \dots x_k) = P(X_{j_1} = x_1, \dots, X_{j_k} = x_k)$ of the random variables. This interpretation is valid since cylinder sets $[x_i]_{j_i}^{\mathbb{T}}$ and events $(X_{j_i} = x_i)$ are exactly the same objects, so we have

$$[x_1]_{j_1}^{\mathbb{T}} \cap \dots \cap [x_k]_{j_k}^{\mathbb{T}} = (X_{j_1} = x_1, \dots, X_{j_k} = x_k) \qquad (2.18)$$

The reader may wonder whether introducing the obscure notation for cylinder sets was necessary at all. As we will see further in this book, some variations of this notation can be still useful, see Sections 3.1 and 7.5.

To conclude this section, we note that stochastic processes can be effectively defined in general via Theorem 2.7 or as measurable functions of other stochastic processes. In this book, we will mostly start from a discrete process $(X_t)_{t \in \mathbb{T}}$ and we will construct other random variables as measurable functions thereof.

2.4 From IID to Finite-State Processes

So far, we have discussed quite abstract foundations of measure-theoretic probability. It is high time to give some examples of probability spaces or, equivalently, of some stochastic processes living on these spaces. In this section, we will define simple examples of discrete stochastic processes. These examples are IID processes, Markov processes, and hidden Markov processes, less formally introduced in Section 1.4. These three classes of processes correspond to an increasing complexity of allowed dependence among the process variables. Although they are too simple to model natural language adequately, they are important building blocks for further constructions.

An IID process is simply a collection of independent random variables sharing the same marginal distribution.

Definition 2.12 (*IID process*): A discrete process $(X_i)_{i \in \mathbb{T}}$, where $X_i : \Omega \rightarrow \mathbb{X}$, is called an IID process or a collection of IID random variables if

$$P(X_{j_1} = x_1, \dots, X_{j_k} = x_k) = \prod_{i=1}^{k} P(X_{j_i} = x_i) \qquad (2.19)$$

where $P(X_{j_i} = x_i) = p(x_i)$, for all indices $j_i \in \mathbb{T}$, all values $x_i \in \mathbb{X}$, and a certain function $p(x)$.

IID processes play a fundamental role in mathematical statistics, where it is usually assumed that random experiments constitute an IID process with an unknown distribution, and the goal is to infer this distribution or its parameters from sample $(X_i)_{i \in \mathbb{T}}$ under various scenarios. IID processes are an important probability model of data where the order of individual observations does not matter, but they cannot model dependent time series such as texts in natural language.

In contrast, when dealing with data such as texts, we have a strong feeling that the order of text symbols matters so texts cannot be modeled by an IID process. In the first approximation, we can then consider another class of simple processes, called Markov processes. The general idea of a Markov process, introduced by Markov (1913) exactly for modeling texts in natural language, is that to predict a random variable from the process, it is sufficient to know only the preceding random variable. For simplicity, we will restrict ourselves to stationary processes of this type. Let us write blocks of random variables $X_i^k := (X_i, X_{i+1}, \ldots, X_k)$ and strings of symbols $x_i^k := (x_i, x_{i+1}, \ldots, x_k)$. From the elementary probability calculus, we will use the notation for conditional probability

$$P(A|B) := \frac{P(A \cap B)}{P(B)} \tag{2.20}$$

for events A and B such that $P(B) > 0$. In particular, for any discrete process $(X_i)_{i \in \mathbb{T}}$, where $\mathbb{T} = \mathbb{N}$ or $\mathbb{T} = \mathbb{Z}$, we may write

$$P(X_{t+1}^{t+n} = x_1^n) = P(X_{t+1} = x_1) \prod_{i=2}^{n} P(X_{t+i} = x_i | X_{t+1}^{t+i-1} = x_1^{i-1}) \tag{2.21}$$

A Markov process is a stochastic process where we can restrict the conditioning in formula (2.21) to the directly preceding random variable.

Definition 2.13 *(Markov process)*: A discrete process $(X_i)_{i \in \mathbb{T}}$, where $\mathbb{T} = \mathbb{N}$ or $\mathbb{T} = \mathbb{Z}$ and $X_i : \Omega \to \mathbb{X}$, is called a (time-homogeneous) Markov process if

$$P(X_{t+n+1} = x_{n+1} | X_{t+1}^{t+n} = x_1^n) = P(X_{t+n+1} = x_{n+1} | X_{t+n} = x_n)$$
$$= p(x_{n+1} | x_n) \tag{2.22}$$

for all indices $t \in \mathbb{T} \cup \{0\}$, all sequences $x_1^{n+1} \in \mathbb{X}^*$, and a certain function $p(x_2|x_1)$. Function $p(x_2|x_1)$ is called the transition matrix.

For a Markov process $(X_i)_{i \in \mathbb{T}}$, elements of set $\mathbb{X}$ are called states. The Markov process is called finite if the set of states $\mathbb{X}$ is finite. In the following chapters, phrase "time-homogeneous" will be omitted.

It follows from representation (2.21) that the distribution of a Markov process can be expressed as

$$P(X_{t+1}^{t+n} = x_1^n) = P(X_{t+1} = x_1) \prod_{i=2}^{n} P(X_{t+i} = x_i | X_{t+i-1} = x_{i-1})$$

$$= P(X_{t+1} = x_1) \prod_{i=2}^{n} p(x_i | x_{i-1}) \qquad (2.23)$$

Thus, we have a simple recipe for computing joint distributions $P(X_{t+1}^{t+n} = x_1^n)$.

The theory of Markov processes is well developed. For an elementary and thorough introduction, we refer the reader to Norris (1997). Two particular properties play a crucial role in this theory.

Definition 2.14 *(Irreducible Markov process)*: A (time-homogeneous) Markov process $(X_i)_{i \in \mathbb{T}}$, where $\mathbb{T} = \mathbb{N}$ or $\mathbb{T} = \mathbb{Z}$ and $X_i : \Omega \to \mathbb{X}$, is called irreducible if for all $x_0, x_n \in \mathbb{X}$ there exists an $n \in \mathbb{N}$ such that $P(X_{t+n} = x_n | X_t = x_0) > 0$.

Definition 2.15 *(Aperiodic Markov process)*: A (time-homogeneous) Markov process $(X_i)_{i \in \mathbb{T}}$, where $\mathbb{T} = \mathbb{N}$ or $\mathbb{T} = \mathbb{Z}$ and $X_i : \Omega \to \mathbb{X}$, is called aperiodic if for each $x_0 \in \mathbb{X}$ there exists an $N \in \mathbb{N}$ such that for all $n \geq N$ we have $P(X_{t+n} = x_0 | X_t = x_0) > 0$.

The above formal definitions correspond to quite simple intuitions. Namely, a Markov process is irreducible if the set of states cannot be decomposed into two disjoint subsets between which transitions are impossible. In contrast, a Markov process is aperiodic if there are no cycles in transitions. Irreducibility and aperiodicity can be related to the process being ergodic and mixing, respectively, which are two important properties of more general stationary stochastic processes. These topics will be discussed in Section 4.3. In this context, another important property of a Markov process is the existence of a stationary distribution.

Definition 2.16 *(Stationary distribution)*: For a Markov process with a transition matrix $p(x|y)$, a stationary distribution $\pi(x)$ is a vector that satisfies conditions $\pi(x) \geq 0$, $\sum_{y \in \mathbb{X}} \pi(y) = 1$ and

$$\sum_{y \in \mathbb{X}} p(x|y)\pi(y) = \pi(x) \qquad (2.24)$$

for all $x \in \mathbb{X}$.

Existence and uniqueness of a stationary distribution for a given Markov process is not trivial. We have the following results.

Theorem 2.8: *For an irreducible Markov process, the stationary distribution is unique and strictly positive, i.e. $\pi(x) > 0$ for all $x \in \mathbb{X}$, if it exists.*

Proof: See Theorems 1.7.7 of Norris (1997). □

Theorem 2.9: *For a finite irreducible Markov process, the stationary distribution exists.*

Proof: See Theorems 1.5.5, 1.5.6, and 1.7.6 of Norris (1997). □

As we can see easily, stationary distributions need not exist if the set of states is infinite, see Problem 2.6.

Once we know whether a stationary distribution exists, we can formulate two important theorems, the ergodic theorem and the convergence to equilibrium. These propositions state that if the Markov process is irreducible and aperiodic or just irreducible, then the appropriate transition probabilities or their averages converge to the stationary distribution.

Theorem 2.10 *(Ergodic theorem):* *For an irreducible Markov process $(X_i)_{i \in \mathbb{T}}$, where $\mathbb{T} = \mathbb{N}$ or $\mathbb{T} = \mathbb{Z}$, with a stationary distribution $\pi(x)$, for all $x, y \in \mathbb{X}$, we have*

$$\lim_{n \to \infty} \frac{1}{n} \sum_{i=1}^{n} P(X_{t+i} = x | X_t = y) = \pi(x) \tag{2.25}$$

Proof: See Theorem 1.10.2 of Norris (1997). We apply the expectation to both sides of the original almost sure statement and we use the dominated convergence (Theorem 3.10), to be discussed in Chapter 3. □

Theorem 2.11 *(Convergence to equilibrium):* *For an irreducible and aperiodic Markov process $(X_i)_{i \in \mathbb{T}}$, where $\mathbb{T} = \mathbb{N}$ or $\mathbb{T} = \mathbb{Z}$, with a stationary distribution $\pi(x)$, for all $x, y \in \mathbb{X}$, we have*

$$\lim_{n \to \infty} P(X_{t+n} = x | X_t = y) = \pi(x) \tag{2.26}$$

Proof: See Theorem 1.8.3 of Norris (1997). □

The above theorems will be used to prove ergodicity or mixing of irreducible or irreducible and aperiodic Markov processes in Section 4.3.

For a Markov process, the conditional probability of a random variable given an arbitrary past equals the conditional probability of this random variable given only the directly preceding random variable. This condition can be also considered too restrictive as for modeling natural language. We can reasonably suspect that the value of the next letter of a text depends on many more previous letters than just one. There are two basic ways of generalizing the concept of a Markov process. In the first approach, we extend the length of the context on which a given random variable effectively depends. As a result, we obtain kth order Markov processes.

Definition 2.17 *(kth order Markov process)*:   A discrete process $(X_i)_{i \in \mathbb{T}}$, where $\mathbb{T} = \mathbb{N}$ or $\mathbb{T} = \mathbb{Z}$ and $X_i : \Omega \to \mathbb{X}$, is called a (time-homogeneous) kth order Markov process if

$$P(X_{t+n+1} = x_{n+1} | X_{t+1}^{t+n} = x_1^n) = P(X_{t+n+1} = x_{n+1} | X_{t+n-k+1}^{t+n} = x_{n-k+1}^n)$$
$$= p(x_{n+1} | x_{n-k+1}^n) \tag{2.27}$$

for all indices $t \in \mathbb{T} \cup \{0\}$, all sequences $x_1^{n+1} \in \mathbb{X}^*$, and a certain function $p(x_{k+1} | x_1^k)$. Function $p(x_{k+1} | x_1^k)$ is also called the transition matrix.

In particular, first-order Markov processes are simply Markov processes. For a kth-order Markov processes, we obtain

$$P(X_{t+1}^{t+n} = x_1^n) = P(X_{t+1}^{t+k} = x_1^k) \prod_{i=k+1}^{n} P(X_{t+i} = x_i | X_{t+i-k}^{t+i-1} = x_{i-k}^{i-1})$$
$$= P(X_{t+1}^{t+k} = x_1^k) \prod_{i=k+1}^{n} p(x_i | x_{i-k}^{i-1}) \tag{2.28}$$

Thus, taking $Y_i = X_{i-k}^{i-1}$ for a kth order Markov process $(X_i)_{i \in \mathbb{T}}$, we obtain a first-order Markov process $(Y_i)_{i \in \mathbb{T}}$. The kth order Markov processes still seem to assume a too rigid model of dependence to model texts in natural language adequately.

The second approach to generalizing the concept of a Markov process assumes that the considered process is a stochastic function of a certain Markov process. This case is called a hidden Markov process, and it proved to be quite useful for practical applications in computational linguistics.

Definition 2.18 *(Hidden Markov process)*:   A discrete process $(X_i)_{i \in \mathbb{T}}$, where $\mathbb{T} = \mathbb{N}$ or $\mathbb{T} = \mathbb{Z}$ and $X_i : \Omega \to \mathbb{X}$, is called a (time-homogeneous) hidden Markov process if there exists a discrete (time-homogeneous) Markov process $(Y_i)_{i \in \mathbb{T}}$, where $Y_i : \Omega \to \mathbb{Y}$, such that

$$P(X_{t+1}^{t+n} = x_1^n | Y_{t+1}^{t+n} = y_1^n) = \prod_{i=1}^{n} P(X_{t+i} = x_i | Y_{t+i} = y_i) \tag{2.29}$$

where $P(X_{t+i} = x_i | Y_{t+i} = y_i) = e(x_i | y_i)$, for all indices $t \in \mathbb{T} \cup \{0\}$, all sequences $y_1^n \in \mathbb{Y}^*$ and $x_1^n \in \mathbb{X}^*$, and a certain function $e(x|y)$. Function $e(x|y)$ is called the emission probability distribution.

For the hidden Markov process $(X_i)_{i \in \mathbb{T}}$, elements of set $\mathbb{Y}$ are called hidden states in contrast to elements of set $\mathbb{X}$ being called emitted symbols. In general, the set of hidden states $\mathbb{Y}$ may be countably infinite. If set $\mathbb{Y}$ is finite, then process $(X_i)_{i \in \mathbb{T}}$ is called a finite-state process. The number of hidden states $\#\mathbb{Y}$ is called the order of the hidden Markov process $(X_i)_{i \in \mathbb{T}}$.

As we have commented in Section 1.4, finite-state processes had been an important tool in computational linguistics until they were replaced by artificial neural networks, which proved to be more accurate. Finite-state processes can model nontrivial statistical dependencies, whereas their parameters can be estimated given enough training data. For practical algorithms, the reader is referred to Jelinek (1997), Manning and Schütze (1999), Ephraim and Merhav (2002), Upper (1997), and Lehéricy (2019). There is also a series of works concerning expressibility of hidden Markov processes: Shalizi and Crutchfield (2001), Crutchfield and Feldman (2003), Löhr (2009), Travers and Crutchfield (2014), and Dębowski (2014). In contrast, in this book, we will argue that finite-state processes may be insufficient as models of natural language since they cannot account for certain phenomena likely exhibited by texts in natural language. Whereas the general insufficiency of finite-state processes was noticed by Chomsky (1956, 1957, 1959) in a nonprobabilistic setting, in this book, we will provide some probabilistic arguments. Thus, we will show that typical finite-state processes cannot satisfy the power-law growth of block mutual information (1.132) and the power-law logarithmic growth of maximal repetition (1.133), which are likely satisfied by natural texts. In the chapters to follow, we will build a mathematical framework to understand these problems formally.

Problems

2.1 In this Problem, we will construct an example of a nonmeasurable set. Let $\mathbb{X}$ be a finite set. The uniform probability measure λ on the space $(\mathbb{X}^{\mathbb{Z}}, \mathcal{X}^{\mathbb{Z}})$ of doubly infinite sequences is defined as

$$\lambda(\{(x_i)_{i \in \mathbb{Z}} : x_{j+1}^{j+k} = y_1^k\}) := (\#\mathbb{X})^{-k} \tag{2.30}$$

for $j \in \mathbb{Z}$, $k \in \mathbb{N}$, and $y_1^k \in \mathbb{X}^k$.

(a) Consider the shift operations $T^p : \mathbb{X}^{\mathbb{Z}} \ni (x_i)_{i \in \mathbb{Z}} \mapsto (x_{i+p})_{i \in \mathbb{Z}} \in \mathbb{X}^{\mathbb{Z}}$ for $p \in \mathbb{Z}$. Show that for every event $A \in \mathcal{X}^{\mathbb{Z}}$ we have $\lambda(T^p(A)) = \lambda(A)$.

(b) A sequence is called periodic if $(x_i)_{i \in \mathbb{Z}} = T^p((x_i)_{i \in \mathbb{Z}})$ for some $p \neq 0$. Let $Q \subseteq \mathbb{X}^{\mathbb{Z}}$ be the set of periodic sequences. Show that $\lambda(Q) = 0$.

(c) Let $N \subseteq \mathbb{X}^{\mathbb{Z}}$ be a certain subset of sequences such that $(x_i)_{i \in \mathbb{Z}} \in N$ if and only if $T^p((x_i)_{i \in \mathbb{Z}}) \notin N$ for all $p \neq 0$, i.e. set N contains exactly one sequence from each class of mutually shifted sequences. Show that sets $T^p(N)$ are disjoint and their union is $\bigcup_{p \in \mathbb{Z}} T^p(N) = \mathbb{X}^{\mathbb{Z}} \backslash Q$.

(d) Show that $T^p(N) \notin \mathcal{X}^{\mathbb{Z}}$ and probabilities $\lambda(T^p(N))$ cannot be defined.

2.2 For $\mathcal{X} = \sigma(\{\{x\} : x \in \mathbb{X}\})$, where $\mathbb{X}$ is countable, consider measurable space $(\mathbb{X}^{\mathbb{N}}, \mathcal{X}^{\mathbb{N}})$. Let $\mathcal{A} = \{[A]_k^{\mathbb{N}} : A \in \mathcal{X}, k \in \mathbb{N}\}$ be the class of cylinder sets. Show that $\mathcal{A} \subsetneq \pi(\mathcal{A}) \subsetneq \sigma_0(\mathcal{A}) \subsetneq \sigma(\mathcal{A}) = \mathcal{X}^{\mathbb{N}} \subsetneq 2^{(\mathbb{X}^{\mathbb{N}})}$. Exhibit some examples of events by which these classes differ.

2.3 Prove the following inequalities, called the Bonferroni inequalities:

$$0 \leq \sum_{1 \leq i \leq n} P(A_i) - P\left(\bigcup_{1 \leq i \leq n} A_i\right) \leq \sum_{1 \leq i < j \leq n} P(A_i \cap A_j) \qquad (2.31)$$

2.4 Let us denote the symmetric difference

$$A \Delta B := (A \backslash B) \cup (B \backslash A) = (A \cup B) \backslash (A \cap B) \qquad (2.32)$$

Show that symmetric difference satisfies the following identities:

$$A^c \Delta B^c = A \Delta B \qquad (2.33)$$

$$A \Delta B \subseteq (A \Delta C) \cup (C \Delta B) \qquad (2.34)$$

$$(A \backslash C) \Delta B \subseteq (A \Delta B) \cup (C \cap B) \qquad (2.35)$$

$$\left(\bigcup_{i \in C} A_i\right) \Delta \left(\bigcup_{i \in C} B_i\right) \subseteq \bigcup_{i \in C} (A_i \Delta B_i) \qquad (2.36)$$

Prove also inequality $|P(A) - P(B)| \leq P(A \Delta B)$.

2.5 Verify that processes defined in Definitions 2.12, 2.13, 2.17, and 2.18 exist, i.e. their distributions satisfy the conditions prescribed in the Kolmogorov process theorem (Theorem 2.7).

2.6 Construct an example of an irreducible aperiodic Markov process such that the stationary distribution does not exist.

2.7 Let $\mathbb{T} = \mathbb{N}$ or $\mathbb{T} = \mathbb{Z}$. Anticipating further developments, an arbitrary discrete process $(X_i)_{i \in \mathbb{T}}$ is called stationary when probabilities $P(X_{t+1}^{t+n} = x_1^n)$ do not depend on index t for any block x_1^n. Show that a Markov process $(X_i)_{i\in\mathbb{N}}$ is stationary if and only if $P(X_t = x) = \pi(x)$ for the stationary distribution $\pi(x)$.

2.8 Without invoking Theorem 2.11 show that convergence (2.26) holds if the transition matrix satisfies inequality $p(x|y) \geq \epsilon$ for all $x, y \in \mathbb{X}$ and an $\epsilon > 0$.

2.9 Let $\mathbb{T} = \mathbb{N}$ or $\mathbb{T} = \mathbb{Z}$. A process $(X_i)_{i\in\mathbb{T}}$ is called a function of a Markov process if there exists a function $f : \mathbb{Y} \to \mathbb{X}$ and a Markov process $(Y_i)_{i\in\mathbb{T}}$ such that $X_i = f(Y_i)$. Show that if a process is a function of a Markov process, then it is a hidden Markov process.

3

Probabilistic Toolbox

In this chapter, we will review some important technical results in probability theory, which concern mostly real random variables and related mathematical concepts. Formal linguists may rightly ask what the utility of real numbers is in language research. Indeed, the language structure is largely about discrete symbols and hierarchical categorical distinctions (Adger, 2019). Yet rational and computable real numbers arise naturally when we count the frequencies of those symbols and categories or compute other kinds of statistics for linguistic data, such as fractions, means, or variances. One can debate whether it is more difficult to see some patterns in large tables of numbers than in collections of categorical data, but we cannot deny that for quantitative and computational linguists fractional numbers in language matter as well.

Moreover, we should be aware that it is mathematically possible to encode an infinite sequence of binary digits $(z_k)_{k \in \mathbb{N}}$ as a single real number $y = 0.z_1 z_2 z_3 \ldots := \sum_{k=1}^{\infty} 2^{-k} z_k$. Except for countably many cases such as binary numbers $0.011\,11\ldots = 0.100\,00\ldots$, this mapping between binary sequences and real numbers is one-to-one. In particular, the text of any novel or even a collection of large encyclopedias can be represented as a single real number in range $[0, 1]$. The "minor" physical problem of this representation is that to extract useful information from such a real number, we need to represent it and read it with a very high precision, which is unachievable in a physical model of the real line such as a wooden ruler. Mathematically speaking, however, real numbers are defined with an infinite precision and, when sampled uniformly at random, they usually contain an infinite amount of information.

The aforementioned observations lead us to a conclusion that even when we are studying natural language through the lens of discrete stochastic processes, we should be aware of exact mathematical constructions and tools for real random variables. The minimal toolbox should cover the concepts of the Lebesgue measure and the Lebesgue integral, the Markov and Jensen inequalities, weak and strong convergence of random variables, conditional distributions and martingales. It is

Information Theory Meets Power Laws: Stochastic Processes and Language Models, First Edition.
Łukasz Dębowski.
© 2021 John Wiley & Sons, Inc. Published 2021 by John Wiley & Sons, Inc.

quite a load of ideas and technicalities that we would like to convey in this chapter. For some readers, it may be advisable to skip thorough reading of this chapter and catch up with its contents later once the relevant material is actually needed. Intentionally, we did not make this chapter an appendix to preserve the linear flow of formal ideas.

More detailed contents of this chapter can be described as follows: In Section 3.1, first we will introduce Borel σ-fields and the Lebesgue measure. The Borel σ-fields are some natural σ-fields for the event space being the real line, whereas the Lebesgue measure is the natural measure on the real line generalizing the concept of the length of an interval. Having these, we will also introduce real random variables and nonatomic measures. It turns out as well that the Lebesgue measure on the interval $[0, 1]$ is the probability distribution of an infinite sequence of fairly tossed binary digits $(Z_k)_{k \in \mathbb{N}}$ encoded as the real number $Y = \sum_{k=1}^{\infty} 2^{-k} Z_k$.

Once we have real random variables, it is natural to ask about their expectations. In the modern probability theory, the expectation of an arbitrary real random variable is defined as the Lebesgue integral with respect to the probability measure. The Lebesgue integral itself is a generalization of the Riemann integral, which is defined as the area under the graph of a continuous function. In contrast, the Lebesgue integral exists also for highly discontinuous functions. Lebesgue integrals and their properties will be discussed in Section 3.2.

In Section 3.3, we will present a few useful propositions such as the Borel–Cantelli lemma as well as the Markov, the Jensen, the Cauchy–Schwarz, the Paley–Zygmund, and the median–mean inequalities. The three first propositions will be often used in the book. The Markov inequality allows to bound probability of some events in terms of expectations of real random variables that construct these events. Subsequently, the Borel–Cantelli lemma states that, for an infinite sequence of events, finitely many of them occur almost surely if the sum of probabilities of all events is finite. Eventually, the Jensen inequality allows to bound expectations of some random variables in terms of expectations of other random variables.

As an example, in Section 3.4, we will discuss some important basic results in information theory: the Gibbs and the Barron inequalities. These results combine the Jensen inequality, the Markov inequality, and the Borel–Cantelli lemma together. Basically, the Gibbs and the Barron inequalities say that if we have two probability distributions, the real one and a model thereof, then the real probability cannot be essentially much less than the probability under the model.

Section 3.5 will be devoted to conditional probability – a concept which distinguishes probability theory from the ordinary measure theory. In the elementary probability theory, the conditional probability of an event A given an event B is defined as $P(A|B) := P(A \cap B)/P(B)$. This definition can be applied only if the probability of event *B* is strictly positive. The measure-theoretic probability calculus provides some means to define the conditional probability for probabilities of conditioning events tending to zero. Under certain assumptions, we can speak of random conditional measures and convergence of conditional expectations, generalized as the martingale convergence theorem.

In Section 3.6, we will be occupied with convergence of arbitrary real random variables. There are different modes of probabilistic convergence such as convergence in probability and almost sure convergence. It turns out that the almost sure convergence is stronger than the convergence in probability, but not too much. According to the Riesz theorem, if a sequence of random variables converges in probability, then there exists a subsequence thereof which converges almost surely.

Eventually, Section 3.7 states a few technical but also useful propositions, which may seem quite hair-splitting upon the first reading. The first statement concerns an approximation of σ-fields by their generating fields. Second, we discuss completion of σ-fields with events of outer measure zero. This discussion leads us to finally show that for each sub-σ-field of a countably generated probability space there exists a countably generated sub-σ-field which is almost surely equal.

3.1 Borel σ-Fields and a Fair Coin

In this section, we will construct Borel σ-fields, which are generated by classes of intervals on the real line. This construction is useful for two further movements: a definition of the Lebesgue measure, which generalizes the concept of the length of an interval, and a definition of real random variables, which gives rise to real stochastic processes. Since real numbers and real random variables can be represented via their binary expansions, hence the Lebesgue measure on the unit interval can be also regarded as the probability distribution of a sequence of independent uniformly distributed binary digits, which we will call a fair-coin process. Subsequently, we will show that the Lebesgue measure is nonatomic, and we will characterize nonatomic probability measures in terms of fair-coin processes in a simple way.

Let $\mathbb{R}$ be the set of real numbers. Let us observe that the class of closed intervals

$$[a, b] := \{r \in \mathbb{R} : a \leq r \leq b\} \tag{3.1}$$

where $a, b \in \mathbb{R}$ and $a \leq b$, is a π-system. The Borel σ-field is a σ-field generated by this π-system.

Definition 3.1 *(Borel σ-field):* The Borel σ-field on the real line $\mathbb{R}$ is defined as

$$\mathcal{R} := \sigma(\{[a,b] : a,b \in \mathbb{R}\}) \qquad\qquad (3.2)$$

In particular, all singletons $\{r\}$, where $r \in \mathbb{R}$, belong to the Borel σ-field $\mathcal{R}$.

It is also convenient to consider the extended set of real numbers $\mathbb{R}_\infty := \mathbb{R} \cup \{-\infty, \infty\}$ and the unit interval $\mathbb{U} := [0,1]$. For these sets, we can define respective Borel σ-fields as well.

Definition 3.2 *(Borel σ-fields):* The Borel σ-fields on sets $\mathbb{R}_\infty$ and $\mathbb{U}$ are defined as

$$\mathcal{R}_\infty := \sigma(\{[a,b] : a,b \in \mathbb{R}_\infty\}) \qquad\qquad (3.3)$$

$$\mathcal{U} := \sigma(\{[a,b] : a,b \in \mathbb{U}\}) \qquad\qquad (3.4)$$

The length of an interval can be defined as $\lambda([a,b]) := b - a$. In fact, the length of intervals can be extended to a measure. This measure, also denoted as λ, is called the Lebesgue measure.

Theorem 3.1 *(Lebesgue measure):* *There is a unique measure λ on the measurable space $(\mathbb{R}, \mathcal{R})$ such that $\lambda([a,b]) := b - a$ for all real numbers $b \geq a$.*

Proof: See Billingsley (1979, sections 2 and 3). □

The Lebesgue measure λ is the first measure ever constructed. Let us note that this measure cannot be extended to the σ-field $2^{\mathbb{R}}$ and the Borel σ-field $\mathcal{R}$ is its natural domain (Billingsley, 1979, section 3), see also Problem 2.1.

Let us also observe that the Lebesgue measure on the unit interval is a probability measure, which is a simple but an important observation. Namely, the proposition below gives us a geometric interpretation of the concept of a probability measure.

Theorem 3.2 *(Lebesgue measure):* *The Lebesgue measure λ restricted to the measurable space $(\mathbb{U}, \mathcal{U})$ is a probability measure.*

Proof: Clearly, we have $\lambda(\mathbb{U}) = \lambda([0,1]) = 1 - 0 = 1$. □

Now, we can also define (extended) real random variables.

Definition 3.3 *(Real random variable):* A real random variable Y is a random variable with an image space $(\mathbb{R}_\infty, \mathcal{R}_\infty)$.

Some simple example of a random variable on the probability space $(\mathbb{U}, \mathcal{U}, \lambda)$ can be defined as $Y(r) := r$ for $r \in \mathbb{U}$. However, real random variables may be also discrete, such as real functions of discrete random variables. In our applications, real random variables will arise as functions of discrete random variables or as their limits. In the latter case, they need not be discrete. Let us also note that all continuous functions $\mathbb{R} \to \mathbb{R}$ are measurable from $(\mathbb{R}, \mathcal{R})$ to $(\mathbb{R}, \mathcal{R})$ (Billingsley, 1979, section 13). So continuous functions of real random variables are also real random variables.

Analogously to discrete-valued stochastic processes, we define real-valued stochastic processes. For our applications, an important special case of real stochastic processes will be random measures. They will be defined in Section 3.5.

Definition 3.4 *(Real stochastic process)*: A real stochastic process $(Y_t)_{t \in \mathbb{T}}$ is a collection of real random variables Y_t indexed by certain indices $t \in \mathbb{T}$ sharing the same image space $(\mathbb{R}_\infty, \mathcal{R}_\infty)$. The image space of process $(Y_t)_{t \in \mathbb{T}}$ is pair $(\mathbb{R}_\infty^{\mathbb{T}}, \mathcal{R}_\infty^{\mathbb{T}})$, where $\mathcal{R}_\infty^{\mathbb{T}}$ is the σ-field generated by cylinder sets.

Now, we will show how a stochastic process consisting of simple random variables can be regarded as a single real random variable. Consider a discrete stochastic process $(Z_k)_{k \in \mathbb{N}}$ on a probability space $(\Omega, \mathcal{J}, P)$, consisting of simple random variables Z_k with an image space $(\{0, 1\}, 2^{\{0,1\}})$. Since binary digits $z_k \in \{0, 1\}$ are called bits, random variables Z_k will be called random bits.

Definition 3.5 *(Random bit)*: A simple random variable $Z : \Omega \to \{0, 1\}$ will be called a random bit.

A sequence of random bits is called a fair-coin process when the probability measure of sequence $(Z_k)_{k \in \mathbb{N}}$ is the product measure of the probability distributions of random bits Z_k distributed uniformly on set $\{0, 1\}$.

Definition 3.6 *(Fair-coin process)*: A sequence of random bits $(Z_k)_{k \in \mathbb{N}}$ is called a fair-coin process when

$$P(Z_1^k = z_1^k) = 2^{-k} \tag{3.5}$$

for all strings $z_1^k \in \{0, 1\}^*$.

Let us introduce now function f that maps infinite binary expansions of real numbers to the corresponding real numbers. Formally, we have

$$f((z_k)_{k \in \mathbb{N}}) := 0.z_1 z_2 z_3 \ldots := \sum_{k=1}^{\infty} 2^{-k} z_k \tag{3.6}$$

for $z_k \in \{0, 1\}$. It can be shown that function f is measurable from the measurable space of infinite sequences $(\{0, 1\}^{\mathbb{N}}, (2^{\{0,1\}})^{\mathbb{N}})$ to the unit interval space $(\mathbb{U}, \mathcal{U})$. Let $\mathbb{Q}$ be the set of rational numbers. When $y \in \mathbb{U} \setminus \mathbb{Q}$, i.e. when y is an irrational number, then its binary expansion is unique, i.e. there exists a unique sequence $(z_k)_{k \in \mathbb{N}}$ such that $y = \sum_{k=1}^{\infty} 2^{-k} z_k$. We will write this functional dependence as $(z_k)_{k \in \mathbb{N}} = f^{-1}(y)$.

Consequently, we may ask how the distributions of a real random variable $Y : \Omega \to \mathbb{U}$ and its binary expansion $(Z_k)_{k \in \mathbb{N}} = f^{-1}(Y)$ can be linked in some simple cases. It turns out that random variable Y is distributed according to the Lebesgue measure if and only if its binary expansion $(Z_k)_{k \in \mathbb{N}}$ is a fair-coin process.

Theorem 3.3: *A sequence of random bits $(Z_k)_{k \in \mathbb{N}}$ is a fair-coin process if and only if the probability distribution of random variable $Y = f((Z_k)_{k \in \mathbb{N}})$, where f is defined in (3.6), is the Lebesgue measure on space $(\mathbb{U}, \mathcal{U})$.*

Proof: Consider cylinder sets $[z_1^k] \in \mathcal{J}$ defined as $[z_1^k] := (Z_1 = z_1, \ldots, Z_k = z_k)$. These cylinder sets form a π-system which generates the preimage space of process $(Z_k)_{k \in \mathbb{N}}$, which contains the preimage space of random variable Y. By the π-λ theorem (Theorem 2.3), it is sufficient to show that probability $P([z_1^k])$ for $(Z_k)_{k \in \mathbb{N}}$ being a fair-coin process equals to probability $P([z_1^k])$ if the distribution of Y is the Lebesgue measure. In fact, if the distribution of random variable Y is the Lebesgue measure, then $P(Y \in \mathbb{Q}) = 0$ since the set of rational numbers $\mathbb{Q}$ is countable. Then the unique binary expansion $(Z_k)_{k \in \mathbb{N}} = f^{-1}(Y)$ exists almost surely and since $Y([z_1^k]) \setminus \mathbb{Q} = [0.z_1^k 000 \ldots, 0.z_1^k 111 \ldots] \setminus \mathbb{Q}$, we can compute that

$$P([z_1^k] \setminus (Y \in \mathbb{Q})) = 0.z_1^k 111 \ldots - 0.z_1^k 000 \ldots = 0.0^k 111 \ldots = 2^{-k} \tag{3.7}$$

Hence $P([z_1^k]) = 2^{-k}$, which is the same as for the fair-coin process $(Z_k)_{k \in \mathbb{N}}$. $\qquad \square$

Measurable space $(\mathbb{U}, \mathcal{U})$ with the Lebesgue measure λ, which we will call the Lebesgue probability space $(\mathbb{U}, \mathcal{U}, \lambda)$, is an important example of a probability space which is not discrete. This concept is captured by the following definition of a nonatomic probability space. In plain words, a probability space is nonatomic if we can shatter any event into events of an arbitrarily small strictly positive measure.

Definition 3.7 *(Nonatomic probability space):* A probability space $(\Omega, \mathcal{J}, P)$ is called nonatomic if for each set $A \in \mathcal{J}$ and $x \in [0, P(A)]$, there exists a set $B \in \mathcal{J}$ such that $B \subseteq A$ and $P(B) = x$.

It can be easily checked that if probability space $(\Omega, \mathcal{J}, P)$ is nonatomic and we have $\{\omega\} \in \mathcal{J}$ for a certain point $\omega \in \Omega$, then $P(\{\omega\}) = 0$. Moreover, if the probability space is nonatomic, then σ-field $\mathcal{J}$ cannot be finite. It turns out that the Lebesgue probability space has this property.

Theorem 3.4: *The Lebesgue probability space* $(\mathbb{U}, \mathcal{U}, \lambda)$ *is nonatomic.*

Proof: See Billingsley (1979, problem 2.17). □

Sometimes it is convenient to regard nonatomicity as a property of a particular sub-σ-field rather than of the whole probability space.

Definition 3.8 *(Nonatomic σ-field)*: A σ-field $\mathcal{A}$ is called nonatomic with respect to a probability measure P if probability space $(\Omega, \mathcal{A}, P)$ is nonatomic.

In fact, Theorem 3.3 can be used to provide a general characterization of nonatomic σ-fields. Namely, an arbitrary σ-field contains a nonatomic sub-σ-field if and only if there lives some fair-coin process on it.

Theorem 3.5: *A σ-field $\mathcal{A}$ contains a nonatomic sub-σ-field if and only if there exists an $\mathcal{A}$-measurable fair-coin process* $(Z_k)_{k\in\mathbb{N}}$.

Proof: First, let $\mathcal{A}_0$ be a nonatomic sub-σ-field of $\mathcal{A}$. For any $A \in \mathcal{A}_0$ and $x \in [0, P(A)]$, there exists $B \in \mathcal{A}_0$ such that $B \subseteq A$ and $P(B) = x$. This property can be used to construct a family of nested sets $[u] \in \mathcal{A}_0$ indexed by binary words $u \in \{0, 1\}^*$ such that $[\lambda] = \Omega$ for the empty string λ, $[uz] \subseteq [u]$, and $P([u0]) = P([u1]) = P([u])/2$. Next, for each $k \in \mathbb{N}$, we define $Z_k(\omega) := \mathbf{1}\{\omega \in B_k\}$, where $B_k := \bigcup_{u\in\{0,1\}^{k-1}} [u1]$. Process $(Z_k)_{k\in\mathbb{N}}$ is a fair-coin process.

As for the converse, let $(Z_k)_{k\in\mathbb{N}}$ be an $\mathcal{A}$-measurable fair-coin process. We construct an $\mathcal{A}$-measurable random variable $Y := \sum_{k=1}^{\infty} 2^{-k} Z_k$. By Theorem 3.3, the distribution of Y is the Lebesgue measure on interval $[0, 1]$. Hence, by Theorem 3.4, σ-field $Y^{-1}(\mathcal{U})$ is nonatomic and contained in σ-field $\mathcal{A}$. □

Theorem3.5 will be used to prove Theorems 5.39 and 8.12.

3.2 Integral and Expectation

Now, we can consider the concepts of the integral and the expectation. In the high school calculus, the integral is defined as the area under the graph of a continuous real function $f(x)$, which is known as the Riemann integral $\int_a^b f(x)dx$. Measure theory builds on this idea but extends the concept of the integral to arbitrary measurable functions, such as the Dirichlet function

$$f(x) = \begin{cases} 1, & \text{if } x \text{ is irrational} \\ 0, & \text{if } x \text{ is rational} \end{cases} \tag{3.8}$$

In fact, we can define the integral of this function so that $\int_a^b f(x)dx = b - a$. The respective general concept is called the Lebesgue integral. To be honest, the Lebesgue integral is so useful not only because it can be used to integrate quite exotic functions but also because it enjoys a number of interesting properties.

To define the Lebesgue integral, first we need the concepts of the supremum and the infimum of a set. The supremum $\sup_{a \in A} f(a)$ is defined as the least real number $r \in \mathbb{R}_\infty$ such that $r \geq f(a)$ for all $a \in A$. In contrast, infimum $\inf_{a \in A} f(a)$ is the largest real number $r \in \mathbb{R}_\infty$ such that $r \leq f(a)$ for all $a \in A$.

Definition 3.9 *(Lebesgue integral):* For a real random variable Y on a finite measure space $(\Omega, \mathcal{J}, \mu)$, we define the Lebesgue integral $\int Y\, d\mu$ as follows:

1. If $Y \geq 0$ and Y is a simple random variable:

$$\int Y\, d\mu := \sum_{y:\mu(Y=y)>0} y\mu(Y = y) \tag{3.9}$$

2. If $Y \geq 0$ and Y is not a simple random variable:

$$\int Y\, d\mu := \sup_{X \leq Y} \int X\, d\mu \tag{3.10}$$

   where the supremum ranges over all real-valued simple random variables X such that $X \leq Y$.

3. If $Y = Y_+ - Y_-$, where $Y_+, Y_- \geq 0$ and either $\int Y_+ d\mu < \infty$ or $\int Y_- d\mu < \infty$:

$$\int Y\, d\mu := \int Y_+ d\mu - \int Y_- d\mu \tag{3.11}$$

Some comments to the above definition are due. For a simple random variable, if $\mu(|Y| = \infty) = 0$, then the Lebesgue integral is finite, $|\int Y\, d\mu| < \infty$. If $Y \geq 0$ and $\mu(Y = \infty) > 0$, then the Lebesgue integral is infinite, $\int Y\, d\mu = \infty$. In contrast, if both $\mu(Y = \infty) > 0$ and $\mu(Y = -\infty) > 0$, then the Lebesgue integral $\int Y\, d\mu$ is not defined.

A real random variable Y is called integrable (with respect to a measure μ) if $\int |Y| d\mu < \infty$. It can be shown that the Lebesgue integral is a linear function of integrable random variables, namely,

$$\int (aY + bX)d\mu = a \int Y\, d\mu + b \int X\, d\mu \tag{3.12}$$

for $a, b \in \mathbb{R}$ and integrable random variables X and Y.

Now, we will introduce a few additional conventions and notations for the Lebesgue integral. The characteristic function of a subset $A \subseteq \Omega$ is defined as

$$I_A(\omega) := \mathbf{1}\{\omega \in A\} \tag{3.13}$$

Subsequently, we denote the integral over a measurable subset $A \in \mathcal{J}$ as

$$\int_A Y \, d\mu := \int Y I_A d\mu \tag{3.14}$$

Traditionally, the integral with respect to the Lebesgue measure is denoted like the Riemann integral,

$$\int_a^b f(x)dx := \int_{[a,b]} f \, d\lambda \tag{3.15}$$

In the context of measure-theoretic probability, the Lebesgue integral of a random variable with respect to a probability measure is called the expectation.

Definition 3.10 *(Expectation)*: For a probability measure P, we will write

$$\mathbf{E}Y := \mathbf{E}_P Y := \int Y \, dP \tag{3.16}$$

and we will call this integral the expectation of the random variable Y.

It is easy to see that the above definition generalizes the concept of expectation from the elementary probability calculus, which is simply defined as (3.9).

For a random proposition $\varphi : \Omega \to \{\text{true, false}\}$, we will say that φ holds μ-almost everywhere (or simply – almost surely – for a probability measure $P = \mu$) if

$$\mu(\neg\varphi) := \mu(\{\omega \in \Omega : \varphi(\omega) \text{ is false}\}) = 0 \tag{3.17}$$

In this case, we also say synonimically that "$\varphi(\omega)$ holds for μ-almost all ω".

Let us make this simple observation.

Theorem 3.6: *If $Y \geq 0$ and $\int Y \, d\mu = 0$, then $Y = 0$ holds μ-almost everywhere.*

Proof: We will prove the converse. Suppose that $\mu(Y > 0) > 0$ for $Y \geq 0$. By the countable additivity of μ, we have $\mu(Y > 0) = \sup_{n \in \mathbb{N}} \mu(Y > 1/n)$. Thus, there exists an $\epsilon > 0$ such that $\mu(Y \geq \epsilon) > 0$. Hence, $\int Y \, d\mu \geq \epsilon\mu(Y \geq \epsilon) > 0$. ☐

In the above proof, we have used in fact the Markov inequality $\int Y \, d\mu \geq \epsilon\mu(Y \geq \epsilon)$, which will be formally stated in Theorem 3.11.

Subsequently, we would like to state three important results about convergence of sequences of integrals. For this goal, we need to introduce generalized limits of sequences of real numbers.

Definition 3.11 *(Limits of sequences)*: The upper limit of a sequence of real numbers $(a_n)_{n \in \mathbb{N}}$ is defined as

$$\limsup_{n \to \infty} a_n := \inf_{n \in \mathbb{N}} \sup_{m \geq n} a_m \tag{3.18}$$

whereas the lower limit of this sequence is defined as

$$\liminf_{n \to \infty} a_n := \sup_{n \in \mathbb{N}} \inf_{m \geq n} a_m \tag{3.19}$$

If the upper and the lower limit are equal, $\limsup_{n \to \infty} a_n = \liminf_{n \to \infty} a_n = a$, then we say that the (proper) limit $\lim_{n \to \infty} a_n$ exists, and we define $\lim_{n \to \infty} a_n := a$, whereas otherwise limit $\lim_{n \to \infty} a_n$ is not defined, and we say that it does not exist.

The upper and the lower limits exist for any sequence, but they may be different. For example, we obtain two different values for $a_n = (-1)^n$,

$$\limsup_{n \to \infty} (-1)^n = 1 \tag{3.20}$$

$$\liminf_{n \to \infty} (-1)^n = -1 \tag{3.21}$$

It turns out that limits of sequences of real random variables are measurable.

Theorem 3.7: *Let $(Y_i)_{i \in \mathbb{N}}$ be a sequence of real random variables. Then supremum $\sup_{n \in \mathbb{N}} Y_n$, infimum $\inf_{n \in \mathbb{N}} Y_n$, upper limit $\limsup_{n \to \infty} Y_n$, and lower limit $\liminf_{n \to \infty} Y_n$ are also real random variables.*

Proof: See Billingsley (1979, theorem 13.4). □

Now, we can state the three theorems about sequences of integrals. These are the monotone convergence, the Fatou lemma, and the dominated convergence.

Theorem 3.8 *(Monotone convergence)*: *Let $(Y_n)_{n \in \mathbb{N}}$ be a sequence of positive, $Y_n \geq 0$, and growing, $Y_{n+1} \geq Y_n$, real random variables. Then*

$$\sup_{n \in \mathbb{N}} \int Y_n d\mu = \int \sup_{n \in \mathbb{N}} Y_n d\mu \tag{3.22}$$

Proof: See Billingsley (1979, theorem 16.2). □

By the above result, if $Y \geq 0$ and $\int Y \, d\mu < \infty$, then function $\nu(A) = \int_A Y \, d\mu$ satisfies countable additivity and hence it is also a finite measure.

Theorem 3.9 *(Fatou lemma):* *Let* $(Y_n)_{n\in\mathbb{N}}$ *be a sequence of positive,* $Y_n \geq 0$, *real random variables. Then*

$$\liminf_{n\to\infty} \int Y_n d\mu \geq \int \liminf_{n\to\infty} Y_n d\mu \tag{3.23}$$

Proof: Denote $X_n := \inf_{k\geq n} Y_k \leq Y_n$. We have $X_{n+1} \geq X_n$ and $\liminf_{n\to\infty} Y_n = \sup_{n\in\mathbb{N}} X_n$. Hence, by Theorem 3.8, we have

$$\liminf_{n\to\infty} \int Y_n d\mu \geq \lim_{n\to\infty} \int X_n d\mu = \int \liminf_{n\to\infty} Y_n d\mu \tag{3.24}$$

$\square$

Inequality (3.23) can be strict. For example, for the measurable space $(\mathbb{N}, 2^{\mathbb{N}})$ and the counting measure $\mu(A) := \#A$, where $\#A$ is the cardinality of set A, let us put $Y_n(m) := \mathbf{1}\{n = m\}$. Then we obtain $\int Y_n d\mu = 1$ and $Y(m) = \liminf_{n\to\infty} Y_n(m) = 0$ so $\int Y d\mu = 0$.

Theorem 3.10 *(Lebesgue dominated convergence):* *Let* $(Y_n)_{n\in\mathbb{N}}$ *be a sequence of real random variables which satisfy* $|Y_n| \leq Z$, *where* $\int Z d\mu < \infty$. *If there exists limit* $\lim_{n\to\infty} Y_n$, *then*

$$\lim_{n\to\infty} \int Y_n d\mu = \int \lim_{n\to\infty} Y_n d\mu \tag{3.25}$$

Proof: Let $X_m := |Y_m - \lim_{n\to\infty} Y_n|$. We have $0 \leq X_n \leq 2Z$. Hence, by Theorem 3.9, we obtain

$$\int 2Z d\mu = \int \liminf_{m\to\infty} (2Z - X_m) d\mu$$

$$\leq \liminf_{m\to\infty} \int (2Z - X_m) d\mu = \int 2Z d\mu - \limsup_{m\to\infty} \int X_m d\mu \tag{3.26}$$

Thus,

$$0 = \limsup_{m\to\infty} \int X_m d\mu \geq \limsup_{m\to\infty} \left| \int Y_m d\mu - \lim_{n\to\infty} Y_n d\mu \right| \tag{3.27}$$

and hence the claim follows. $\square$

All three above theorems will be applied in the book.

3.3 Inequalities and Corollaries

To prove many theorems in probability theory, three propositions are particularly useful: the Markov inequality, the Borel–Cantelli lemma, and the Jensen inequality. The Markov inequality bounds the probability of an unlikely event by the expectation of a related random variable. The Borel–Cantelli lemma states that, for an infinite sequence of events, finitely many of them happen almost surely if the sum of probabilities of all events in the sequence is finite. The Jensen inequality compares the expectation of a convex function with the convex function of the expectation. We will derive these results in this section. Additionally, we will show other important inequalities such as the Cauchy–Schwarz, the Paley–Zygmund, and the median–mean inequalities.

First, we will demonstrate the Markov inequality. The Markov inequality upper bounds the probability that a positive random variable Y exceeds a certain threshold in terms of the expectation of Y. This inequality is very simple but surprisingly powerful, since it is often much easier to upper bound the expectation than the probability of some special event.

Theorem 3.11 (*Markov inequality*): *For a real random variable Y such that $Y \geq 0$ and any $\epsilon > 0$, we have*

$$P(Y \geq \epsilon) \leq \frac{\mathbf{E}Y}{\epsilon} \tag{3.28}$$

Proof: We observe that $\epsilon \mathbf{1}\{Y \geq \epsilon\} \leq Y$. Taking the expectations of both sides of this inequality, we obtain the claim. □

A few special cases of the Markov inequality have their own names, such as Chernoff bounds, to be used in Theorem 11.18.

Using the Markov inequality, we may demonstrate the Borel–Cantelli lemma. The Borel–Cantelli lemma states that if the sum of probabilities of some events $A_1, A_2, \ldots$ is finite, then almost surely only finitely many of these events happen simultaneously. That is, we have $\sum_{n=1}^{\infty} I_{A_n} < \infty$ almost surely, where $I_A(\omega) := \mathbf{1}\{\omega \in A\}$ are the characteristic functions.

Theorem 3.12 (*Borel-Cantelli lemma*): *If $\sum_{n=1}^{\infty} P(A_n) < \infty$ for a family of events $A_1, A_2, \ldots$ then*

$$P\left(\sum_{n=1}^{\infty} I_{A_n} < \infty\right) = 1 \tag{3.29}$$

Proof: Fix an $m > 0$. By the Markov inequality (Theorem 3.11) and the monotone convergence (Theorem 3.8), we obtain

$$P\left(\sum_{n=1}^{\infty} I_{A_n} = \infty\right) \leq P\left(\sum_{n=1}^{\infty} I_{A_n} \geq m\right) \leq \frac{\mathbf{E}\sum_{n=1}^{\infty} I_{A_n}}{m} = \frac{\sum_{n=1}^{\infty} P(A_n)}{m} \tag{3.30}$$

Since m can be chosen arbitrarily large and the sum $\sum_{n=1}^{\infty} P(A_n)$ is finite, hence, we obtain $P\left(\sum_{n=1}^{\infty} I_{A_n} = \infty\right) = 0$, whence the claim follows. $\qquad\square$

The Borel–Cantelli lemma is highly useful for proving that some limiting events hold almost surely. We note that some important almost sure propositions probably cannot be proved using the Borel–Cantelli lemma, such as the martingale convergence theorem (Theorem 3.23) and the Birkhoff ergodic theorem (Theorem 4.6).

The next fact to be presented is the Jensen inequality for convex functions. For this aim, let us recall the definition of convex and concave functions.

Definition 3.12 *(Convex and concave functions)*: For an open interval $(a, b) := \{c \in \mathbb{R} : a < c < b\} \subseteq \mathbb{R}$, a real function $f : (a, b) \to \mathbb{R}$ is called convex if

$$p_1 f(x_1) + p_2 f(x_2) \geq f(p_1 x_1 + p_2 x_2) \tag{3.31}$$

for $p_i \geq 0$, $i = 1, 2$, and $p_1 + p_2 = 1$. Moreover, f is called strictly convex if

$$p_1 f(x_1) + p_2 f(x_2) > f(p_1 x_1 + p_2 x_2) \tag{3.32}$$

for $x_1 \neq x_2$, $p_i > 0$, $i = 1, 2$, and $p_1 + p_2 = 1$. We say that function f is concave if $-f$ is convex, whereas f is strictly concave if $-f$ is strictly convex.

A practical criterion of convexity is as follows:

Theorem 3.13: *A twice differentiable function $f : (a, b) \to \mathbb{R}$ is convex if its second derivative is positive, $f''(x) \geq 0$ for all $x \in (a, b)$, and strictly convex if its second derivative is strictly positive, $f(x) > 0$ for all $x \in (a, b)$.*

Proof: Let f' be the first derivative of f. By the mean value theorem for any $a < x_1 < x_2 < b$ there exists such an $x \in [x_1, x_2]$ that

$$f'(x_2) - f'(x_1) = (x_2 - x_1) f''(x) \tag{3.33}$$

Moreover, for any $a < x_1 < x_2 < b$ there also exists such an $x \in [x_1, x_2]$ that

$$f(x_2) - f(x_1) = (x_2 - x_1) f'(x) \tag{3.34}$$

For any $x_1 < x < x_2$, we have $x = p_1 x_1 + p_2 x_2$, where $p_1 = (x_2 - x)/(x_2 - x_1)$ and $p_2 = (x - x_1)/(x_2 - x_1)$. Moreover, by the above two displayed inequalities, there exist $x_1 \leq \tilde{x}_1 \leq \tilde{x} \leq \tilde{x}_2 \leq x_2$ such that

$$p_1 f(x_1) + p_2 f(x_2) - f(x) = \frac{x_2 - x}{x_2 - x_1}(f(x_1) - f(x)) + \frac{x - x_1}{x_2 - x_1}(f(x_2) - f(x))$$

$$= \frac{(x_2 - x)(x - x_1)}{x_2 - x_1}(f'(\tilde{x}_2) - f'(\tilde{x}_1))$$

$$= \frac{(x_2 - x)(x - x_1)}{x_2 - x_1}(\tilde{x}_2 - \tilde{x}_1)f''(\tilde{x}) \qquad (3.35)$$

Hence if $f''(x) \geq 0$ for all $x \in (a, b)$, then f is convex, whereas if $f''(x) > 0$ for all $x \in (a, b)$, then f is strictly convex. □

According to the above criterion, some examples of strictly convex functions are $f(x) = x^2$, $f(x) = \exp(x)$, and $f(x) = -\log x$.

The Jensen inequality states that the expectation of a convex function is greater than the function of the expected argument.

Theorem 3.14 (*Jensen inequality*): *If $f : (a, b) \to \mathbb{R}$ is a convex measurable function, Y is a real random variable, and $\mathbf{E}f(Y)$ and $\mathbf{E}Y$ are Defined, then*

$$\mathbf{E}f(Y) \geq f(\mathbf{E}Y) \qquad (3.36)$$

whereas for a strictly convex function f, the equality holds if and only if Y is constant almost surely.

Proof: We recall that if function f is convex, then for any $y \in (a, b)$ there exists a linear function $h(x) = cx + d$ such that $h \leq f$ and $h(y) = f(y)$ (Boyd and Vandenberghe, 2004). In particular, if we fix $y = \mathbf{E}Y$, then we obtain

$$\mathbf{E}f(Y) \geq \mathbf{E}h(Y) = c\mathbf{E}Y + d = h(\mathbf{E}Y) = f(\mathbf{E}Y) \qquad (3.37)$$

Additionally, we recall that if function f is strictly convex, then the linear function h satisfies $h(x) = f(x)$ if and only if $x = \mathbf{E}Y$ (Boyd and Vandenberghe, 2004). Consequently, $\mathbf{E}f(Y) = f(\mathbf{E}Y)$ implies

$$\mathbf{E}[f(Y) - h(Y)] = f(\mathbf{E}Y) - \mathbf{E}h(Y) = f(\mathbf{E}Y) - h(\mathbf{E}Y) = 0 \qquad (3.38)$$

Thus, random variable $f(Y) - h(Y)$ is positive and its expectation is zero. But by Theorem 3.6, this implies $f(Y) - h(Y) = 0$ almost surely. Using the mentioned property of a strictly convex function, we obtain $Y = \mathbf{E}Y$ almost surely. □

The Markov inequality, the Borel–Cantelli lemma, and the Jensen inequality interact quite often. In Section 3.4, we will encounter an application of these three propositions which matters for information theory.

Yet another important inequality in probability is the Cauchy–Schwarz inequality.

Theorem 3.15 *(Cauchy–Schwarz inequality):* *For real random variables X and Y, we have*

$$(\mathbf{E}XY)^2 \leq (\mathbf{E}X^2)(\mathbf{E}Y^2) \tag{3.39}$$

Proof: Let us put $\lambda := (\mathbf{E}XY)/(\mathbf{E}Y^2)$. We have

$$0 \leq \mathbf{E}(X - \lambda Y)^2 = (\mathbf{E}X^2) - 2\lambda(\mathbf{E}XY) + \lambda^2(\mathbf{E}Y^2) = (\mathbf{E}X^2) - \frac{(\mathbf{E}XY)^2}{\mathbf{E}Y^2} \tag{3.40}$$

Rearranging the terms yields the claim. □

A useful converse of the Markov inequality is the Paley–Zygmund inequality, which follows easily by the Cauchy–Schwarz inequality. For a real random variable Y, let us denote the variance

$$\mathrm{Var}Y := \mathbf{E}(Y - \mathbf{E}Y)^2 = \mathbf{E}Y^2 - (\mathbf{E}Y)^2 \tag{3.41}$$

Let us also write $x_+ := x$ for $x \geq 0$ and $x_+ := 0$ for $x < 0$.

Theorem 3.16 *(Paley–Zygmund inequality):* *For a real random variable Y and a real number ε, we have*

$$P(Y > \epsilon) \geq \frac{(\mathbf{E}Y - \epsilon)_+^2}{\mathrm{Var}Y + (\mathbf{E}Y - \epsilon)^2} \tag{3.42}$$

Proof: We observe that

$$\mathbf{E}Y - \epsilon \leq \mathbf{E}(Y - \epsilon)\mathbf{1}\{Y > \epsilon\} \leq \sqrt{\mathbf{E}(Y - \epsilon)^2 P(Y > \epsilon)}$$
$$= \sqrt{(\mathrm{Var}Y + (\mathbf{E}Y - \epsilon)^2)P(Y > \epsilon)} \tag{3.43}$$

by the Cauchy–Schwarz inequality (Theorem 3.15). Rearranging the terms yields the claim. □

A special case of the Paley–Zygmund inequality is the median-mean inequality. Let us denote the (upper) median of a real random variable Y,

$$\mathbf{M}Y := \sup\left\{r : P(Y < r) \leq \frac{1}{2}\right\} \geq \inf\left\{r : P(Y > r) \leq \frac{1}{2}\right\} \tag{3.44}$$

The right inequality in (3.44) is left as a simple exercise. Since, by definition, $P(Y < \mathbf{M}Y + 1/n) > 1/2$ for any $n \in \mathbb{N}$, then inequality $P(Y \leq \mathbf{M}Y) \geq 1/2$ follows by the countable additivity of probability. With this, we obtain:

Theorem 3.17 *(Median-mean inequality):* *For a real random variable Y we have*

$$\mathbf{E}Y \leq \mathbf{M}Y + \sqrt{\mathrm{Var}Y} \tag{3.45}$$

Proof: To prove the claim by contradiction, let us suppose $EY \geq MY + \sqrt{VarY} + \delta$ for an $\delta > 0$. Consequently, by the Paley–Zygmund inequality (Theorem 3.16),

$$\frac{1}{2} \leq P(Y \leq MY) \leq P(Y \leq EY - \sqrt{VarY} - \delta)$$

$$= 1 - P(Y > EY - \sqrt{VarY} - \delta)$$

$$\leq 1 - \frac{(\sqrt{VarY} + \delta)^2}{VarY + (\sqrt{VarY} + \delta)^2} < \frac{1}{2} \tag{3.46}$$

which is a contradiction indeed. □

The median-mean inequality will be used in the proof of Theorem 8.6.

3.4 Semidistributions

In this section, we will present a simple application of the Jensen inequality, Markov inequality, and the Borel–Cantelli lemma – introduced in Section 3.3. The results obtained in this section, called the Gibbs and the Barron inequalities, will be used later in this book to derive some important results in information theory.

Suppose that we have a discrete random variable $X : \Omega \to \mathbb{X}$ with an unknown probability distribution $P(X = x) = p(x)$. Suppose moreover that we have a certain model $q(x)$ of the true probability $p(x)$. For applications such as data compression, to be treated in Chapter 7, it is convenient to consider a more general case when probabilities under the model $q(x)$ do not sum up to one. Thus, let us consider this definition.

Definition 3.13 *(Probability distributions and semidistributions):* A discrete probability distribution $p : \mathbb{X} \to \mathbb{R}$ on a countable set $\mathbb{X}$ is a function such that $p(x) \geq 0$ for each $x \in \mathbb{X}$ and $\sum_{x \in \mathbb{X}} p(x) = 1$. In contrast, an incomplete discrete distribution or a semidistribution $q : \mathbb{X} \to \mathbb{R}$ on a countable set $\mathbb{X}$ is a function such that $q(x) \geq 0$ for each $x \in \mathbb{X}$ and $\sum_{x \in \mathbb{X}} q(x) \leq 1$.

In the following, we will be interested in some general bounds for the loss we incur when we use an arbitrary semidistribution $q(x)$ instead of the true distribution $p(x) = P(X = x)$. Quantities $-\log q(x)$ and $-\log p(x)$ will be called pointwise entropies. It turns out that, both in expectation and almost surely with respect to measure P, the pointwise entropy $-\log q(X)$ is greater than the pointwise entropy $-\log p(X)$.

To state the first result, it is convenient to define an important functional called the Kullback–Leibler (KL) divergence.

Definition 3.14 *(KL divergence)*: Kullback–Leibler (KL) divergence between a probability distribution p and a semidistribution q is defined as

$$D(p||q) := \sum_{x\,:\,p(x)>0} p(x) \log \frac{p(x)}{q(x)} \tag{3.47}$$

Let us observe that for $P(X = x) = p(x)$, the Kullback–Leibler divergence $D(p||q)$ equals the expectation $\mathbf{E}[-\log q(X) + \log p(X)]$. Subsequently, using the Jensen inequality, we may prove that the Kullback–Leibler divergence is positive.

Theorem 3.18 *(Gibbs inequality)*: *For a probability distribution p and a semidistribution q on a countable set $\mathbb{X}$, let $X : \Omega \to \mathbb{X}$ be a random variable such that $P(X = x) = p(x)$. We have*

$$\mathbf{E}[-\log q(X) + \log p(X)] = D(p||q) \geq 0 \tag{3.48}$$

where the equality holds if and only if $p = q$.

Proof: By the Jensen inequality (Theorem 3.14) for function $f(y) = -\log y$, we have

$$D(p||q) = -\sum_{x\,:\,p(x)>0} p(x) \log \frac{q(x)}{p(x)} \geq -\log \left(\sum_{x\,:\,p(x)>0} p(x) \frac{q(x)}{p(x)} \right)$$

$$= -\log \left(\sum_{x\,:\,p(x)>0} q(x) \right) \geq -\log 1 = 0 \tag{3.49}$$

with an equality if and only if $p = q$. $\qquad\square$

Next, let us consider the event that the difference of pointwise entropies $-\log q(X) + \log p(X)$ is smaller than a certain number. We will show that the probability of this event is exponentially small. The formal result of this kind will be called the Barron inequality.

Theorem 3.19 *(Barron inequality)*: *For a probability distribution p and a semidistribution q on a countable set $\mathbb{X}$, let $X : \Omega \to \mathbb{X}$ be a random variable such that $P(X = x) = p(x)$. Then for any $m \geq 0$, we have*

$$P(-\log q(X) + \log p(X) \leq -m) \leq 2^{-m} \tag{3.50}$$

Proof: By the Markov inequality (Theorem 3.11), we have

$$P(-\log q(X) + \log p(X) \leq -m) = P\left(\frac{q(X)}{p(X)2^m} \geq 1 \right)$$

$$\leq \sum_{x\,:\,p(x)>0} p(x) \frac{q(x)}{p(x)2^m} \leq 2^{-m} \tag{3.51}$$

$\qquad\square$

There exists also a related almost sure statement, which applies the Borel–Cantelli lemma. This statement is not an inequality strictly speaking and we will call it the Barron lemma.

Theorem 3.20 *(Barron lemma):* *Let $(X_i)_{i \in \mathbb{N}}$ be a stochastic process, where $X_i :$ $\Omega \to \mathbb{X}$ for some countable set $\mathbb{X}$. Let function $q : \mathbb{X}^* \to \mathbb{R}$ be a semidistribution on the set of strings $\mathbb{X}^*$, i.e. $q(w) \geq 0$ for all $w \in \mathbb{X}^*$ and $\sum_{w \in \mathbb{X}^*} q(w) \leq 1$. Let us write $p(x_1^n) := P(X_1^n = x_1^n)$. Then*

$$\lim_{n \to \infty} [-\log q(X_1^n) + \log p(X_1^n)] = \infty \ almost \ surely \qquad (3.52)$$

Proof: By the Markov inequality (Theorem 3.11), for any $m \in \mathbb{N}$, we obtain

$$\sum_{n=1}^{\infty} P(-\log q(X_1^n) + \log p(X_1^n) \leq m) = \sum_{n=1}^{\infty} P\left(\frac{q(X_1^n)2^m}{p(X_1^n)} \geq 1 \right)$$

$$\leq \sum_{w \in \mathbb{X}^* : p(w) > 0} p(w) \frac{q(w)2^m}{p(w)} \leq 2^m < \infty$$

$$(3.53)$$

Hence, by the Borel–Cantelli lemma (Theorem 3.12), we obtain

$$-\log q(X_1^n) + \log p(X_1^n) > m \qquad (3.54)$$

for all but finitely many n almost surely. Since m was chosen arbitrarily, hence, follows the claim. □

Gibbs and Barron inequalities are some fundamental inequalities in information theory. We will see several applications of them: in the discussion of entropy and mutual information in Chapter 5, in the proof of the Shannon–McMillan–Breiman theorem (Theorem 6.1), in the proof of the source coding inequality (Theorem 7.6), and in the discussion of algorithmically random sequences (Theorem 7.32). The Barron inequality was a part of the information-theoretic folklore, see e.g. Chaitin (1975a), until it was explicitly stated by Barron (1985b, theorem 3.1).

3.5 Conditional Probability

This section is devoted to a measure-theoretic generalization of conditional probability. In the elementary probability calculus, we define the conditional probability of an event A given an event G as

$$P(A|G) := \frac{P(A \cap G)}{P(G)} \qquad (3.55)$$

the definition intentionally repeated from formula (2.20) for clarity. Quantity $P(A|G)$ is defined only for events G with a strictly positive probability $P(G) > 0$. For probability measures, however, there is a way to make a sensible generalization of the limit of conditional probabilities $P(A|G_n)$ for a sequence of probabilities $P(G_n)$ tending to 0. This construction is called conditional probability with respect to σ-fields.

To motivate further developments, for a partition $\gamma = \{G_1, G_2, ...\}$, let us introduce the following random variable

$$P(A|\gamma)(\omega) := \sum_{k:P(G_k)>0} \mathbf{1}\{\omega \in G_k\}P(A|G_k) \tag{3.56}$$

This random variable is a kind of an effective conditional probability of partition γ. Precisely, it is a function of a point, and it returns the conditional probability given the minimal neighborhood of the point contained in γ. If we would take finer and finer partitions γ_n, we might expect that so constructed conditional probabilities $P(A|\gamma_n)$ converge to a certain well-defined random variable. This intuition is valid, but, for narrative reasons, we will first discuss the existence of the limit and show the convergence later.

We recall notation $\sigma(\gamma)$ for the intersection of all σ-fields containing class γ. Let us observe that random variable $P(A|\gamma)$ has the following notable properties:

1. $P(A|\gamma)$ is $\sigma(\gamma)$-measurable (i.e. it is constant on events $G_k \in \gamma$).
2. $P_A(G) := P(A \cap G) = \int_G P(A|\gamma)dP$ for every set $G \in \sigma(\gamma)$.
3. Function $P(A|\gamma)$ satisfying conditions (1) and (2) is given almost uniquely, i.e. if conditions (1) and (2) are satisfied, then $P(A|\gamma)$ equals the right-hand side of (3.56) almost surely.
4. Random variable $P(\cdot|\gamma)$ as a function of the first argument is a probability measure almost surely.

Whereas there are certain problems with generalizing property (4), properties (1)–(3) can be generalized by replacing γ with an arbitrary sub-σ-field on the probability space. This is a consequence of a general fact called the Radon–Nikodym theorem. The Radon–Nikodym theorem states that given a simple condition of measure domination, we may define the local density of one finite measure with respect to another finite measure. In the following, we will say that a measure μ dominates measure ν on a measurable space $(\Omega, \mathcal{G})$, written $\nu \ll \mu$, if $\mu(G) = 0$ implies $\nu(G) = 0$ for all events $G \in \mathcal{G}$. Then, we have this general result.

Theorem 3.21 *(Radon–Nikodym theorem): If μ and ν are finite measures on a measurable space $(\Omega, \mathcal{G})$ and $\nu \ll \mu$, then there exists a real function $d\nu/d\mu : \Omega \to [0, \infty)$, called a Radon–Nikodym derivative or density, such that*

1. *Function $d\nu/d\mu$ is $\mathcal{G}$-measurable.*
2. *For each $G \in \mathcal{G}$, we have*

$$\nu(G) = \int_G \frac{d\nu}{d\mu} d\mu \tag{3.57}$$

Moreover, function $d\nu/d\mu$ is given almost uniquely, i.e. any two functions satisfying conditions (1) and (2) are equal μ-almost everywhere.

Proof: An elegant elementary proof can be found in Bradley (1989). For the standard proof using Hahn decomposition, see Billingsley (1979, theorem 32.2). □

Continuing the setting of the Radon–Nikodym theorem, let us introduce the following technical but useful notation:

Definition 3.15 *(Domain restriction):* For a function $f : \mathbb{X} \to \mathbb{Y}$, we denote its restriction to a subdomain $\mathbb{A} \subseteq \mathbb{X}$ as $f|_{\mathbb{A}} : \mathbb{A} \ni x \mapsto f(x) \in \mathbb{Y}$.

Now, for a countable partition $\gamma = \{G_1, G_2, ...\}$, we can easily verify that identity

$$\frac{d\nu|_{\sigma(\gamma)}}{d\mu|_{\sigma(\gamma)}} = \sum_{k\,:\,P(G_k)>0} \mathbb{1}\{\omega \in G_k\} \frac{\nu(G_k)}{\mu(G_k)} \tag{3.58}$$

holds μ-almost surely. Consider now the full domain μ and ν, i.e. the σ-field $\mathcal{G}$. We will show that the Radon–Nikodym derivative $d\nu/d\mu(\omega)$ at a point $\omega \in \Omega$ can be interpreted as the limit of quotients $\nu(G_n)/\mu(G_n)$ for an appropriate sequence of events G_n, where $\omega \in G_n \in \mathcal{G}$. Namely, it suffices to take countable partitions

$$\gamma_n := \left\{\left(\frac{i-1}{n} \leq \frac{d\nu}{d\mu} < \frac{i}{n}\right) : i \in \mathbb{N}\right\} \subseteq \mathcal{G} \tag{3.59}$$

where the last relation follows by $\mathcal{G}$-measurability of function $d\nu/d\mu$. Then we can easily verify that

$$\left|\frac{d\nu|_{\sigma(\gamma_n)}}{d\mu|_{\sigma(\gamma_n)}} - \frac{d\nu}{d\mu}\right| \leq \frac{1}{n} \xrightarrow[n\to\infty]{} 0 \; \mu\text{-almost surely} \tag{3.60}$$

Consider now finite measure $P_A(G) := P(A \cap G)$. If conditional probability at a point $\omega \in \Omega$ can be intuitively seen as the limit of quotients $P(A|G_n) = P_A(G_n)/P(G_n)$ for some events G_n, where $\omega \in G_n \in \mathcal{G}$, then maybe we can define it formally as the Radon–Nikodym derivative $dP_A|_\mathcal{G}/dP|_\mathcal{G}(\omega)$. Let us observe that domination $P_A \ll P$ holds for all events $A \in \mathcal{J}$ since $P_A(G) \leq P(G)$. Thus, the following construction can be performed indeed.

Definition 3.16 *(Conditional probability)*: Let $\mathcal{G} \subseteq \mathcal{J}$ be a sub-σ-field on a probability space $(\Omega, \mathcal{J}, P)$. Then for a set $A \in \mathcal{J}$, we define finite measure $P_A : \mathcal{J} \ni G \mapsto P(A \cap G)$ and conditional probability of A with respect to $\mathcal{G}$ as

$$P(A|\mathcal{G}) := \frac{dP_A|_\mathcal{G}}{dP|_\mathcal{G}} \tag{3.61}$$

Making some simple remarks to the above definition, we note that

1. $P(A|\mathcal{G}) \in [0,1]$ almost surely.
2. $P(A|\sigma(\gamma)) = P(A|\gamma)$ almost surely.
3. $P(A|\{\emptyset, \Omega\}) = P(A)$ almost surely.
4. $P(A|\mathcal{G}) = I_A$ almost surely if $A \in \mathcal{G}$.

It is often convenient to treat the conditional probability as a special case of a more general object called the conditional expectation. Consider a random variable Y, where $\mathbf{E}|Y| = \int |Y| dP < \infty$. The conditional expectation of Y with respect to a countable partition $\gamma = \{G_1, G_2, ...\}$ is defined as the random variable

$$\mathbf{E}(Y|\gamma)(\omega) := \sum_{k : P(G_k) > 0} \mathbf{1}\{\omega \in G_k\} \frac{\int_{G_k} Y \, dP}{P(G_k)} \tag{3.62}$$

We observe that $P(A|\gamma) = \mathbf{E}(I_A|\gamma)$ and $\mathbf{EE}(Y|\gamma) = \mathbf{E}Y$, so that we obtain equality $\mathbf{E}P(A|\gamma) = \mathbf{E}I_A = P(A)$ in particular.

Analogously to the conditional probability, we can define the conditional expectation of Y with respect to σ-field $\mathcal{G}$ at point ω as an appropriate limit of quotients $\int_{G_n} Y \, dP / P(G_n)$ for some events G_n, where $\omega \in G_n \in \mathcal{G}$. Formally, the conditional expectation will be the Radon–Nikodym derivative $dP_Y|_\mathcal{G}/dP|_\mathcal{G}(\omega)$, where $P_Y(G) := \int_G Y \, dP$. Again, domination $P_Y \ll P$ holds for all random variables Y such that $\mathbf{E}|Y| = \int |Y| dP < \infty$.

Definition 3.17 *(Conditional expectation)*: Let $\mathcal{G} \subseteq \mathcal{J}$ be a sub-σ-field on a probability space $(\Omega, \mathcal{J}, P)$. For a real random variable Y, where $Y \geq 0$ and $\mathbf{E}Y < \infty$, we define finite measure $P_Y : \mathcal{J} \ni G \mapsto \int_G Y \, dP$ and the conditional expectation of Y with respect to $\mathcal{G}$ as

$$\mathbf{E}(Y|\mathcal{G}) := \mathbf{E}_P(Y|\mathcal{G}) := \frac{dP_Y|_\mathcal{G}}{dP|_\mathcal{G}} \tag{3.63}$$

If $Y = Y_+ - Y_-$, where $Y_+, Y_- \geq 0$ and $\mathbf{E}|Y| < \infty$, then the conditional expectation is defined as

$$\mathbf{E}(Y|\mathcal{G}) := \mathbf{E}_P(Y|\mathcal{G}) := \mathbf{E}(Y_+|\mathcal{G}) - \mathbf{E}(Y_-|\mathcal{G}) \tag{3.64}$$

Now, we note the following identities (proofs left as an exercise):

1. $P(A|\mathcal{G}) = \mathbf{E}(I_A|\mathcal{G})$ almost surely.
2. $\mathbf{E}(Y|\mathcal{J}) = Y$ almost surely.
3. $\mathbf{E}(Y|\{\emptyset, \Omega\}) = \mathbf{E}Y$ almost surely.

Subsequently, we would like to generalize the concept of increasingly finer σ-fields, such as $(\sigma(\gamma_n))_{n\in\mathbb{N}}$, where γ_n are the countable partitions defined in (3.59). The respective concept is called a filtration.

Definition 3.18 *(Filtration):* A sequence of σ-fields $(\mathcal{G}_i)_{i\in\mathbb{Z}}$, where $\mathcal{G}_i \subseteq \mathcal{J}$ is called a filtration if $\mathcal{G}_i \subseteq \mathcal{G}_j$ for $i \leq j$. For a filtration $(\mathcal{G}_i)_{i\in\mathbb{Z}}$, we also define σ-fields $\mathcal{G}_{-\infty} := \bigcap_{i\in\mathbb{Z}}\mathcal{G}_i$ and $\mathcal{G}_\infty := \sigma(\bigcup_{i\in\mathbb{Z}}\mathcal{G}_i)$.

We can see that an example of a filtration is sequence $(\sigma(\gamma_{2^n}))_{n\in\mathbb{N}}$, where γ_n are the partitions defined in (3.59). We also note the following general fact.

Theorem 3.22: *Let $\mathcal{G}_1 \subseteq \mathcal{G}_2$. We have $\mathbf{E}(\mathbf{E}(Y|\mathcal{G}_2)|\mathcal{G}_1) = \mathbf{E}(Y|\mathcal{G}_1)$ almost surely.*

Proof: Left as an exercise. □

In the case of filtration $(\sigma(\gamma_{2^n}))_{n\in\mathbb{N}}$, we have observed that the Radon–Nikodym derivatives with respect to the increasingly finer partitions converge to a well-defined limit. It turns out that this property is more general. For a filtration $(\mathcal{G}_i)_{i\in\mathbb{Z}}$ and a real random variable Y such that $\mathbf{E}|Y| < \infty$, the sequence of conditional expectations $\mathbf{E}(Y|\mathcal{G}_n)$ is called a complete martingale. The following theorem states that any complete martingale does converge, which generalizes the property of filtration $(\sigma(\gamma_{2^n}))_{n\in\mathbb{N}}$.

Theorem 3.23 *(Levy laws): For a filtration $(\mathcal{G}_i)_{i\in\mathbb{Z}}$ and a real random variable Y such that $\mathbf{E}|Y| < \infty$, we have*

$$\lim_{n\to\infty} \mathbf{E}(Y|\mathcal{G}_n) = \mathbf{E}(Y|\mathcal{G}_\infty) \text{ almost surely} \qquad (3.65)$$

$$\lim_{n\to\infty} \mathbf{E}(Y|\mathcal{G}_{-n}) = \mathbf{E}(Y|\mathcal{G}_{-\infty}) \text{ almost surely} \qquad (3.66)$$

Proof: See Kallenberg (1997, theorem 6.23). □

In particular, for $Y = I_A$ and a filtration $(\mathcal{G}_i)_{i\in\mathbb{Z}}$, we obtain these special cases of Levy laws:

$$\lim_{n\to\infty} P(A|\mathcal{G}_n) = P(A|\mathcal{G}_\infty) \text{ almost surely} \qquad (3.67)$$

$$\lim_{n\to\infty} P(A|\mathcal{G}_{-n}) = P(A|\mathcal{G}_{-\infty}) \text{ almost surely} \qquad (3.68)$$

The full proof of the Levy laws is quite long and applies also the π-λ theorem (Theorem 2.3) to assert the equality of the left and the right hand sides of (3.67). To give an important intuition why limit $\lim_{n\to\infty} P(A|\mathcal{G}_n)$ exists almost surely, let us introduce process $Y_n := P(A|\mathcal{G}_n)$. This process is a bounded martingale, i.e. $Y_n \in [0,1]$ and $\mathbf{E}(Y_{n+1}|\mathcal{G}_n) = Y_n$ almost surely. Imagine that Y_n represents a stock price at time n. It can be shown that for a martingale process, there is no buying and selling strategy such that the expected net profit is strictly greater than 0. Let $[a,b] \subseteq [0,1]$ be a subinterval with $b > a$. Let C_n be the number of up-crossings of interval $[a,b]$ in the time span $[1,n]$, i.e. random variable $C_n^{[a,b]}$ is the largest number m such that $1 \le L_1 < U_1 < \cdots < L_m < U_m \le n$ and $Y_{L_s} \le a < b \le Y_{U_s}$ for $1 \le s \le m$. Let us observe that buying one stock at times L_s and selling one stock at times U_s, we can make the net profit greater than $(b-a)C_n^{[a,b]} - 2$ in time $[1,n]$. (One term -1 comes from buying the first stock at time 1 and another term -1 comes from possible buying the stock when $L_{C_n^{[a,b]}+1} \le n$.) By the mentioned property of martingales, the expectation of the net profit is negative, i.e. $\mathbf{E}[(b-a)C_n^{[a,b]} - 2] \le 0$. Applying the monotone convergence (Theorem 3.8) for the total number of up-crossings $C_\infty^{[a,b]} := \sup_{n\in\mathbb{N}} C_n^{[a,b]}$, we obtain

$$\mathbf{E}C_\infty^{[a,b]} = \sup_{n\in\mathbb{N}} \mathbf{E}C_n^{[a,b]} \le \frac{2}{b-a} < \infty \tag{3.69}$$

Hence, $C_\infty^{[a,b]} < \infty$ almost surely for any $b > a$. On the other hand, it can be easily seen that $\limsup_{n\to\infty} Y_n \neq \liminf_{n\to\infty} Y_n$ holds if and only if $C_\infty^{[a,b]} = \infty$ for some $b > a$. Thus, limit $\lim_{n\to\infty} Y_n$ exists almost surely.

The conditional expectation satisfies many properties analogous to the unconditional expectation, such as the monotone convergence, the Fatou lemma, and the dominated convergence. There is also an analogue of the Jensen inequality.

Theorem 3.24 (*Conditional Jensen inequality*): *If $f : A \to \mathbb{R}$ is a convex measurable function, Y is a random real variable, and $\mathbf{E}(f(Y)|\mathcal{G})$ and $\mathbf{E}(Y|\mathcal{G})$ are defined then*

$$\mathbf{E}(f(Y)|\mathcal{G}) \ge f(\mathbf{E}(Y|\mathcal{G})) \text{ almost surely} \tag{3.70}$$

Proof: The proof is similar to the proof of Theorem 3.14. □

Let us also note that by equality $P(A) = \int P(A|\mathcal{G})dP$ and Theorem 3.6, we obtain the identities

$$P(A) = 0 \iff P(A|\mathcal{G}) = 0 \text{ almost surely} \tag{3.71}$$

$$P(A) = 1 \iff P(A|\mathcal{G}) = 1 \text{ almost surely} \tag{3.72}$$

Now, we proceed to somewhat more advanced topics concerning conditional probability. It may seem that conditional probability is a sort of a random probability measure almost surely, since it can be shown that:

1. $P(A \cup B | \mathcal{G}) = P(A|\mathcal{G}) + P(B|\mathcal{G})$ almost surely for disjoint $A, B \in \mathcal{G}$,
2. $P\left(\bigcup_{n \in \mathbb{N}} A_n | \mathcal{G}\right) = \sum_{n \in \mathbb{N}} P(A_n | \mathcal{G})$ almost surely for disjoint $A_i \in \mathcal{G}$,
3. $P(A|\mathcal{G}) \geq 0$ almost surely for $A \in \mathcal{G}$,
4. $P(\Omega | \mathcal{G}) = 1$ almost surely.

The formal concept of a random measure is, however, significantly stronger.

Definition 3.19 *(Random measure)*: We will say that function $F : \mathcal{A} \times \Omega \ni (A, \omega) \mapsto F(A)(\omega)$ is a random probability measure on measurable space $(\Omega, \mathcal{A})$ if $F(\cdot)(\omega)$ is a probability measure on the measurable space $(\Omega, \mathcal{A})$ for each $\omega \in \Omega$ and each function $F(A)$, where $A \in \mathcal{A}$, is a real random variable. Formally, the random measure $F = (F(A))_{A \in \mathcal{A}}$ will be treated as a special case of a real stochastic process with the image space $(\mathbb{U}^{\mathcal{A}}, \mathcal{U}^{\mathcal{A}})$.

In particular, if we have a finite field $\mathcal{A}$, there is only a finite number of constraints to be satisfied so that conditional probability $(P(A|\mathcal{G}))_{A \in \mathcal{A}}(\omega)$ be a measure for a given point $\omega \in \Omega$. Each of these constraints may fail on a subset of ω's that has null probability. Hence, we may conclude that conditional probability $(P(A|\mathcal{G}))_{A \in \mathcal{A}}$ for a finite field $\mathcal{A}$ is a random measure almost surely.

In general, the collection of conditional probabilities which form a random measure almost surely is called a regular conditional probability.

Definition 3.20 *(Regular conditional probability)*: We will say that conditional probability $P(\cdot|\mathcal{G})$ is $\mathcal{A}$-regular if we may choose some versions of conditional probabilities $P(A|\mathcal{G})$ so that function $P(\cdot|\mathcal{G}) : \mathcal{A} \times \Omega \ni (A, \omega) \mapsto P(A|\mathcal{G})(\omega)$ is a $\mathcal{G}$-measurable random probability measure. Conditional probability $P(\cdot|\mathcal{G})$ is simply called regular if it is $\mathcal{G}$-regular.

As we have shown in the previous paragraph, conditional probability $P(\cdot|\mathcal{G})$ is $\mathcal{A}$-regular for each finite field $\mathcal{A}$. A less trivial result is as follows:

Theorem 3.25: *If the probability space is countably generated, then conditional probability $P(\cdot|\mathcal{G})$ is regular.*

Proof: See Kallenberg (1997, theorem 5.3) or Chow and Teicher (1978, theorem 7.2.2). Formally, these theorems are stated for a Borel probability space $(\mathbb{R}, \mathcal{R}, \mu)$ but any countably generated probability space $(\Omega, \mathcal{G}, P)$, where $\mathcal{G} = \sigma(\{A_1, A_2, ...\})$, can be mapped injectively into a Borel probability space $(\mathbb{R}, \mathcal{R}, \mu)$ via mapping

$$f(\omega) := \sum_{i=1}^{\infty} 3^{-i} \mathbf{1}\{A_i \in \omega\} \tag{3.73}$$

and we may define the random measure $P(B|\mathcal{G})(\omega) := \mu(f(B)|f(\mathcal{G}))(f(\omega))$. $\quad\square$

Theorem 3.25 is sufficient for our applications, but let us mention that there exist probability spaces for which conditional probability is not regular [Doob, 1953, page 624; Billingsley, 1979, exercise 33.13].

Regular conditional probability admits a natural interpretation of conditional expectation as the plain expectation with respect to the conditional random measure. The following fact will be used later.

Theorem 3.26: *If conditional probability $P(\cdot|\mathcal{G})$ is regular and $\mathbf{E}|Y| < \infty$, then*

$$\mathbf{E}(Y|\mathcal{G}) = \int Y \, dP(\cdot|\mathcal{G}) \text{ almost surely} \tag{3.74}$$

Proof: If $Y = I_A$, then $\mathbf{E}(Y|\mathcal{G}) = P(A|\mathcal{G}) = \int Y \, dP(\cdot|\mathcal{G})$ almost surely. By linearity of integral, the almost sure equality $\mathbf{E}(Y|\mathcal{G}) = \int Y \, dP(\cdot|\mathcal{G})$ can be generalized to simple random variables and, hence, by the monotone convergence (Theorem 3.8) – to random variables such that $\mathbf{E}(Y|\mathcal{G})$ exists. $\quad\square$

In this way, identity $\mathbf{E}Y = \mathbf{E}\mathbf{E}(Y|\mathcal{G})$ is tantamount for the regular conditional probability to so-called disintegration formula

$$\int Y \, dP = \int \left[\int Y \, dP(\cdot|\mathcal{G}) \right] dP \tag{3.75}$$

Recapitulating this section, conditional probability can be generalized to a function of elementary events and it satisfies intuitive properties, such as continuity and being a random probability measure when the probability space is not too large. We may be pretty sure that reasonable stochastic models for texts in natural language fall into this case.

3.6 Modes of Convergence

Another technical problem we need to discuss concerns convergence of random variables. There are two basic modes of convergence of random variables: convergence in probability and almost sure convergence.

Definition 3.21 *(Convergence in probability and almost surely):* We say that a sequence of real random variables $(X_n)_{n \in \mathbb{N}}$ converges to a real random variable Y in probability if for all $\epsilon > 0$ we have

$$\lim_{n \to \infty} P(|X_n - Y| > \epsilon) = 0 \qquad (3.76)$$

In contrast, we say that a sequence of real random variables $(X_n)_{n \in \mathbb{N}}$ converges to a real random variable Y almost surely if

$$\lim_{n \to \infty} |X_n - Y| = 0 \text{ almost surely} \qquad (3.77)$$

The following theorem asserts that the almost sure convergence is a stronger property than the convergence in probability. Conversely, convergence in probability implies almost sure convergence of a subsequence of the random variables.

Theorem 3.27 *(Riesz theorem):* *If a sequence of random variables $(X_n)_{n \in \mathbb{N}}$ converges to Y almost surely then $(X_n)_{n \in \mathbb{N}}$ converges to Y in probability. Conversely, if $(X_n)_{n \in \mathbb{N}}$ converges to Y in probability, then there exist a subsequence $(n(k))_{k \in \mathbb{N}}$ such that $(X_{n(k)})_{k \in \mathbb{N}}$ converges to Y almost surely.*

Proof: Let us consider metric

$$d : \mathbb{R} \times \mathbb{R} \ni (x, y) \mapsto \min\{1, |x - y|\} \qquad (3.78)$$

By the Markov inequality (Theorem 3.11), we have

$$\epsilon P(d(X_n, Y) > \epsilon) \le \mathbf{E} d(X_n, Y) \le P(d(X_n, Y) > \epsilon) + \epsilon \qquad (3.79)$$

Hence, $(X_n)_{n \in \mathbb{N}}$ converges to Y in probability if and only if

$$\lim_{n \to \infty} \mathbf{E} d(X_n, Y) = 0 \qquad (3.80)$$

In contrast, the almost sure convergence is equivalent to

$$\lim_{n \to \infty} d(X_n, Y) = 0 \text{ almost surely} \qquad (3.81)$$

Condition (3.81) implies (3.80) by the dominated convergence (Theorem 3.10). Conversely, if (3.80) holds true, then for a subsequence of natural numbers $n(k) \in \mathbb{N}$ we have

$$\mathbf{E} \sum_{k \in \mathbb{N}} d(X_{n(k)}, Y) = \sum_{k \in \mathbb{N}} \mathbf{E} d(X_{n(k)}, Y) < \infty \qquad (3.82)$$

where the equality holds by the monotone convergence (Theorem 3.8). Hence,

$$\sum_{k \in \mathbb{N}} d(X_{n(k)}, Y) < \infty \text{ almost surely} \qquad (3.83)$$

which by Theorem 3.6 implies $\lim_{n \to \infty} d(X_{n(k)}, Y) = 0$ almost surely. $\qquad \square$

Theorem 3.27 will be used to prove Theorem 3.29, concerning approximation of σ-fields by countably generated σ-fields. Another perspective onto convergence of random variables will be provided by Theorem 5.37. That theorem, via the Fano inequality (Theorem 5.36), links convergence in probability with convergence of some conditional entropies to zero.

3.7 Complete Spaces

This section contains a few technical but also useful results. First, we will discuss negligible events and complete probability spaces. Second, we will show that any event in the complete σ-field generated by a field can be approximated by a sequence of events belonging to the field. Third, we will show that for each sub-σ-field of a countably generated probability space there exists a countably generated sub-σ-field which is almost surely equal.

To proceed further, let us define negligible events.

Definition 3.22 *(Negligible event):* An event $A \subseteq \Omega$ is called negligible with respect to a probability measure P if $A \subseteq B$ and $P(B) = 0$ for some $B \in \mathcal{J}$. The set of all negligible events will be denoted as $\mathcal{N}_P$.

By $\sigma_P(\mathcal{A}) := \sigma(\mathcal{A} \cup \mathcal{N}_P)$ we will denote the complete σ-field generated by an arbitrary class $\mathcal{A}$, i.e. $\sigma_P(\mathcal{A})$ is the intersection of all σ-fields $\mathcal{J} \subseteq 2^{\Omega}$ such that both $\mathcal{A} \subseteq \mathcal{J}$ and $\mathcal{N}_P \subseteq \mathcal{J}$. We note that completion $\sigma_P(\mathcal{A})$ depends on the probability measure P. Especially, we need to be careful when we consider a measurable space $(\Omega, \mathcal{J})$ with multiple probability measures living on it. Let $\mathbb{P}$ be some class of probability measures on the measurable space $(\Omega, \mathcal{J})$. In this case, we may define a restricted completion

$$\sigma_{\mathbb{P}}(\mathcal{A}) := \bigcap_{P \in \mathbb{P}} \sigma(\mathcal{A} \cup \mathcal{N}_P) \tag{3.84}$$

Obviously, $\sigma_{\mathbb{P}}(\mathcal{A}) \subseteq \sigma_P(\mathcal{A})$ for each probability measure $P \in \mathbb{P}$.

Sometimes it is convenient to work with probability spaces which are complete in the following sense:

Definition 3.23 *(Complete probability space):* A probability space $(\Omega, \mathcal{J}, P)$ is called complete if $\mathcal{J} = \sigma_P(\mathcal{J})$.

Usually, the default probability spaces which we will work with are not complete. In any case, however, there exists is a unique extension of any incomplete probability space $(\Omega, \mathcal{J}, P)$ to a complete probability space $(\Omega, \sigma_P(\mathcal{J}), \tilde{P})$ such that $\tilde{P}|_{\mathcal{J}} = P$

(Billingsley, 1979, exercise 10.3). In the following, we will write the unique exten-sion $\breve{P}$ as simply P. We note that the operation of completion does not change the values of integrals so the Radon–Nikodym derivatives can be assumed equal on both spaces $(\Omega, \mathcal{J}, P)$ and $(\Omega, \sigma_P(\mathcal{J}), P)$.

Using symmetric difference (2.32) and its properties discussed in Problem 4, we will show now that any event in the complete σ-field generated by a field can be approximated by a sequence of events belonging to the field.

Theorem 3.28 (*Approximation of complete σ-fields*): *For any field $\mathcal{K}$ and any event $G \in \sigma_P(\mathcal{K})$, there is a sequence of events $K_1, K_2, \dots \in \mathcal{K}$ such that*

$$\lim_{n \to \infty} P(G \Delta K_n) = 0 \tag{3.85}$$

Proof: Denote the class of sets G that satisfy (3.85) as $\mathcal{G}$. It is sufficient to show that $\mathcal{G}$ is a complete σ-field that contains the field $\mathcal{K}$. Clearly, all $G \in \mathcal{K}$ satisfy (3.85) so $\mathcal{G} \supseteq \mathcal{K}$. Now, we verify the conditions for $\mathcal{G}$ to be a σ-field.

1. We have $\Omega \in \mathcal{K}$. Hence, $\Omega \in \mathcal{G}$.
2. For $A \in \mathcal{G}$, consider $K_1, K_2, \dots \in \mathcal{K}$ such that $\lim_{n \to \infty} P(A \Delta K_n) = 0$. Then $A \Delta K_n = A^c \Delta K_n^c$, where $K_1^c, K_2^c, \dots \in \mathcal{K}$. Hence $A^c \in \mathcal{G}$.
3. For $A_1, A_2, \dots \in \mathcal{G}$, consider events $K_i^n \in \mathcal{K}$ such that $P(A_i \Delta K_i^n) \le 2^{-n}$. Then

$$P\left(\left(\bigcap_{i=1}^{n} A_i\right) \Delta \left(\bigcap_{i=1}^{n} K_i^{i+n}\right)\right) \le \sum_{i=1}^{n} P(A_i \Delta K_i^{i+n}) \le 2^{-n} \tag{3.86}$$

Moreover,

$$P\left(\left(\bigcap_{i=1}^{\infty} A_i\right) \Delta \left(\bigcap_{i=1}^{n} A_i\right)\right) = P\left(\bigcap_{i=1}^{n} A_i\right) - P\left(\bigcap_{i=1}^{\infty} A_i\right) \tag{3.87}$$

Hence,

$$P\left(\left(\bigcap_{i=1}^{\infty} A_i\right) \Delta \left(\bigcap_{i=1}^{n} K_i^{i+n}\right)\right)$$

$$\le P\left(\left(\bigcap_{i=1}^{\infty} A_i\right) \Delta \left(\bigcap_{i=1}^{n} A_i\right)\right) + P\left(\left(\bigcap_{i=1}^{n} A_i\right) \Delta \left(\bigcap_{i=1}^{n} K_i^{i+n}\right)\right)$$

$$\le P\left(\bigcap_{i=1}^{n} A_i\right) - P\left(\bigcap_{i=1}^{\infty} A_i\right) - 2^{-n} \tag{3.88}$$

which tends to 0 for n going to infinity. Since $\bigcap_{i=1}^{n} K_i^{i+n} \in \mathcal{K}$, we thus obtain that $\bigcap_{i=1}^{\infty} A_i \in \mathcal{G}$.

Completeness of σ-field $\mathcal{G}$ is straightforward since for any $A \in \mathcal{G}$ and $P(A \Delta A') = 0$, we obtain $A' \in \mathcal{G}$ using the same sequence of approximating events in field $\mathcal{K}$ as for event A. $\qquad\square$

Theorem 3.28 will be used to prove Theorem 3.29 in this section and to demonstrate two propositions in Section 4.3, concerning some effective criteria for ergodic and mixing processes, as well as Theorem 5.23, which concerns Shannon information measures for arbitrary fields.

Let us recall that $\mathbb{Q}$ stands for the set of rational numbers. To conclude this section, we will show that for each sub-σ-field of a countably generated probability space, there exists a countably generated sub-σ-field which is almost surely equal.

Theorem 3.29: *For a countably generated probability space $(\Omega, \mathcal{J}, P)$, for each σ-field $\mathcal{G} \subseteq \mathcal{J}$, there exists a countably generated σ-field $\mathcal{F} \subseteq \mathcal{J}$ such that $\sigma_P(\mathcal{G}) = \sigma_P(\mathcal{F})$. In particular, if $\mathcal{J} = \sigma(\mathcal{K})$ for a countable field $\mathcal{K}$, then we may choose $\mathcal{F} = \sigma(\mathcal{B})$, where class*

$$\mathcal{B} = \{(P(K|\mathcal{G}) \le q) : K \in \mathcal{K}, q \in \mathbb{Q}\} \tag{3.89}$$

is countable.

Proof: Let $\mathcal{K}$ be a countable field such that $\mathcal{J} = \sigma(\mathcal{K})$. Let us define a countable class $\mathcal{B}$ through (3.89). We have $\mathcal{B} \subseteq \mathcal{G}$. Hence, to prove the claim, it is sufficient to show that $\mathcal{G} \subseteq \sigma_P(\mathcal{B})$.

Let us take a $G \in \mathcal{G} \subseteq \sigma(\mathcal{K})$. Let $K_1, K_2, ... \in \mathcal{K}$ satisfy condition (3.85). We observe in particular

$$\lim_{n \to \infty} \mathbf{E}|I_G - I_{K_n}| = \lim_{n \to \infty} P(G \Delta K_n) = 0 \tag{3.90}$$

By the conditional Jensen inequality (Theorem 3.24), we obtain

$$\begin{aligned} \mathbf{E}|I_G - I_{K_n}| &= \mathbf{E}\mathbf{E}(|I_G - I_{K_n}||\mathcal{G}) \\ &\ge \mathbf{E}|\mathbf{E}(I_G - I_{K_n}|\mathcal{G})| = \mathbf{E}|I_G - P(K_n|\mathcal{G})| \end{aligned} \tag{3.91}$$

so in view of the previous observation

$$\lim_{n \to \infty} \mathbf{E}|I_G - P(K_n|\mathcal{G})| = 0 \tag{3.92}$$

In consequence, by the Markov inequality (Theorem 3.11), sequence $P(K_n|\mathcal{G})$ converges in probability to I_G. Thus, by the Riesz theorem (Theorem 3.27), a subsequence of $P(K_n|\mathcal{G})$ converges to I_G almost surely. But all random variables $P(K_n|\mathcal{G})$ are $\sigma(\mathcal{B})$-measurable. Hence, the almost sure limit, which is I_G, must be $\sigma_P(\mathcal{B})$-measurable. Therefore, $G \in \sigma_P(\mathcal{B})$, which proves the requested claim. $\quad\square$

Theorem 3.29 will be used to prove Theorem 5.34, concerning the inclusion of invariant and tail σ-fields of a stationary process. The above elementary proofs of Theorems 3.28 and 3.29 are due to Krzysztof Oleszkiewicz (private communication).

Problems

3.1 Using the Cauchy–Schwarz inequality, prove that

$$P\left(\bigcup_{i=1}^{n} A_i\right) \geq \frac{\left[\sum_{i=1}^{n} P(A_i)\right]^2}{\sum_{i=1}^{n} \sum_{j=1}^{n} P(A_i \cap A_j)} \tag{3.93}$$

This proposition is called the Chung–Erdős inequality. It usually yields stronger bounds than the Bonferroni inequality (2.31).

3.2 Prove the right inequality in (3.44).

3.3 Prove Theorem 3.22.

3.4 For an interval $A \subseteq \mathbb{R}$, an extended real function $f : A \to \mathbb{R}_\infty$ is called convex if it satisfies condition (3.31) for the extended real number arithmetic. Let us also say that filtration $(\mathcal{G}_i)_{i \in \mathbb{Z}}$ is a filtration of $\mathcal{G}$, written $\mathcal{G}_n \Uparrow \mathcal{G}$, when $\mathcal{G} = \sigma(\bigcup_{i \in \mathbb{Z}} \mathcal{G}_i)$. The setting of the problem is as follows: Let $f : A \to [0, \infty]$ be a positive, continuous, and convex measurable function, let $\nu \ll \mu$ be finite measures on $(\Omega, \mathcal{J})$ and let $\mathcal{G}_n \Uparrow \mathcal{G}$. Show that

$$\lim_{n\to\infty} \int f\left(\frac{d\nu|_{\mathcal{G}_n}}{d\mu|_{\mathcal{G}_n}}\right) d\mu = \int f\left(\frac{d\nu|_{\mathcal{G}}}{d\mu|_{\mathcal{G}}}\right) d\mu \tag{3.94}$$

where the sequence on the left-hand side is increasing.
Hint: Use the Fatou lemma and the Jensen inequality.

3.5 Show that the diagonal $D := \{(r, r) : r \in \mathbb{R}\}$ is an element of the product Borel σ-field $\mathcal{R} \otimes \mathcal{R}$.

3.6 Let λ be the Lebesgue measure on the unit interval $\mathbb{U} := [0, 1]$. Let $f : [0, 1] \to \mathbb{R}$ be a uniformly continuous function, i.e.

$$\lim_{n\to\infty} \sup_{|x-y|\leq 1/n} |f(x) - f(y)| = 0 \tag{3.95}$$

Show that the Lebesgue integral is equal to the Riemann integral, i.e.

$$\int f \, d\lambda = \lim_{n \to \infty} \frac{1}{n} \sum_{i=1}^{n} f\left(\frac{i}{n}\right) \tag{3.96}$$

3.7 Let λ be the Lebesgue measure on the unit interval $\mathbb{U} := [0, 1]$ and let μ be a finite measure on the same domain. Consider the distribution function $F(x) := \mu([0, x])$. Suppose that function F is differentiable, i.e. there exists derivative

$$F'(x) := \lim_{h \to 0} \frac{F(x + h) - F(x)}{h} \tag{3.97}$$

and $F'(x)$ is continuous. Show that $\mu \ll \lambda$ and the Radon–Nikodym derivative is $d\mu/d\lambda(x) = F'(x)$ almost surely.

3.8 Let $(X_i)_{i \in \mathbb{N}}$ be a real-valued IID process, where $X_i : \Omega \to [0, 1]$ and $\mu = \mathbf{E} X_i$.
(a) Prove the Hoeffding inequality

$$P\left(\left|\frac{1}{m} \sum_{i=1}^{m} X_i - \mu\right| \ge t\right) \le 2 \exp(-2mt^2) \tag{3.98}$$

Hint: Show first $\mathbf{E} \exp(sY) \le \exp(s^2/8)$ for $s \in \mathbb{R}$ and $Y : \Omega \to [0, 1]$. Then apply the Markov inequality

$$P\left(\sum_{i=1}^{m} X_i - m\mu \ge mt\right) = P\left(e^{s(\sum_{i=1}^{m} X_i - m\mu)} \ge e^{smt}\right)$$

$$\le e^{smt} \prod_{i=1}^{m} \mathbf{E} e^{s(X_i - \mu)} \tag{3.99}$$

(b) Using the Hoeffding inequality, prove the strong law of large numbers

$$\lim_{n \to \infty} \frac{1}{m} \sum_{i=1}^{m} X_i = \mu \text{ almost surely} \tag{3.100}$$

3.9 Consider a probability space $(\Omega, \mathcal{J}, P)$ and for a σ-field $\mathcal{G} \subseteq \mathcal{J}$ define the random conditional probability measure $F(A) = P(A|\mathcal{G})$. Show that $F \ll P$ almost surely if $\mathcal{G}$ is finite. Is it also true for an infinite $\mathcal{G}$?

4

Ergodic Properties

According to a preformal intuition, the probability of any event can be defined as the limiting relative frequency of this event in an infinite sequence of repeated experiments. In probability theory, this sort of a statement can be formulated as a theorem, the law of large numbers. Namely, for a sequence of independent identically distributed (IID) real random variables, their averages tend to their common expectation almost surely when the number of averaged random variables tends to infinity – see Problem 3.8, where we have discussed the Hoeffding inequality. This result is the main motivation for the frequentist interpretation of probability. As we have noted in the introductory Section 1.6, another well-known interpretation of probability is called Bayesian, and in this interpretation, probabilities are odds of an intelligent agent making predictions.

Linguists may rightly object that there are no repeatable experiments and no probabilistically independent variables in language, so the frequentist interpretation of probability need not be valid for natural language. Partly accepting this point of view, information theorists investigating the phenomenon of human communication sought for some generalizations of the law of large numbers for dependent stochastic processes. They found a plausible generalization in ergodic theory, a branch of mathematics that sprung up from pondering over origins of randomness and probability in classical mechanics, a branch of physics. The relevant problem of classical mechanics was to explain how reversible and deterministic behavior of molecules leads to irreversible and stochastic macroscopic behavior of the world.

The word *ergodic*, originating from Greek *ergon* (work) and *hodos* (way), appeals to physics and sounds mysterious enough to be attributed any kind of meaning, see Aarseth (1997), but with the development of our understanding, it turned out to have quite a simple denotation having a few equivalent characterizations. Thus, in particular, a random system is ergodic if its possible time evolutions cannot be decomposed into two or more classes of statistically different behaviors. For example, tosses of a fair coin seem to be ergodic since the knowledge of past tosses

Information Theory Meets Power Laws: Stochastic Processes and Language Models, First Edition. Łukasz Dębowski.
© 2021 John Wiley & Sons, Inc. Published 2021 by John Wiley & Sons, Inc.

cannot influence the knowledge of future tosses. In contrast, if we have a hidden random variable, which selects whether we continuously generate a sequence of 1s or a sequence of 0s, then such a system is not ergodic. Moreover, if a human being has a plenty of such hidden random variables like a description of a particular topic on which she is going to write some novel text, then, in some sense, she constitutes a strongly nonergodic system.

Thus, we can see that the setting of ergodic theory is more general and invites considerations relevant for linguistics. As we have sketched in the introductory Sections 1.5, 1.6, and 1.11, some concepts from ergodic theory like the distinction between Bayesian and frequentist interpretations of probability and the idea of strong nonergodicity have a clear linguistic or even semantic interpretation. Indeed, probabilities in our minds are not the relative frequencies of events in the physical world, and this distinction arises naturally if the natural language describes an infinitely complex reality. We should add as well that these important interpretations have not been seriously touched by linguists or mathematicians so far. Consequently, we will develop these ideas formally in Chapter 8, whereas in this chapter, we will present more basic concepts from ergodic theory. Subsequently, these concepts will interact with fundamental ideas from information theory – to be exposed in Chapters 5–7.

The contents of this chapter are as follows. The central notion of ergodic theory is a measurable operation on points of a probability space. This operation corresponds to letting the random system evolve freely until the next experiment. In Section 4.1, we will be studying what happens to the random variables on average when we repeat this operation. It turns out that a regular behavior can be observed only for some special points which are typical of some special probability measures. In particular, we will see that the relative frequency interpretation of probability fails if we take it verbatim for any events in the σ-field and if we assume that the probability measure is nonatomic. That is, probability measures usually do not have a completely general frequentist interpretation.

However, the relative frequency interpretation of probability is sound enough to survive in a reasonably restricted form, as we will show in Section 4.2. If the probability measure is stationary and ergodic, then we can generalize the law of large numbers as the Birkhoff ergodic theorem. That is, the relative frequency of any event in an infinite sequence of repeated experiments tends to the probability of this event almost surely. This convergence holds for each event almost surely separately, and since there are uncountably many different events, the convergence for all events simultaneously may fail almost surely as well. It is notable that the limit of relative frequencies exists almost surely also for stationary and nonergodic processes but is equal to a conditional probability with respect to the invariant σ-field.

From this point of view, it is important to decide which known probability measures are stationary and ergodic. In Section 4.3, we will derive several practical criteria. These criteria allow to inspect the algebraic properties of the probability measure only on the π-system of cylinder sets, which is quite feasible. Consequently, it turns out that all so-called mixing stochastic processes are ergodic, whereas among stationary Markov processes, those irreducible are ergodic, and those irreducible and aperiodic are mixing. We can also state some theorems how taking products or functions of stochastic processes affects their ergodicity and mixing.

In Section 4.4, we will discuss the ergodic decomposition theorems, another fundamental result concerning stationary processes. The ergodic decomposition theorems allow us to represent any stationary measure as an integral over the space of stationary ergodic measures. This representation is unique, i.e. no stationary measure can be decomposed in two different ways. A proper understanding of the ergodic decomposition is crucial for language modeling since we may suppose that good statistical models of natural language should be strongly nonergodic. In plain words, the latter condition – to be formally defined in Section 5.6 – roughly models the intuition that we can potentially write texts on infinitely many different topics. This semantic interpretation of strong nonergodicity will be, however, relegated to Section 8.3.

4.1 Plain Relative Frequency

In this section, we will investigate the frequency interpretation of probability. According to a naive intuition, the probability of any event can be defined as the limiting relative frequency of this event in an infinite sequence of repeated experiments. We will see that this interpretation fails if we take it quite verbatim for any events in the σ-field and if we assume that the probability measure is nonatomic. As it will be explained in the subsequent sections, however, there is a way to save the frequency interpretation of probability if we only confine ourselves to events in a countable generating field.

Let us begin with a general concept as follows:

Definition 4.1 *(Dynamical system)*: A dynamical system $(\Omega, \mathcal{J}, T)$ is a triple that consists of a measurable space $(\Omega, \mathcal{J})$ and a measurable operation $T : \Omega \to \Omega$, i.e. $T^{-1}(A) \in \mathcal{J}$ for all $A \in \mathcal{J}$.

Usually, a dynamical system is defined as a quadruple, with an additional probability measure. Since we will consider different probability measures for the same measurable space and a measurable operation, here we will consider dynamical systems without a specified measure.

For example, if we have the space of one-sided or two-sided infinite sequences $(\mathbb{X}^{\mathbb{T}}, \mathcal{X}^{\mathbb{T}})$, where $\mathbb{T} = \mathbb{N}$ or $\mathbb{T} = \mathbb{Z}$, we can consider the shift operation

$$T : \mathbb{X}^{\mathbb{T}} \ni (x_i)_{i \in \mathbb{T}} \mapsto (x_{i+1})_{i \in \mathbb{T}} \in \mathbb{X}^{\mathbb{T}} \tag{4.1}$$

For $\mathbb{T} = \mathbb{Z}$, operation T is invertible, i.e. the preimage $T^{-1}(\{(x_i)_{i \in \mathbb{Z}}\})$ contains exactly one sequence and there exists the inverse operation

$$T^{-1} : \mathbb{X}^{\mathbb{Z}} \ni (x_i)_{i \in \mathbb{Z}} \mapsto (x_{i-1})_{i \in \mathbb{T}} \in \mathbb{X}^{\mathbb{Z}} \tag{4.2}$$

whereas for $\mathbb{T} = \mathbb{N}$, operation T is not invertible, i.e. the preimage of a single sequence is $T^{-1}(\{(x_i)_{i \in \mathbb{N}}\}) = \{(y_i)_{i \in \mathbb{N}} : y_1 \in \mathbb{X}, y_{j+1} = x_j, j \geq 1\}$, so it contains more than one sequence. For both $\mathbb{T} = \mathbb{N}$ or $\mathbb{T} = \mathbb{Z}$, the shift operation T defined by (4.1) is measurable.

Consider a dynamical system $(\Omega, \mathcal{J}, T)$. For a real random variable Y on the measurable space $(\Omega, \mathcal{J})$, let us introduce a stochastic process

$$Y \circ T^i : \Omega \ni \omega \mapsto Y(T^i(\omega)) \in \mathbb{R} \tag{4.3}$$

where $i = 0, 1, 2, \ldots$ and T^i is T iterated i times, i.e. $T^0(\omega) := \omega$, $T^{i+1}(\omega) := T(T^i(\omega))$, and $T^{i-1}(\omega) := T^{-1}(T^i(\omega))$ if T is invertible. For example, for the shift operation (4.1) and projections $Y((x_j)_{j \in \mathbb{T}}) = x_0$, we have $Y \circ T^i = Y_i$, where $Y_i((x_j)_{j \in \mathbb{T}}) = x_i$. In general, if operation T describes a certain discrete-time transformation of the probability space then $Y \circ T^i$ is a value of a certain observed statistic of the dynamical system after i steps.

If the random variable Y takes numerical values, then it is quite natural to ask what happens to the average of observations $Y \circ T^i$ in the long run. If $Y \circ T^i$ are bounded IID random variables, then the average of $Y \circ T^i$ converges almost surely to the expectation of Y by the strong law of large numbers (3.101). Here, however, we want to abstract for a while from assuming a prior probability distribution on Y, and we are simply interested whether limit

$$\lim_{n \to \infty} \frac{1}{n} \sum_{i=0}^{n-1} Y \circ T^i \tag{4.4}$$

exists in general and whether it has some interesting properties. In particular, for a while, let us confine ourselves to the characteristic functions $Y = I_A$, where $I_A(\omega) = \mathbf{1}\{\omega \in A\}$ and $A \in \mathcal{J}$. Then limit (4.4) is the relative frequency of event A in a series of possibly dependent experiments performed on a dynamical system that evolves in discrete time steps according to operation T. It is natural to ask in this case if limit (4.4) is equal to a probability measure for some points $\omega \in \Omega$.

Let $\bot$ stand for an undefined value. For the unit interval $\mathbb{U} := [0, 1]$, let us denote the extended unit interval $\mathbb{U}_\bot := \mathbb{U} \cup \{\bot\}$ and the corresponding Borel σ-field as

$$\mathcal{U}_\bot := \sigma(\{[a, b] : a, b \in \mathbb{U}\} \cup \{\{\bot\}\}) \tag{4.5}$$

We will define the following function of sets and points:

Definition 4.2 *(Relative frequency)*: For a dynamical system $(\Omega, \mathcal{J}, T)$, the relative frequency $\Phi : \mathcal{J} \times \Omega \ni (A, \omega) \mapsto \Phi(A)(\omega) \in \mathbb{U}_\bot$ is defined through

$$\Phi_n(A)(\omega) := \frac{1}{n} \sum_{i=0}^{n-1} I_A \circ T^i(\omega) \tag{4.6}$$

$$\Phi(A)(\omega) := \begin{cases} \lim_{n \to \infty} \Phi_n(A)(\omega) & \text{if the limit exists} \\ \bot & \text{else} \end{cases} \tag{4.7}$$

We will say that $\Phi(A)(\omega)$ exists if $\Phi(A)(\omega) \neq \bot$.

Let us stress that whereas functions Φ_n are random probability measures for any $\omega \in \Omega$, the relative frequency Φ need not be necessarily a random probability measure, for the simple reason that limits $\Phi(A)(\omega)$ need not exist. Still, the relative frequency Φ enjoys several nice properties. Two properties are so important that they deserve special names.

Definition 4.3 *(Stationary and ergodic set functions)*: For a dynamical system $(\Omega, \mathcal{J}, T)$, an arbitrary set function $S : \mathcal{J} \to \mathbb{U}_\bot$ is called stationary on $(\Omega, \mathcal{J}, T)$ if $S(T^{-1}(A)) = S(A)$ for any event $A \in \mathcal{J}$. The set function S is called ergodic on $(\Omega, \mathcal{J}, T)$ if $S(A) \in \{0, 1\}$ for any event $A \in \mathcal{I}$, where class

$$\mathcal{I} := \{A \in \mathcal{J} : A = T^{-1}(A)\} \tag{4.8}$$

is called the invariant σ-field.

It can be easily verified that class $\mathcal{I}$ is a σ-field. To give some more intuition, consider the real random variables $Y \circ T^i$ on $(\Omega, \mathcal{J}, T)$. The following events belong to the invariant σ-field $\mathcal{I}$:

$$(Y \circ T^i = a \text{ for all } i \in \mathbb{Z}) = \bigcap_{i=-\infty}^{\infty} (Y \circ T^i = a) \tag{4.9}$$

$$(Y \circ T^i = a \text{ for infinitely many } i \geq 1) = \bigcap_{i=1}^{\infty} \bigcup_{j=i}^{\infty} (Y \circ T^i = a) \tag{4.10}$$

$$\left(\lim_{n \to \infty} \frac{1}{n} \sum_{i=1}^{n} Y \circ T^i = a \right) = \bigcap_{p=1}^{\infty} \bigcup_{N=1}^{\infty} \bigcap_{n=N}^{\infty} \left(\left| \frac{1}{n} \sum_{i=1}^{n} Y \circ T^i - a \right| \leq \frac{1}{p} \right) \tag{4.11}$$

We observe the following simple facts:

Theorem 4.1: *For a dynamical system* $(\Omega, \mathcal{J}, T)$ *and for each* $\omega \in \Omega$, *we have:*

1. $\Phi(\cdot)(\omega)$ *is stationary.*
2. $\Phi(\cdot)(\omega)$ *is ergodic.*
3. Φ *is invariant, i.e.* $\Phi(\cdot)(T(\omega)) = \Phi(\cdot)(\omega)$.

Proof: Left as an easy exercise. □

The next theorem is some prerequisite to decide when the relative frequency is a probability measure.

Theorem 4.2 *(Vitali–Hahn–Saks theorem):* *If we have a sequence of probability measures* $(P_n)_{n \in \mathbb{N}}$ *on a field* $\mathcal{K}$ *such that limits*

$$P(A) = \lim_{n \to \infty} P_n(A) \tag{4.12}$$

exist for all $A \in \mathcal{K}$, *then function* P *is a probability measure on* $\mathcal{K}$.

Proof: See Gray (2009, lemma 7.4). □

Now, we may state when the relative frequency Φ is a probability measure.

Theorem 4.3: *For a dynamical system* $(\Omega, \mathcal{J}, T)$ *and for each* $\omega \in \Omega$, *if* $\Phi(A)(\omega) \neq \perp$ *for all* $A \in \mathcal{J}$, *then* $\Phi(\cdot)(\omega)$ *is a probability measure. Moreover, for each* $A \in \mathcal{J}$ *function* $\Phi(A)$ *is measurable from* $(\Omega, \mathcal{J})$ *to* $(\mathbb{U}_\perp, \mathcal{U}_\perp)$.

Proof: The first claim follows by the Vitali–Hahn–Saks theorem. In our case, we have $\Phi(A)(\omega) = \lim_{n \to \infty} \Phi_n(A)(\omega)$, where functions $\Phi_n(\cdot)(\omega)$ are probability measures. As for the second claim, measurability of $\Phi(A)$ from $(\Omega, \mathcal{J})$ to $(\mathbb{U}_\perp, \mathcal{U}_\perp)$ follows by measurability of $\Phi_n(A)$ from $(\Omega, \mathcal{J})$ to $(\mathbb{U}_\perp, \mathcal{U}_\perp)$. In the following, we apply claim (3) of Theorem 4.1 to infer that $\Phi(A)$ is also measurable from $(\Omega, \mathcal{J})$ to $(\mathbb{U}_\perp, \mathcal{U}_\perp)$. □

The points for which the relative frequency is a probability measure will be called regular. In contrast, periodic points are those that are mapped onto themselves for some iteration of operation T.

Definition 4.4 *(Regular and periodic points):* A point $\omega \in \Omega$ is called regular if $\Phi(A)(\omega) \neq \perp$ for all $A \in \mathcal{J}$. The point $\omega \in \Omega$ is called periodic if $\omega = T^n(\omega)$ for some $n \geq 0$.

It turns out that in a typical case, the regular points must be periodic. The condition is rather weak, namely, the singleton sets must be measurable. This holds in particular for measurable spaces $(\mathbb{X}^{\mathbb{T}}, \mathcal{X}^{\mathbb{T}})$ with the shift operation (4.1).

Theorem 4.4: *If $\{\omega\} \in \mathcal{J}$ for all $\omega \in \Omega$, then a point $\omega \in \Omega$ is regular if and only if it is periodic.*

Proof: Clearly, if a point is periodic, then it is regular, since $\Phi(A)(\omega) = n^{-1} \sum_{i=0}^{n-1} I_A \circ T^i(\omega)$ for $\omega = T^n(\omega)$. Now, assume that the point is not periodic. Consider a subset of natural numbers

$$K = \{i \in \mathbb{N} : 3^{2k-1} < i \le 3^{2k}, k \in \mathbb{N}\} \tag{4.13}$$

Then we obtain

$$\liminf_{n \to \infty} \frac{1}{n} \sum_{i=1}^{n} \mathbf{1}\{i \in K\} \le \frac{1}{3} \tag{4.14}$$

$$\limsup_{n \to \infty} \frac{1}{n} \sum_{i=1}^{n} \mathbf{1}\{i \in K\} \ge \frac{2}{3} \tag{4.15}$$

Since sets $\{T^{i-1}(\omega)\}$ belong to $\mathcal{J}$ for $i \in \mathbb{N}$, they are all distinct, and there are only countably many of them, then we may consider an event $A := \bigcup_{i \in K}\{T^{i-1}(\omega)\} \in \mathcal{J}$ to see that the sequence of averages

$$\frac{1}{n} \sum_{i=0}^{n-1} I_A \circ T^i(\omega) = \frac{1}{n} \sum_{i=1}^{n} \mathbf{1}\{i \in K\} \tag{4.16}$$

obviously does not converge. $\square$

A probability measure P will be called periodic if $P = \Phi(\cdot)(\omega)$ for a certain periodic point ω. Thus, contrary to a naive intuition, Theorem 4.4 states that, in some typical cases, the relative frequency Φ can be a probability measure if and only if it is periodic. Moreover, it can be easily shown that if probability measure is periodic, then it is not nonatomic. In other words, nonatomic probability measures, such as the Lebesgue measure, strictly speaking are not the asymptotic frequencies of events in some infinite series of repeated experiments. In the construction of nonatomic probability measures, there is an inherent smoothing which excludes a naive frequency interpretation of these probability measures.

Let us stress that measures which we will work in this book with are mostly nonatomic. It is important to ask whether the relative frequencies exist for them in some relaxed sense. In the next section, we will show that the relative frequency $\Phi(A)$ exists almost surely for each event A separately with respect to any stationary measure P and equals almost surely to conditional probability $P(A|\mathcal{J})$. This fact is called the Birkhoff ergodic theorem. Since conditional probability $P(\cdot|\mathcal{J})$

can be chosen to be a random probability measure, the Birkhoff ergodic theorem asserts that the relative frequency Φ is almost surely a probability measure when restricted to a countable field, which necessarily does not contain elementary events $\{\omega\}$. Thus, in this relaxed sense, the frequency interpretation of certain nonatomic probability measures holds true.

4.2 Birkhoff Ergodic Theorem

In this section, we will show that a restricted frequency interpretation of probability holds true in some important cases. The respective result is called the Birkhoff ergodic theorem. The Birkhoff ergodic theorem has been already mentioned half-formally in Section 1.5. Here we will present a fully formal exposition.

Analogously to stationary and ergodic set functions, let us define stationary and ergodic probability measures and stochastic processes.

Definition 4.5 *(Stationary and ergodic measures):* A stationary probability measure is a set function which is both a probability measure and stationary, whereas an ergodic probability measure is a set function which is both a probability measure and ergodic.

Definition 4.6 *(Stationary and ergodic processes):* A stochastic process $(X_i)_{i\in\mathbb{T}}$, where $\mathbb{T} = \mathbb{N}$ or $\mathbb{T} = \mathbb{Z}$, on a probability space $(\Omega, \mathcal{J}, P)$, is called stationary or ergodic, respectively, when the distribution $\mu(A) = P((X_i)_{i\in\mathbb{T}} \in A)$ of the process is stationary or ergodic with respect to the shift operation (4.1).

For measurable spaces of infinite sequences, stationary probability measures can be easily constructed, as guaranteed by the following version of the Kolmogorov process theorem (see Theorem 2.7).

Theorem 4.5 *(Stationary process theorem):* *For a countable set $\mathbb{X}$, let function $p : \mathbb{X}^* \to \mathbb{R}$ satisfy conditions:*

1. *$p(x_1 \dots x_k) \geq 0$ for all $x_i \in \mathbb{X}$;*
2. *$\sum_{a\in\mathbb{X}} p(ax_1 \dots x_k) = p(x_1 \dots x_k) = \sum_{a\in\mathbb{X}} p(x_1 \dots x_k a)$ for all $x_i \in \mathbb{X}$;*
3. *$p(\lambda) = 1$ for the empty string λ.*

Then there exists a unique probability measure μ on the measurable space $(\mathbb{X}^\mathbb{T}, \mathcal{X}^\mathbb{T})$, where $\mathbb{T} = \mathbb{N}$ or $\mathbb{T} = \mathbb{Z}$, which is stationary for the shift operation (4.1) and satisfies $\mu([x_1]_{j+1}^\mathbb{T} \cap \dots \cap [x_k]_{j+k}^\mathbb{T}) = p(x_1 \dots x_k)$. Conversely, if there exists a probability measure μ on the measurable space $(\mathbb{X}^\mathbb{T}, \mathcal{X}^\mathbb{T})$ which is stationary for operation (4.1), then function $p(x_1 \dots x_k) = \mu([x_1]_{j+1}^\mathbb{T} \cap \dots \cap [x_k]_{j+k}^\mathbb{T})$ satisfies conditions (1)–(3).

Proof: All claims but stationarity follow from the Kolmogorov process theorem (Theorem 2.7). Now, let $\mathcal{A}$ be the class of cylinder sets $[x_1]_{j+1}^{\top} \cap \cdots \cap [x_k]_{j+k}^{\top}$. Then $\mathcal{A}$ is a π-system that generates $\mathcal{X}^{\top}$, i.e. $\mathcal{X}^{\top} = \sigma(\mathcal{A})$, and $\mu(T^{-1}(A)) = \mu(A)$ for all $A \in \mathcal{A}$. Thus, applying the π-λ theorem (Theorem 2.3), we conclude that $\mu(T^{-1}(A)) = \mu(A)$ must hold for all $A \in \mathcal{X}^{\top}$. □

By Theorem 4.5, we can easily verify in particular that IID processes are stationary. Similarly, Markov processes with the marginal distribution equal to the stationary distribution, if it exists, are stationary – see Problem 2.7. Another, somewhat abstract example of a stationary measure is the Lebesgue measure λ on the dynamical system $(\mathbb{U}, \mathcal{U}, \phi_a)$, where $a \in (0, 1)$ and

$$\phi_a(x) := \begin{cases} x + a, & x + a \leq 1 \\ x + a - 1, & x + a > 1 \end{cases} \tag{4.17}$$

is a rotation of the interval $\mathbb{U} = [0, 1]$. It can be shown that this measure is ergodic for an irrational rotation, i.e. if $a \in (0, 1) \backslash \mathbb{Q}$. Moreover, using so-called cutting and stacking method (Shields, 1991), it can be shown that any dynamical system with a stationary measure is isomorphic with the dynamical system $(\mathbb{U}, \mathcal{U}, \phi)$ with the Lebesgue measure λ and some operation ϕ. In general, this operation ϕ can be quite complicated.

Now we will present the Birkhoff ergodic theorem – the central point of this section. The proposition states that, for any stationary measure and finite expectation $\mathbf{E}|Y| < \infty$, the time average of process $(Y \circ T^i)_{i \in \mathbb{N}}$ converges to the conditional expectation of random variable Y with respect to the invariant σ-field.

Theorem 4.6 *(Birkhoff ergodic theorem):* *Let P be a stationary probability measure with respect to a dynamical system $(\Omega, \mathcal{I}, T)$ and let Y be a real random variable such that $\mathbf{E}|Y| < \infty$. Then*

$$\lim_{n \to \infty} \frac{1}{n} \sum_{i=0}^{n-1} Y \circ T^i = \mathbf{E}(Y|\mathcal{I}) \text{ almost surely} \tag{4.18}$$

Before we prove the Birkhoff ergodic theorem, let us make some important remarks. First of all, if P is both stationary and ergodic, then $\mathbf{E}(Y|\mathcal{I}) = \mathbf{E}Y$ almost surely. In that case, the time average of process $(Y \circ T^i)_{i \in \mathbb{N}}$ converges almost surely to the expectation of Y. Plugging in $Y = I_A$ for an event $A \in \mathcal{I}$ and a stationary probability measure P, the Birkhoff ergodic theorem may be formulated as

$$\Phi(A) = P(A|\mathcal{I}) \text{ almost surely} \tag{4.19}$$

where Φ is the relative frequency discussed in Section 4.1, whereas if P is stationary and ergodic, then

$$\Phi(A) = P(A) \text{ almost surely} \tag{4.20}$$

Hence, if we restrict ourselves to events $A \in \mathcal{K}$, where class $\mathcal{K}$ is countable, then the relative frequency $A \mapsto \Phi(A)$ is a probability measure almost surely. Thus, the Birkhoff ergodic theorem asserts a restricted frequency interpretation of stationary ergodic probability measures – and therefore, it is an important milestone in philosophy of probability. Moreover, as we will see in Problem 4.7, the Birkhoff ergodic theorem allows to single out some well-behaved points $\omega \in \Omega$ for which the restricted relative frequency $\Phi(\cdot)(\omega)$ exists as a random measure.

Now, we will present a simple proof of the Birkhoff ergodic theorem. First, we will demonstrate an auxiliary fact called the maximal ergodic theorem.

Theorem 4.7 *(Maximal ergodic theorem)*: *Let P be a stationary measure with respect to a dynamical system $(\Omega, \mathcal{J}, T)$ and let Y be a real random variable such that $E|Y| < \infty$. Define $S_k := \sum_{i=0}^{k-1} Y \circ T^i$ and $M_n := \max(0, S_1, S_2, \ldots, S_n)$. We have*

$$\int_{M_n > 0} Y \, dP \geq 0 \tag{4.21}$$

Proof: For $1 \leq k \leq n$, we have $M_n \circ T \geq S_k \circ T$. Hence,

$$Y + M_n \circ T \geq Y + S_k \circ T = S_{k+1} \tag{4.22}$$

Let us write it as

$$Y \geq S_{k+1} - M_n \circ T, \quad k = 1, \ldots, n \tag{4.23}$$

But we also have

$$Y = S_1 \geq S_1 - M_n \circ T \tag{4.24}$$

Both inequalities yield $Y \geq \max(S_1, S_2, \ldots, S_n) - M_n \circ T$. Hence

$$\int_{M_n > 0} Y \, dP \geq \int_{M_n > 0} [M_n - M_n \circ T] \, dP \tag{4.25}$$

$$= \int_{M_n > 0} M_n \, dP - \int_{M_n > 0} M_n \circ T \, dP. \tag{4.26}$$

Now, we observe that $\int_{M_n > 0} M_n \, dP = \int M_n \, dP$, whereas $\int_{M_n > 0} M_n \circ T \, dP \leq \int M_n \circ T \, dP$ since $M_n \circ T \geq 0$. Moreover, by stationarity $\int M_n \, dP = \int M_n \circ T \, dP$. Hence $\int_{M_n > 0} Y \, dP \geq \int M_n \, dP - \int M_n \circ T \, dP \geq 0$. $\qquad\square$

In the next step, we will prove the Birkhoff ergodic theorem.

Proof of Theorem 4.6:   We use the notation from Theorem 4.7. Without loss of generality, let us assume $\mathbf{E}(Y|\mathcal{I}) = 0$. Statement (4.18) can be derived applying the proof below to random variable $Y - \mathbf{E}(Y|\mathcal{I})$. For a fixed $\epsilon > 0$, denote the event

$$G = (\limsup_{n\to\infty} S_n/n > \epsilon) \tag{4.27}$$

We introduce random variable $Y^* = (Y - \epsilon)I_G$ and, by analogy, we define S_k^* and M_n^* as in the statement of the maximal ergodic theorem. Events

$$(M_n^* > 0) = (\max_{1\le k\le n} S_k^* > 0) \tag{4.28}$$

converge to

$$(\sup_{k\ge 1} S_k^* > 0) = (\sup_{k\ge 1} S_k^*/k > 0) = (\sup_{k\ge 1} S_k/k > \epsilon) \cap G = G \tag{4.29}$$

Inequality $\mathbf{E}|Y^*| \le \mathbf{E}|Y| + \epsilon < \infty$ allows to use the Lebesgue dominated convergence theorem (Theorem 3.10), which yields

$$\int_G Y^* dP = \lim_{n\to\infty} \int_{M_n^* > 0} Y^* dP \ge 0 \tag{4.30}$$

by the maximal ergodic theorem. But $G \in \mathcal{I}$ so $\int_G Y dP = \int_G \mathbf{E}(Y|\mathcal{I})dP = 0$. Hence,

$$\int_G Y^* dP = \int_G Y dP - \epsilon P(G) = -\epsilon P(G) \tag{4.31}$$

and thus $P(G) = 0$. By the arbitrary choice of ϵ, we hence obtain that

$$\limsup_{n\to\infty} S_n/n \le 0 \text{ almost surely} \tag{4.32}$$

Applying the analogous reasoning to random variable $-Y$ yields

$$\liminf_{n\to\infty} S_n/n \le 0 \text{ almost surely} \tag{4.33}$$

Hence, $\lim_{n\to\infty} S_n/n = 0$. □

The original proof of the ergodic theorem given by Birkhoff (1932) was much longer. It was considerably shortened by Garsia (1965), whose proof we have reproduced here. For more background in ergodic theory and ergodic theorems, an interested reader is referred to Cornfeld et al. (1982), Kallenberg (1997), and Gray (2009). The Birkhoff ergodic theorem will be used throughout this book.

4.3   Ergodic and Mixing Criteria

In view of the Birkhoff ergodic theorem, a question arises which particular stationary measures or processes are ergodic. Several effective criteria can be proposed, as we have mentioned them half-formally in Section 1.5. The following theorem is the first step to carve them out.

Theorem 4.8:   *A stationary measure P on a dynamical system* $(\Omega, \mathcal{J}, T)$ *is ergodic if and only if for all* $A, B \in \mathcal{J}$ *we have*

$$\lim_{n \to \infty} \frac{1}{n} \sum_{i=0}^{n-1} P(T^{-i}(A) \cap B) = P(A)P(B) \tag{4.34}$$

Proof: If (4.34) holds for all $A, B \in \mathcal{J}$ then putting $A = B$ we obtain $P(A) = [P(A)]^2 \in \{0, 1\}$ and P is ergodic. Conversely, if P is ergodic, then the Birkhoff ergodic theorem (Theorem 4.6) and the dominated convergence (Theorem 3.10) yield

$$P(A)P(B) = \int_B P(A)dP = \int_B \left[\lim_{n \to \infty} \frac{1}{n} \sum_{i=0}^{n-1} I_A \circ T^i \right] dP$$

$$= \lim_{n \to \infty} \frac{1}{n} \sum_{i=0}^{n-1} P(T^{-i}(A) \cap B) \tag{4.35}$$

for all $A, B \in \mathcal{J}$. $\square$

Subsequently, we will observe that condition (4.34) need not be tested for all $A, B \in \mathcal{J}$. Instead of testing it for all elements of the σ-field, it suffices to check the condition for the elements of the generating field.

Theorem 4.9   (*Ergodic criterion*):   *A stationary measure P is ergodic if and only if condition (4.34) is satisfied for all* $A, B \in \mathcal{K}$, *where* $\mathcal{K}$ *is a field such that* $\mathcal{J} = \sigma(\mathcal{K})$.

Proof: We will show that (4.34) for $A, B \in \mathcal{K}$ implies the same for $A, B \in \mathcal{J}$. By Theorem 3.28, for any $\epsilon > 0$ and any $A, B \in \mathcal{J}$ there exist $A', B' \in \mathcal{K}$ such that $P(A \triangle A'), P(B \triangle B') < \epsilon$. We will show then that

$$D := \limsup_{n \to \infty} \left| \frac{1}{n} \sum_{i=0}^{n-1} P(T^{-i}(A) \cap B) - P(A)P(B) \right| < 4\epsilon \tag{4.36}$$

which proves the claim. In fact, using inequalities $|P(A) - P(B)| \leq P(A \triangle B)$ and $P(A \triangle C) \leq P(A \triangle B) + P(B \triangle C)$, we obtain

$$D \leq \sup_{i \in \mathbb{N}} |P(T^{-i}(A) \cap B) - P(T^{-i}(A') \cap B')| + |P(A)P(B) - [P(A')P(B')|$$

$$\leq \sup_{i \in \mathbb{N}} |P(T^{-i}(A \triangle A')) + P(B \triangle B')| + P(A \triangle A') + P(B \triangle B') < 4\epsilon$$

$$\tag{4.37}$$

$\square$

Some version of Theorem 4.9 can be found in Gray (2009, lemma 7.15).

For countably generated spaces, we can further narrow down the class of sets for which we test condition (4.34). It is sufficient to test this condition for the generated π-system, which sounds feasible enough.

Theorem 4.10 *(Ergodic criterion):* *Let $\mathcal{J} = \sigma(\mathcal{A})$ for a countable class $\mathcal{A}$. A stationary measure P is ergodic if and only if condition (4.34) is satisfied for all $A, B \in \pi(\mathcal{A})$.*

Proof: By Theorem 2.1, any $B \in \sigma_0(\mathcal{A})$ can be written as sum $B = \bigcup_{i=1}^{n} B_i$, where sets B_i are disjoint and $B_i = C_{i1} \cap \cdots \cap C_{ik}$, where $C_{ij} = D_j$ or $C_{ij} = D_j^c$ and $D_j \in \mathcal{A}$. Consequently, we may show that condition (4.34) is satisfied for all $A, B \in \sigma_0(\mathcal{A})$ if and only if it is satisfied for all $A, B \in \pi(\mathcal{A}, \mathcal{A}^c)$, where $\mathcal{A}^c = \{A^c : A \in \mathcal{A}\}$. In the next step, we can show that condition (4.34) is satisfied for all $A, B \in \pi(\mathcal{A}, \mathcal{A}^c)$ if and only if it is satisfied for all $A, B \in \pi(\mathcal{A})$. □

Ergodicity is a more fundamental property, but it is usually easier to show a related stronger condition, called mixing. In fact, there is a whole hierarchy of various mixing conditions (Bradley, 2005). Here we will discuss the simplest of them.

Definition 4.7 *(Mixing measure):* A stationary probability measure P on a dynamical system $(\Omega, \mathcal{J}, T)$ is called mixing when for all $A, B \in \mathcal{J}$ we have

$$\lim_{n \to \infty} P(T^{-n}(A) \cap B) = P(A)P(B) \tag{4.38}$$

Definition 4.8 *(Mixing process):* A stochastic process $(X_i)_{i \in \mathbb{T}}$, where $\mathbb{T} = \mathbb{N}$ or $\mathbb{T} = \mathbb{Z}$, on a probability space $(\Omega, \mathcal{J}, P)$ is called mixing when the distribution $\mu(A) = P((X_i)_{i \in \mathbb{T}} \in A)$ of the process is mixing with respect to the shift operation (4.1).

Let us show that mixing is a stronger condition than ergodicity.

Theorem 4.11: *If a stationary measure P is mixing, then it is ergodic.*

Proof: Condition (4.38) for $A = B \in \mathcal{J}$ yields $P(A) = [P(A)]^2$. Hence $P(A) \in \{0, 1\}$ and P is ergodic. □

Applying the technique from the proof of Theorem 4.9, we may demonstrate this analogical statement.

Theorem 4.12 *(Mixing criterion):* *A stationary measure P is mixing if and only if condition (4.38) is satisfied for all $A, B \in \mathcal{K}$, where $\mathcal{K}$ is a field such that $\mathcal{J} = \sigma(\mathcal{K})$.*

Proof: It suffices to demonstrate that (4.38) holds for $A, B \in \mathcal{K}$ implies the same for $A, B \in \mathcal{J}$. By Theorem 3.28, for any $\epsilon > 0$ and any $A, B \in \mathcal{J}$ there exist $A', B' \in \mathcal{K}$ such that $P(A \triangle A'), P(B \triangle B') < \epsilon$. Consider

$$D := \limsup_{n \to \infty} |P(T^{-n}(A) \cap B) - P(A)P(B)| \tag{4.39}$$

In fact, for this D bound (4.37) remains valid, which proves the claim. $\qquad\square$

Similarly, we prove the counterpart of Theorem 4.10.

Theorem 4.13 (*Mixing criterion*): *Let $\mathcal{J} = \sigma(\mathcal{A})$ for a countable class $\mathcal{A}$. A stationary measure P is mixing if and only if condition (4.38) is satisfied for all $A, B \in \pi(\mathcal{A})$.*

Proof: By Theorem 2.1, any $B \in \sigma_0(\mathcal{A})$ can be written as sum $B = \bigcup_{i=1}^{n} B_i$, where sets B_i are disjoint and $B_i = C_{i1} \cap \cdots \cap C_{ik}$, where $C_{ij} = D_j$ or $C_{ij} = D_j^c$ and $D_j \in \mathcal{A}$. Consequently, we may show that condition (4.38) is satisfied for all $A, B \in \sigma_0(\mathcal{A})$ if and only if it is satisfied for all $A, B \in \pi(\mathcal{A}, \mathcal{A}^c)$, where $\mathcal{A}^c = \{A^c : A \in \mathcal{A}\}$. In the next step, we can show that condition (4.38) is satisfied for all $A, B \in \pi(\mathcal{A}, \mathcal{A}^c)$ if and only if it is satisfied for all $A, B \in \pi(\mathcal{A})$. $\qquad\square$

Although mixing condition (4.38) is stronger than ergodicity, usually it is easier to demonstrate it for particular processes, such as IID processes and Markov processes. First, we note that in the case of an IID process, there are no preconditions.

Theorem 4.14: *A discrete IID process $(X_i)_{i \in \mathbb{T}}$, where $\mathbb{T} = \mathbb{N}$ or $\mathbb{T} = \mathbb{Z}$, is mixing.*

Proof: By Theorem 4.13, it is sufficient to prove mixing condition (4.38) for $A = (X_1 = x_1, \ldots, X_k = x_k)$ and $B = (X_1 = y_1, \ldots, X_k = y_k)$. In fact, for such A and B, we have (4.38) since events $T^{-n}(A)$ and B are independent for sufficiently large n. $\qquad\square$

The next theorem states necessary and sufficient conditions for a stationary Markov process to be ergodic or mixing. These conditions can be simply expressed in terms of irreducibility and aperiodicity of the Markov process, defined in Section 2.4.

Theorem 4.15: *Consider a discrete stationary Markov process $(X_i)_{i \in \mathbb{T}}$, where $\mathbb{T} = \mathbb{N}$ or $\mathbb{T} = \mathbb{Z}$ and $X_i : \Omega \to \mathbb{X}$.*

1. *The process is ergodic with $P(X_i = x) > 0$ for all $x \in \mathbb{X}$ if and only if it is irreducible.*

2. *The process is mixing with $P(X_i = x) > 0$ for all $x \in \mathbb{X}$ if and only if it is irreducible and aperiodic.*

Proof:
1. Suppose that the Markov process is stationary ergodic with $P(X_i = x) > 0$ for all $x \in \mathbb{X}$. Then

$$\lim_{n \to \infty} \frac{1}{n} \sum_{i=1}^{n} P(X_{t+i} = x | X_t = y) = P(X_i = x) > 0 \qquad (4.40)$$

and consequently there exists an $n \in \mathbb{N}$ such that $P(X_{t+n} = x | X_t = y) > 0$. Hence, the process must be irreducible. Conversely assume that the process is irreducible. Then by Theorems 2.9 and 2.10, we have (4.40). But since the process is Markov, we hence obtain (4.34) for $A = (X_1 = x_1, \dots, X_k = x_k)$ and $B = (X_1 = y_1, \dots, X_k = y_k)$. Thus, by Theorem 4.10, the process is ergodic.
2. Suppose that the Markov process is stationary mixing with $P(X_i = x) > 0$ for all $x \in \mathbb{X}$. Then

$$\lim_{n \to \infty} P(X_{t+n} = x | X_t = y) = P(X_i = x) > 0 \qquad (4.41)$$

and consequently for all sufficiently large $n \in \mathbb{N}$ we have $P(X_{t+n} = x | X_t = y) > 0$. Hence, the process must be irreducible and aperiodic. Conversely assume that the process is irreducible and aperiodic. Then by Theorems 2.9 and 2.11, we have (4.41). But since the process is Markov, we hence obtain (4.38) for $A = (X_1 = x_1, \dots, X_k = x_k)$ and $B = (X_1 = y_1, \dots, X_k = y_k)$. Thus, by Theorem 4.13, the process is mixing. $\qquad \square$

To inspect ergodicity and mixing of a hidden Markov process, let us consider extended dynamical systems of form $(\Omega, \mathcal{J}, T, P)$ with adjoined stationary measures P and let us say that an extended dynamical system $(\Omega, \mathcal{J}, T, P)$ is ergodic or mixing when the respective condition holds for measure P. We have some general propositions about the mixing property for projections and products of extended dynamical systems, which we will consider now. The results are traditionally stated in terms of probability measures, but we will briefly comment on their applications to processes.

A projection of an extended dynamical system is an appropriately measurable function of the system.

Definition 4.9 *(Projection of a dynamical system):* Let $(\Omega_1, \mathcal{J}_1, T_1, P_1)$ and $(\Omega_2, \mathcal{J}_2, T_2, P_2)$ be two dynamical systems with stationary measures. We say that $(\Omega_2, \mathcal{J}_2, T_2, P_2)$ is a projection of $(\Omega_1, \mathcal{J}_1, T_1, P_1)$ if there exists a function f measurable from $(\Omega_1, \mathcal{J}_1)$ to $(\Omega_2, \mathcal{J}_2)$ such that $T_1^{-1}(f^{-1}(A_2)) = f^{-1}(T_2^{-1}(A_2))$ and $P_1(f^{-1}(A_2)) = P_2(A_2)$ for all $A \in \mathcal{J}_2$.

Theorem 4.16: *A projection* $(\Omega_2, \mathcal{J}_2, T_2, P_2)$ *of a dynamical system* $(\Omega_1, \mathcal{J}_1, T_1, P_1)$ *is mixing if* $(\Omega_1, \mathcal{J}_1, T_1, P_1)$ *is mixing.*

Proof: By induction, we have $T_1^{-n}(f^{-1}(A_2)) = f^{-1}(T_2^{-n}(A_2))$. Hence,

$$P_2(T_2^{-n}(A_2) \cap B_2) = P_1(f^{-1}(T_2^{-n}(A_2) \cap B_2))$$
$$= P_1(f^{-1}(T_2^{-n}(A_2)) \cap f^{-1}(B_2)) = P_1(T_1^{-n}(f^{-1}(A_2)) \cap f^{-1}(B_2))$$
$$\xrightarrow[n \to \infty]{} P_1(f^{-1}(A_2)) P_1(f^{-1}(B_2)) = P_2(A_2) P_2(B_2) \qquad (4.42)$$

Thus, the projection is mixing as well. □

Let $(X_i)_{i \in \mathbb{T}}$ be a function of a discrete stationary Markov process $(Y_i)_{i \in \mathbb{T}}$, i.e. $X_i = f(Y_i)$. We can see that ergodicity and mixing of the hidden Markov process $(X_i)_{i \in \mathbb{T}}$ depend partly on the respective properties of the underlying Markov process $(Y_i)_{i \in \mathbb{T}}$. By Theorem 4.16, process $(X_i)_{i \in \mathbb{T}}$ is mixing if process $(Y_i)_{i \in \mathbb{T}}$ is so. Similarly, it can be shown that $(X_i)_{i \in \mathbb{T}}$ is ergodic if process $(Y_i)_{i \in \mathbb{T}}$ is ergodic. These are only sufficient conditions. It can be easily shown that they are not necessary. To see it, consider, for example $Y_i = (X_i, Z_i)$ where $(X_i)_{i \in \mathbb{T}}$ is mixing and $(Z_i)_{i \in \mathbb{T}}$ is nonergodic. Then obviously process $(Y_i)_{i \in \mathbb{T}}$ is not ergodic.

Coming back to more general facts, the product of extended dynamical systems is the respective extended dynamical system for the product measure.

Definition 4.10 *(Product of dynamical systems):* Let $(\Omega_i, \mathcal{J}_i, T_i, P_i)$, where $i \in \mathbb{T}$, be some dynamical systems with stationary measures. For $\omega = (\omega_i)_{i \in \mathbb{T}}$, where $\omega_i \in \Omega_i$, define $(\prod_{i \in \mathbb{T}} T)(\omega) := (T_i(\omega_i))_{i \in \mathbb{T}}$. The product of systems $(\Omega_i, \mathcal{J}_i, T_i, P_i)$ is defined as $(\prod_{i \in \mathbb{T}} \Omega_i, \bigotimes_{i \in \mathbb{T}} \mathcal{J}_i, \prod_{i \in \mathbb{T}} T_i, \prod_{i \in \mathbb{T}} P_i)$.

Theorem 4.17: *Let* $\mathbb{T}$ *be a countable set and let probability spaces* $(\Omega_i, \mathcal{J}_i, P_i)$ *be countably generated for all* $i \in \mathbb{T}$. *The extended product dynamical system* $(\prod_{i \in \mathbb{T}} \Omega_i, \bigotimes_{i \in \mathbb{T}} \mathcal{J}_i, \prod_{i \in \mathbb{T}} T_i, \prod_{i \in \mathbb{T}} P_i)$ *is mixing if the dynamical systems* $(\Omega_i, \mathcal{J}_i, T_i, P_i)$ *are mixing for all* $i \in \mathbb{T}$.

Proof: The class of cylinders $\mathcal{A} = \{[A]_k^{\mathbb{T}} : A \in \mathcal{J}_k, k \in \mathbb{T}\}$ is countable by the assumption and generates the product σ-field $\bigotimes_{i \in \mathbb{T}} \mathcal{J}_i$. By Theorem 4.13 it is sufficient to show the mixing property (4.38) for $A, B \in \pi(\mathcal{A})$. Without loss of generality, we may write $A = [A_1]_{j_1}^{\mathbb{T}} \cap \cdots \cap [A_m]_{j_m}^{\mathbb{T}}$, $B = [B_1]_{j_1}^{\mathbb{T}} \cap \cdots \cap [B_m]_{j_m}^{\mathbb{T}}$. Hence, for $P = \prod_{i \in \mathbb{T}} P_i$, we have

$$P(T^{-n}(A) \cap B) = P_{j_1}(T_{j_1}^{-n}(A_1) \cap B_1) \times \cdots \times P_{j_m}(T_{j_m}^{-n}(A_m) \cap B_m)$$
$$\xrightarrow[n \to \infty]{} P_{j_1}(A_1) P_{j_1}(B_1) \times \cdots \times P_{j_m}(A_m) P_{j_m}(B_m)$$
$$= P_{j_1}(A_1) \times \cdots \times P_{j_m}(A_m) P_{j_1}(B_1) \times \cdots \times P_{j_m}(B_m) = P(A) P(B) \qquad (4.43)$$

Thus, the extended product dynamical system is mixing. □

The above theorem asserts in particular that a process $(W_i)_{i \in \mathbb{T}}$, where $W_i = (X_{ik})_{k \in \mathbb{N}}$, is mixing if processes $(X_{ik})_{i \in \mathbb{T}}$ are all mixing and probabilistically independent from each other. For ergodicity, there is no such characterization. Theorems 4.16 and 4.17 slightly generalize the results from Cornfeld et al. (1982, Section 10.§1). Theorems 4.16 and 4.17 will be applied in Section 11.2 for the construction of some linguistically motivated mixing processes.

4.4 Ergodic Decomposition

In this section, we will discuss another fundamental result concerning stationary processes, which is called the ergodic decomposition and will be stated in two separate theorems. The ergodic decomposition theorems combined allow to represent any stationary measure as an integral over the space of stationary ergodic measures. A proper understanding of the ergodic decomposition theorems is crucial for the later developments in this book – since as we have informally argued in Section 1.11 and as it will become clear formally from Sections 5.4, 5.6, 8.3, and 8.4, a reasonable statistical model of natural language should be either nonergodic or noncomputable, depending whether we adopt a Bayesian or a frequentist interpretation of probability.

Let us define measurable spaces of stationary and ergodic measures.

Definition 4.11 *(Spaces of stationary and ergodic measures)*: The set of stationary probability measures on a dynamical system $(\Omega, \mathcal{J}, T)$ will be denoted as $\mathbb{S} \subseteq \mathbb{U}^{\mathcal{J}}$, whereas the set of stationary ergodic probability measures on $(\Omega, \mathcal{J}, T)$ will be denoted as $\mathbb{E} \subseteq \mathbb{S}$. We define the σ-fields of stationary and stationary ergodic probability measures as

$$\mathcal{S} := \{B \cap \mathbb{S} : B \in \mathcal{U}^{\mathcal{J}}\} \tag{4.44}$$

$$\mathcal{E} := \{B \cap \mathbb{E} : B \in \mathcal{U}^{\mathcal{J}}\} \tag{4.45}$$

We note this technical corollary of the Birkhoff ergodic theorem.

Theorem 4.18: *For a dynamical system $(\Omega, \mathcal{J}, T)$, where $\mathcal{J} = \sigma(\mathcal{K})$ for a certain countable π-system $\mathcal{K}$, we have*

$$f \in \mathbb{E} \implies f(\Phi|_{\mathcal{K}} = f|_{\mathcal{K}}) = 1 \tag{4.46}$$

$$f \in \mathbb{S} \implies f(\Phi|_{\mathcal{K}} = f(\cdot|\mathcal{J})|_{\mathcal{K}}) = 1 \tag{4.47}$$

$$f \in \mathbb{S} \implies [f \in \mathbb{E} \iff f(f|_{\mathcal{K}} = f(\cdot|\mathcal{J})|_{\mathcal{K}}) = 1] \tag{4.48}$$

Proof: Implications (4.46) and (4.47) follow by the Birkhoff ergodic theorem for (Theorem 4.6) since $(\Phi|_{\mathcal{K}} = f|_{\mathcal{K}}), (\Phi|_{\mathcal{K}} = f(\cdot|\mathcal{I})|_{\mathcal{K}}) \in \mathcal{I}$ by the assumed countability of $\mathcal{K}$. As for implication (4.48), by implications (4.46) and (4.47) it is sufficient to show that conditions $f \in \mathbb{S}$ and $f(f|_{\mathcal{K}} = f(\cdot|\mathcal{I})|_{\mathcal{K}}) = 1$ imply condition $f \in \mathbb{E}$. Assume that the first two conditions hold. Then Theorem 3.25 and the π–λ theorem (Theorem 2.3) imply that $f = f(\cdot|\mathcal{I})$ holds f-almost surely. Hence, $f(A) = f(A|\mathcal{I}) \in \{0,1\}$ holds almost surely for $A \in \mathcal{I}$, which in turn implies that f is ergodic. $\qquad\square$

In particular, the above theorem and the following two ones can be applied to dynamical systems of infinite sequences $(\mathbb{X}^{\mathbb{T}}, \mathcal{X}^{\mathbb{T}}, T)$ where $\mathbb{T} = \mathbb{N}$ or $\mathbb{T} = \mathbb{Z}$, the alphabet $\mathbb{X}$ is countable, and T is the shift operation (4.1), since σ-field $\mathcal{X}^{\mathbb{T}}$ is generated by a countable π-system in this case.

Now we proceed to stating the first ergodic decomposition theorem. Let $P \in \mathbb{S}$ be a stationary probability measure. We note that if the probability space is countably generated, then conditional probability

$$F(A) := P(A|\mathcal{I}), \quad A \in \mathcal{I} \tag{4.49}$$

is a probability measure almost surely, where $F(A) \in \{0,1\}$ almost surely for each $A \in \mathcal{I}$. A question arises whether some version of the random measure F is stationary and ergodic almost surely, i.e. whether $F \in \mathbb{E}$ almost surely. The answer to this question is positive for a stationary probability measure P on a countably generated space and is formally captured by the first ergodic decomposition theorem.

Theorem 4.19 *(Ergodic decomposition): For a dynamical system $(\Omega, \mathcal{I}, T)$, where $\mathcal{I} = \sigma(\mathcal{K})$ for a certain countable π-system $\mathcal{K}$, consider a stationary measure $P \in \mathbb{S}$. We have $P(F \in \mathbb{E}) = 1$ for some version of the random set function $F(A) := P(A|\mathcal{I})$, where $A \in \mathcal{I}$.*

Proof: By Theorem 3.25, function F is a probability measure almost surely. To show that $F \in \mathbb{S}$ almost surely, first we check countably many conditions

$$F(A) = \Phi(A) = \Phi(T^{-1}A) = F(T^{-1}A) \tag{4.50}$$

for $A \in \mathcal{K}$, which hold all simultaneously almost surely by the Birkhoff ergodic theorem written as (4.19), and we apply the π–λ theorem (Theorem 2.3) to infer $F(A) = F(T^{-1}A)$ simultaneously for all $A \in \mathcal{I}$ almost surely. Hence, $P(F \in \mathbb{S}) = 1$.

Having reached this point, we prove $P(F \in \mathbb{E}) = 1$. Using the Birkhoff ergodic theorem for P (Theorem 4.6) and identity (3.73) produces equality $F(\Phi|_{\mathcal{K}} = F|_{\mathcal{K}}) = 1$ P-almost surely, whereas using the Birkhoff ergodic theorem for F yields equality $F(\Phi|_{\mathcal{K}} = F(\cdot|\mathcal{I})|_{\mathcal{K}}) = 1$ holding P-almost surely. Hence,

$F(F(\cdot|\mathcal{I})|_{\mathcal{K}} = F|_{\mathcal{K}}) = 1$ holds P-almost surely, which by (4.48) implies that $F \in \mathbb{E}$ holds P-almost surely. ☐

Subsequently, we will formulate the second ergodic decomposition theorem. Let $P \in \mathbb{S}$ be a stationary measure on a countably generated space. By the first ergodic decomposition theorem, for the random ergodic measure $F(A) := P(A|\mathcal{I})$, we may write

$$P(A) = \int F(A)dP = \int f(A)d\nu(f), \quad A \in \mathcal{I} \tag{4.51}$$

where $\nu(B) = P(F \in B)$ for $B \in \mathcal{E}$ is a probability measure on the space of stationary ergodic measures $(\mathbb{E}, \mathcal{E})$. The second ergodic decomposition theorem states that measure ν not only exists but also is the unique probability measure on space $(\mathbb{E}, \mathcal{E})$ which satisfies condition (4.51). In other words, each stationary measure is a unique convex combination of stationary ergodic probability measures. Here is the formal statement.

Theorem 4.20 *(Ergodic decomposition):* For a dynamical system $(\Omega, \mathcal{I}, T)$, where $\mathcal{I} = \sigma(\mathcal{K})$ for a certain countable π-system $\mathcal{K}$, consider a stationary measure $P \in \mathbb{S}$. Let $F(A) := P(A|\mathcal{I})$. Measure $\nu(B) = P(F \in B)$ for $B \in \mathcal{E}$ is the unique probability measure on the measurable space $(\mathbb{E}, \mathcal{E})$ such that

$$P(A) = \int f(A)d\nu(f), \quad A \in \mathcal{I} \tag{4.52}$$

Proof: First, we will show the existence of ν. Denoting $\nu(B) = P(F \in B)$ for $B \in \mathcal{E}$, for any set $A \in \mathcal{I}$, we obtain

$$P(A) = \int F(A)dP = \int_{F \in \mathbb{E}} F(A)dP = \int f(A)d\nu(f) \tag{4.53}$$

by Theorem 4.19.

Now, we will show the uniqueness of ν. Suppose that (4.52) holds. Then by (4.46) and (4.47) we have $f(\Phi|_{\mathcal{K}} = f|_{\mathcal{K}}) = 1$ for ν-almost all f and $P(\Phi|_{\mathcal{K}} = F|_{\mathcal{K}}) = 1$. Hence, we derive $f(F|_{\mathcal{K}} = f|_{\mathcal{K}}) = 1$ for ν-almost all f, which by the π–λ theorem (Theorem 2.3) implies $f(F = f) = 1$ for ν-almost all f. Now, let $B \in \mathcal{E}$. Applying identity $f(F = f) = 1$, we obtain

$$P(F \in B) = \int f(F \in B)d\nu(f) = \int f(F \in B, F = f)d\nu(f)$$
$$= \int f(f \in B)d\nu(f) = \int \mathbf{1}\{f \in B\}d\nu(f) = \nu(B) \tag{4.54}$$

which proves uniqueness of measure ν. ☐

Let us note the following geometric interpretation of stationary ergodic measures, which is a consequence of the ergodic decomposition theorems. The set of stationary measures $\mathbb{S}$ is convex, i.e. $pf_1 + (1-p)f_2 \in \mathbb{S}$ for any $p \in (0,1)$ and any $f_1, f_2 \in \mathbb{S}$. An element f of a convex set $\mathbb{S}$ is called its extremal point if $f = pf_1 + (1-p)f_2$ for any $p \in (0,1)$ and any $f_1, f_2 \in \mathbb{S}$ implies $f_1 = f_2 = f$. Thus, a simple consequence of the second ergodic decomposition theorem is that a measure $f \in \mathbb{S}$ is ergodic if and only if it is an extremal point.

The ergodic decomposition theorems were discovered by Rokhlin (1962) and then popularized among information-theorists by Gray and Davisson (1974b). Some modern accounts of these results can be found in Kallenberg (1997, theorems 9.10–9.12) and in Gray (2009, theorem 7.6).

Problems

4.1 Show that the invariant σ-field is actually a σ-field.

4.2 Prove Theorem 4.1.

4.3 Prove that any random variable Y such that $Y = Y \circ T$ almost surely is almost surely constant and measurable with respect to the completion of the invariant σ-field if the probability measure is stationary and ergodic.

4.4 Consider the shift operation $T : \mathbb{X}^\top \ni (x_i)_{i \in \mathbb{N}} \mapsto (x_{i+1})_{i \in \mathbb{N}} \in \mathbb{X}^\mathbb{N}$ on the space of one-sided infinite sequences $(\mathbb{X}^\mathbb{N}, \mathcal{X}^\mathbb{N})$. For each random variable $Y_i((x_j)_{j \in \mathbb{N}}) = x_i$, let $\mathcal{G}_i$ be the smallest σ-field against which Y_i is measurable. The σ-fields of the semi-infinite futures are $\mathcal{G}_i^\infty := \sigma\left(\bigcup_{j \geq i} \mathcal{G}_j\right)$. The tail σ-field of the future is $\mathcal{G}_\infty := \bigcap_{i \in \mathbb{N}} \mathcal{G}_i^\infty$. Show that the invariant σ-field satisfies $\mathcal{J} \subseteq \mathcal{G}_\infty$. Exhibit a few events that belong to $\mathcal{G}_\infty$ but not to $\mathcal{J}$. Does a similar result hold for the space of two-sided infinite sequences $(\mathbb{X}^\mathbb{Z}, \mathcal{X}^\mathbb{Z})$?

4.5 Let $\phi_a(x)$ be the rotation defined in (4.17) and let λ be the Lebesgue measure restricted to the unit interval $\mathbb{U}$. Show that the extended dynamical system $(\mathbb{U}, \mathcal{U}, \phi_a, \lambda)$ is ergodic but not mixing for an irrational rotation, i.e. if $a \in (0,1) \backslash \mathbb{Q}$.

4.6 Here we will show an alternative path to proving the Birkhoff ergodic theorem, which was used by Franklin et al. (2012) to demonstrate an effective version of this theorem for Martin-Löf random points, see Section 7.5. In the following, let $(\Omega, \mathcal{J}, T, P)$ be a stationary ergodic dynamical system.

(a) Assume that $P(E) > 0$ for some event $E \in \mathcal{J}$. Without applying the Birkhoff ergodic theorem show this result, called the Poincaré recurrence theorem,

$$\sum_{i=0}^{\infty} I_E \circ T^i = \infty \text{ almost surely} \tag{4.55}$$

(b) Consider an event $B \in \mathcal{J}$ and a number $q \in \mathbb{Q}$. Define the event

$$F := \left(\exists_{n \in \mathbb{N}} \frac{1}{n} \sum_{i=0}^{n-1} (q - I_B) \circ T^i > 0 \right) \tag{4.56}$$

Applying the maximal ergodic theorem, show that

$$\int_F (q - I_B) dP \geq 0 \tag{4.57}$$

(c) Let $P(B) > q > 0$ for some event $B \in \mathcal{J}$ and a number $q \in \mathbb{Q}$. Using (4.57) show that $P(F^c) > 0$ and consequently using (4.55) show that

$$\liminf_{n \to \infty} \frac{1}{n} \sum_{i=0}^{n-1} I_B \circ T^i \geq q \text{ almost surely} \tag{4.58}$$

(d) Finally, argue that for any $B \in \mathcal{J}$, we have

$$\lim_{n \to \infty} \frac{1}{n} \sum_{i=0}^{n-1} I_B \circ T^i = P(B) \text{ almost surely} \tag{4.59}$$

4.7 Consider a dynamical system $(\Omega, \mathcal{J}, T)$, where $\mathcal{J} = \sigma(\mathcal{K})$ for a countable π-system $\mathcal{K}$. Define the set of stationary points

$$\Omega_{\mathbb{S}} = \bigcap_{A \in \mathcal{K}} (\Phi(A) \neq \bot) \tag{4.60}$$

and the set of stationary ergodic points

$$\Omega_{\mathbb{E}} = \Omega_{\mathbb{S}} \cap \bigcap_{A,B \in \mathcal{K}} \left(\lim_{n \to \infty} \frac{1}{n} \sum_{i=0}^{n-1} \Phi(T^{-i}(A) \cap B) = \Phi(A)\Phi(B) \right) \tag{4.61}$$

Let also $\Omega_f := (\Phi|_{\mathcal{K}} = f|_{\mathcal{K}})$ for $f \in \mathbb{E}$. Show that
(a) $\Omega_{\mathbb{S}}, \Omega_{\mathbb{E}} \in \mathcal{J}$.
(b) $\Omega_f \in \mathcal{J}$ and $\Omega_f \cap \Omega_g = \emptyset$ for $f \neq g$ and $f, g \in \mathbb{E}$.
(c) $\Omega_{\mathbb{E}} = \bigcup_{f \in \mathbb{E}} \Omega_f$.
(d) $f(\Omega_f) = 1$ for $f \in \mathbb{E}$.
(e) $f(\Omega_{\mathbb{E}}) = 1$ for $f \in \mathbb{S}$.

4.8 Let $p(x_1^n) = P(X_1^n = x_1^n)$ be the distribution of a discrete stationary process $(X_i)_{i \in \mathbb{Z}}$. Using results of this chapter, complete the proofs of equivalence of the following four statements stated in Chapter 1:

(a) Process $(X_i)_{i \in \mathbb{Z}}$ is ergodic.

(b) For any sequence x_1^k, we have

$$\lim_{n \to \infty} \frac{1}{n} \sum_{i=0}^{n-1} \mathbf{1}\{y_{i+1}^{i+k} = x_1^k\} = p(x_1^k) \text{ for } P\text{-almost all } (y_i)_{i \in \mathbb{Z}} \quad (4.62)$$

(c) For all strings x_1^k and z_1^m, there holds condition

$$\lim_{n \to \infty} \frac{1}{n} \sum_{l=0}^{n-1} \sum_{y_1^l} p(x_1^k y_1^l z_1^m) = p(x_1^k) p(z_1^m) \quad (4.63)$$

holds for all strings x_1^k and z_1^m.

(d) Stationary distribution $p(x_1^n)$ cannot be decomposed as

$$p(x_1^n) = q_1 r_1(x_1^n) + q_2 r_2(x_1^n) \quad (4.64)$$

where $r_1 \neq r_2$ are two different stationary distributions and $0 < q_1 = 1 - q_2 < 1$ are some prior probabilities independent of x_1^n.

4.9 A discrete process $(X_i)_{i \in \mathbb{Z}}$ is called exchangeable if

$$P(X_{j_1} = x_1, \dots, X_{j_n} = x_n) = p(x_1^n) \quad (4.65)$$

for distinct j_i and some function $x_1^n \mapsto p(x_1^n)$. Show that exchangeable processes are stationary. Moreover, show that the ergodic decomposition of an exchangeable process $(X_i)_{i \in \mathbb{Z}}$ consists only of IID processes, i.e.

$$P(X_1^n = x_1^n) = \int \pi(x_1) \cdots \pi(x_n) d\nu(\pi) \quad (4.66)$$

for some probability distributions $\pi(x)$ and a prior probability measure ν. This result was shown by de Finetti (1931).

4.10 Consider measurable space $(\mathbb{X}^{\mathbb{N}}, \mathcal{X}^{\mathbb{N}})$ and the shift operation (4.1) with $\mathbb{T} = \mathbb{N}$ and $\mathbb{X} = \{0, 1\}$. Does every stationary measure P on the dynamical system $(\mathbb{X}^{\mathbb{N}}, \mathcal{X}^{\mathbb{N}}, T)$ satisfy $P(A) = \Phi(A)(\omega)$ for a certain point $\omega \in \mathbb{X}^{\mathbb{N}}$ and all cylinder sets A?

5

Entropy and Information

Language is a system of communication, whereas communication can be regarded as a process of information transmission. From this point of view, while studying language, we should be interested in understanding the notion of information. The standard mathematical theories of information are theories of the amount of information, which turns out to be a more fundamental concept than the structure of information. Namely, the amount of information is preserved by reversible operations on the data, whereas the structure of information can be distorted by them. These different properties of the amount of information and the structure of information are probably the reason for little interaction between linguists, who prefer to study the structure of information, and information theorists, who tend to investigate the amount of information.

There is one more problem with studying the amount of information in empirical data. Namely, in the standard mathematical theories of information, the amount of information is given relative to a probability distribution or a universal computer. Asymptotically, the choice of a concrete probability distribution or a universal computer does not matter much, but, if we are going to perform experiments on finite samples, the choice can influence our estimates. Of course, we can confine ourselves to studying some concrete estimators of the amount of information, but these procedures introduce another level of approximation of the theoretically absolute values. As a result, information theory is not so widely applied in language research, in our opinion, as it should be.

As we have stated in Sections 1.7 and 1.10, nowadays, there are recognized two main mathematical theories of information: one initiated by Claude Elwood Shannon applies probability, whereas another proposed by Andrey Nikolaevich Kolmogorov uses algorithms. In both the approaches, the amount of information contained in an object is roughly equal to the least number of binary digits that are needed to describe the object. Whereas this idea is directly present in Kolmogorov's algorithmic definition, Shannon's probabilistic definition is more abstract. Probably for this reason, Kolmogorov eventually came to believe

Information Theory Meets Power Laws: Stochastic Processes and Language Models, First Edition.
Łukasz Dębowski.
© 2021 John Wiley & Sons, Inc. Published 2021 by John Wiley & Sons, Inc.

that the algorithmic approach to information is more fundamental than the measure-theoretic approach to probability – and for that reason, it should be exposed first, see Cover et al. (1989).

In this chapter and in the next two ones, we will study some ideas from information theory that will be useful to us. However, we will put Kolmogorov's vision of the correct exposition order on its head. We will begin with Shannon's probabilistic theory and some measure-theoretic excursions in this chapter, subsequently going through asymptotic equipartition and universal distributions in Chapter 6, whereas the coding interpretation of Shannon entropy and algorithmic information theory will be relegated to Chapter 7, which will close the cycle of preliminary chapters.

Thus, Section 5.1 is devoted to the concepts of Shannon entropy and mutual information as well as their conditional variants, collectively called Shannon information measures. Usually, these measures are defined for discrete random variables, but, as it is often done in ergodic theory, we will speak of Shannon information measures for partitions of the probability space, which provides a convenient layer of abstraction and stimulates further generalizations. We will see that Shannon information measures satisfy a large number of identities, called chain rules or polymatroid identities, interacting interestingly with the concept of conditional probabilistic independence.

In Section 5.2, we will apply Shannon information measures to discrete stationary processes. We will study the block entropy, which is the entropy of a sequence of consecutive random variables of the process. We will show that block entropy is positive, growing, and concave. As a result, we can define two asymptotic quantities, entropy rate and excess entropy, which measure the limiting unpredictability of the process and the total memory used by the process, respectively. Whereas for natural language, excess entropy seems infinite by Hilberg's hypothesis introduced in Section 1.9, we will show that this quantity is finite for finite-state processes.

Entropy rate and excess entropy are defined as limits of conditional entropy and mutual information for block lengths tending to infinity. We may rightly ask if we can define them plugging infinitely long blocks directly into their definitions. Such operation presumes that we define Shannon information measures for arbitrary fields. This construction will be performed indeed in Section 5.3. It turns out that Shannon information measures for fields inherit all nice properties of Shannon information measures for partitions and additionally, they enjoy continuity, which allows to define entropy rate and excess entropy using infinitely long blocks.

These generalized definitions of entropy rate and excess entropy matter not only for aesthetic reasons. In Section 5.4, we will use them to derive ergodic decomposition formulas for entropy rate and excess entropy. According to these results, a stationary process has the zero entropy rate if all of its ergodic components have the zero entropy rate, whereas the process has the infinite excess entropy if some of its ergodic components have the infinite excess entropy or if there are uncountably many ergodic components. Hence, we arrive at a reason to suspect that natural language, possibly having an infinite excess entropy, can have uncountably many ergodic components.

Section 5.5 is devoted to understanding processes with the zero entropy rate and disproving a strong form of Hilberg's hypothesis, which demands that natural language has a zero entropy rate indeed. A convenient tool for this analysis are Fano inequalities, which are of an independent interest. Using Fano inequalities, we will show that processes with a zero entropy rate are deterministic in the sense that each random variable of the process is a measurable function of its infinite past. This condition seems intuitively implausible as applied to natural language.

In Section 5.6, we will analyze processes that have uncountably many ergodic components. Formally, these processes will be given the name of strongly nonergodic processes. Subsequently, we will show that the mixture Bernoulli process, a simple example of a strongly nonergodic process, does not satisfy the power-law growth of mutual information stated in the relaxed form of Hilberg's hypothesis. The power-law growth of mutual information requires a different construction of the process, which will be postponed till Chapter 8.

5.1 Shannon Measures for Partitions

In this section, we will discuss the basic Shannon information measures such as the Shannon entropy and the mutual information. This section develops formally the informal exposition from Section 1.7. We will begin with Shannon entropy. The Shannon entropy is the founding concept of information theory. It is usually defined as follows, assuming that $\log y$ stands for the logarithm of y to the base 2.

Definition 5.1 *(Shannon entropy):* For a discrete probability distribution p on a countable set $\mathbb{X}$, we define the Shannon entropy of p as

$$H(p) := - \sum_{x : p(x) > 0} p(x) \log p(x) \tag{5.1}$$

Shannon entropy measures the average amount of information in the objects drawn from a discrete distribution. That is, as we will see in Chapter 7, the Shannon entropy is the average least number of binary digits that are needed to describe a random object – if we know its exact probability distribution, in advance. The concept of Shannon entropy gives rise to many interesting considerations and is highly useful. In this chapter, we will study some algebraic properties of Shannon entropy which contributes to the general utility of this concept.

To begin, by the relationship to the KL divergence (Definition 3.14), we obtain that Shannon entropy is less than the following quantity, called Hartley entropy.

Definition 5.2 *(Hartley entropy):* For a discrete probability distribution p on a countable set $\mathbb{X}$, we define the Hartley entropy of p as

$$H_0(p) := \log \#\{x \in \mathbb{X} : p(x) > 0\} \tag{5.2}$$

Theorem 5.1: *We have $0 \leq H(p) \leq H_0(p)$.*

Proof: $H(p) \geq 0$ follows by $p(x) \leq 1$, whereas $H(p) \leq H_0(p)$ follows by Theorem 3.18 for $q(x) = 2^{-H_0(p)}\mathbf{1}\{p(x) > 0\}$. $\qquad\square$

Inparticular $H(p) < \infty$ if the domain of the probability distribution is finite.

In Eq. (5.1), Shannon entropy is defined as a functional of a discrete probability distribution. Here we would like to entertain a different point of view, which leads to a useful generalization. In case of arbitrary probability spaces, the analogue of discrete distributions are discrete random variables and countable partitions of the probability space. In didactic expositions, the Shannon entropy is usually defined for discrete random variables, but speaking of entropy of countable partitions provides a certain layer of abstraction, which turns out to propel some further developments.

Thus, we will consider the definition of entropy for countable partitions as a primary one. In this setting, we will cast both the Shannon and the Hartley entropy.

Definition 5.3 *(Shannon and Hartley entropy):* We define the Shannon entropy of a partition $\alpha = \{A_1, A_2, ...\}$ with respect to probability measure P as

$$H(\alpha) := H_P(\alpha) := - \sum_{i:P(A_i)>0} P(A_i)\log P(A_i) \tag{5.3}$$

Analogously, we also define the Hartley entropy

$$H_0(\alpha) := H_{0,P}(\alpha) := \log \#\{i : P(A_i) > 0\} \tag{5.4}$$

We have inequality $0 \leq H(\alpha) \leq H_0(\alpha)$ by Theorem 5.1. Thus, the Shannon entropy $H(\alpha)$ is finite if the partition α is finite.

What if we have two partitions? The Shannon entropy of one partition is a special case of the Shannon mutual information between two partitions. The definition of Shannon mutual information is as follows.

Definition 5.4 *(Mutual information):* We define the Shannon mutual information between two partitions $\alpha = \{A_1, A_2, ...\}$ and $\beta = \{B_1, B_2, ...\}$ with respect to probability measure P as

$$I(\alpha;\beta) := I_P(\alpha;\beta) := \sum_{i,j:P(A_i \cap B_j)>0} P(A_i \cap B_j) \log \frac{P(A_i \cap B_j)}{P(A_i)P(B_j)} \qquad (5.5)$$

We can immediately see that mutual information is a special case of the KL divergence (Definition 3.14), hence, $I(\alpha;\beta) \geq 0$. Moreover, $I(\alpha;\alpha) = H(\alpha)$.

What if we have three partitions? Let us observe that for $P(C) > 0$, conditional probability $P(\cdot|C) : \mathcal{J} \ni A \mapsto P(A|C)$ is also a probability measure. Therefore, the following concepts of conditional entropy and conditional mutual information also make sense.

Definition 5.5 *(Conditional entropy):* We define the conditional entropy of a partition α given a partition $\gamma = \{C_1, C_2, ...\}$ as

$$H(\alpha|\gamma) := \sum_{k:P(C_k)>0} P(C_k)H_{P(\cdot|C_k)}(\alpha) \qquad (5.6)$$

Definition 5.6 *(Conditional mutual information):* We define the conditional mutual information between two partitions α and β given a partition $\gamma = \{C_1, C_2, ...\}$ as

$$I(\alpha;\beta|\gamma) := \sum_{k:P(C_k)>0} P(C_k)I_{P(\cdot|C_k)}(\alpha;\beta) \qquad (5.7)$$

Now we can specialize the above definitions to discrete random variables. Let us note that discrete random variables $X_i : \Omega \to \mathbb{X}_i$ generate unique partitions $\alpha_i = \{(X_i = x_i) : x_i \in \mathbb{X}_i\}$. In this case, we will define

$$I(X_1; X_2|X_3) := I(\alpha_1; \alpha_2|\alpha_3) \qquad (5.8)$$
$$I(X_1; X_2) := I(\alpha_1; \alpha_2) \qquad (5.9)$$
$$H(X_1|X_2) := H(\alpha_1|\alpha_2) \qquad (5.10)$$
$$H(X_1) := H(\alpha_1) \qquad (5.11)$$

Sometimes we will also use a notation where we mix random variables and classes of events as arguments of these functions.

Entropy and mutual information, both conditional and unconditional, collectively called the Shannon information measures, satisfy a large number of algebraic identities, as it is generally known (Csiszár and Körner, 2011; Cover and Thomas, 2006; Yeung, 2002). To recall, first of all, the Shannon information measures can be all expressed as conditional mutual information or as linear combinations of Shannon entropies. The expression in terms of conditional mutual information is as follows.

Theorem 5.2: *We have:*

1. $I(\alpha_1;\alpha_2) = I(\alpha_1;\alpha_2|\{\Omega\})$.
2. $H(\alpha_1|\alpha_2) = I(\alpha_1;\alpha_1|\alpha_2)$.
3. $H(\alpha_1) = I(\alpha_1;\alpha_1|\{\Omega\})$.

The expression in terms of Shannon entropies requires an additional notation. For two partitions $\alpha = \{A_1, A_2, ...\}$ and $\beta = \{B_1, B_2, ...\}$, let us denote the join partition $\alpha \wedge \beta := \{A_i \cap B_j : i, j = 1, 2, ...\}$.

Theorem 5.3: *If the negative terms are finite, we have:*

1. $I(\alpha_1;\alpha_2|\alpha_3) = H(\alpha_1|\alpha_3) + H(\alpha_2|\alpha_3) - H(\alpha_1 \wedge \alpha_2|\alpha_3)$.
2. $I(\alpha_1;\alpha_2) = H(\alpha_1) - H(\alpha_1|\alpha_2) = H(\alpha_1) + H(\alpha_2) - H(\alpha_1 \wedge \alpha_2)$.
3. $H(\alpha_1|\alpha_2) = H(\alpha_1 \wedge \alpha_2) - H(\alpha_2)$.

The proofs of the two above theorems are left as an easy exercise. Now, we will introduce several relations on partitions. The first three ones have to do with refinements of partitions.

1. We will say that partition β is finer than α, written $\alpha \prec \beta$, if for each event $B \in \beta$, there exists an event $A \in \alpha$ such that $B \subseteq A$. For discrete random variables $Y_i : \Omega \to \mathbb{Y}_i$ and partitions $\beta_i = \{(Y_i = y_i) : y_i \in \mathbb{Y}_i\}$, relation $\beta_1 \prec \beta_2$ holds if and only if $Y_1 = f(Y_2)$ for a certain function f (measurability is not an issue here – it is guaranteed by discreteness of the variables).

2. We will say that partition β is almost surely finer than α, written $\alpha \precsim \beta$, if for each event $B \in \beta$ with $P(B) > 0$ we have $P(A|B) \in \{0, 1\}$ for all $A \in \alpha$. For discrete random variables $Y_i : \Omega \to \mathbb{Y}_i$ and partitions $\beta_i = \{(Y_i = y_i) : y_i \in \mathbb{Y}_i\}$, relation $\beta_1 \precsim \beta_2$ holds if and only if $Y_1 = f(Y_2)$ almost surely for a certain function f.

3. We will say that partitions α and β are almost surely equal, written $\alpha \sim \beta$, if both $\alpha \precsim \beta$ and $\beta \precsim \alpha$. For discrete random variables $Y_i : \Omega \to \mathbb{Y}_i$ and partitions $\beta_i = \{(Y_i = y_i) : y_i \in \mathbb{Y}_i\}$, relation $\beta_1 \sim \beta_2$ holds if and only if $Y_1 = f(Y_2)$ almost surely for a certain bijection f.

The next few relations have to do with probabilistic independence. In general, we say that events A and B are (probabilistically) independent if $P(A \cap B) = P(A)P(B)$, which reduces to $P(A|B) = P(A)$ for $P(B) > 0$. Analogously, we say that events A and B are conditionally independent given C, where $P(C) > 0$, if $P(A \cap B|C) = P(A|C)P(B|C)$. In particular, events A and B are independent if and only if they are conditionally independent given Ω. Let us generalize these concepts to partitions.

1. We will say that partitions α and β are independent, written $\alpha \perp\!\!\!\perp \beta$, if all $A \in \alpha$ and $B \in \beta$ are pairwise independent, i.e. if $P(A \cap B) = P(A)P(B)$.
2. We will say that partitions α and β are conditionally independent given partition γ, written $\alpha \perp\!\!\!\perp \beta|\gamma$, if all $A \in \alpha$ and $B \in \beta$ are pairwise conditionally independent given all $C \in \gamma$ with $P(C) > 0$, i.e. if $P(A \cap B|C) = P(A|C)P(B|C)$. We have $\alpha \perp\!\!\!\perp \beta|\{\Omega\}$ if and only if $\alpha \perp\!\!\!\perp \beta$.

Extending these concepts, we will say that discrete random variables Y_i are (conditionally) independent if the respective partitions $\{(Y_i = y_i) : y_i \in \mathbb{Y}_i\}$ are (conditionally) independent.

The following four results found the algebra of the Shannon information measures. The results apply the above-defined relations of refinement and independence. The first proposition concerns refinements of partitions.

Theorem 5.4: *We have* $\alpha \prec \beta \iff \beta = \alpha \wedge \beta \implies \alpha \precsim \beta \iff \beta \sim \alpha \wedge \beta$.

Proof: We have

1. $\alpha \prec \beta \iff \forall B \in \beta : \exists A \in \alpha : B \subseteq A \implies \forall B \in \beta : [P(B) > 0 \implies \forall A \in \alpha : P(A|B) \in \{0, 1\}] \iff \alpha \precsim \beta$.
2. $\beta = \alpha \wedge \beta \iff [\beta \prec \alpha \wedge \beta$ and $\alpha \wedge \beta \prec \beta] \iff \alpha \wedge \beta \prec \beta \iff \alpha \prec \beta$.
3. $\beta \sim \alpha \wedge \beta \iff [\beta \precsim \alpha \wedge \beta$ and $\alpha \wedge \beta \precsim \beta] \iff \alpha \wedge \beta \precsim \beta \iff \alpha \precsim \beta$. $\quad\square$

The second proposition provides two algebraic properties of conditional mutual information: symmetry and vanishing for conditionally independent partitions.

Theorem 5.5: *We have*

1. $I(\alpha; \beta|\gamma) = I(\beta; \alpha|\gamma)$.
2. $I(\alpha; \beta|\gamma) \geq 0$ *with equality if and only if* $\alpha \perp\!\!\!\perp \beta|\gamma$.

Proof:
1. Trivial.
2. By Theorem 3.18, $I(\alpha; \beta) \geq 0$ with equality if and only if $\alpha \perp\!\!\!\perp \beta$. Hence, $I(\alpha; \beta|\gamma) \geq 0$ with equality if and only if $\alpha \perp\!\!\!\perp \beta|\gamma$. $\quad\square$

The third proposition characterizes partly the conditional independence in terms of partition refinements.

Theorem 5.6: *We have* $\alpha \precsim \beta \iff \alpha \perp\!\!\!\perp \alpha|\beta \implies \alpha \perp\!\!\!\perp \gamma|\beta$.

Proof: For each $B \in \beta$ such that $P(B) > 0$, we have $P(A|B)P(A'|B) = P(A \cap A'|B)$ for all $A, A' \in \alpha$ if and only if $P(A|B) \in \{0, 1\}$ for all $A \in \alpha$. Hence,

$\alpha \precsim \beta \iff \alpha \perp\!\!\!\perp \alpha|\beta$. In contrast, if for each $B \in \beta$ such that $P(B) > 0$, we have $P(A|B) \in \{0,1\}$ for all $A \in \alpha$, then $P(A|B)P(C|B) = P(A \cap C|B)$ for all $A \in \alpha$ and $C \in \gamma$. Hence, $\alpha \precsim \beta \implies \alpha \perp\!\!\!\perp \gamma|\beta$. □

The fourth proposition is called the chain rule for conditional information.

Theorem 5.7 *(Chain rule):* *We have* $I(\alpha; \beta \wedge \gamma|\delta) = I(\alpha; \beta|\delta) + I(\alpha; \gamma|\beta \wedge \delta)$.

Proof: The claim follows by

$$\frac{P(A_i \cap B_j \cap C_k|D_l)}{P(A_i|D_l)P(B_j \cap C_k|D_l)}$$
$$= \frac{P(A_i \cap B_j|D_l)}{P(A_i|D_l)P(B_j|D_l)} \times \frac{P(A_i \cap C_k|B_j \cap D_l)}{P(A_i|B_j \cap D_l)P(C_k|B_j \cap D_l)} \tag{5.12}$$
□

The above four theorems yield a large number of other useful identities. We will say that partition α is trivial if $P(A) \in \{0,1\}$ for all $A \in \alpha$. A handful of useful corollaries of Theorems 5.2 and 5.4–5.7 is as follows:

Theorem 5.8: *We have*

1. $I(\alpha; \beta|\gamma) = I(\alpha; \beta)$ *if* $\alpha \wedge \beta \perp\!\!\!\perp \gamma$.
2. $H(\alpha|\beta) = 0$ *if and only if* $\alpha \precsim \beta$.
3. $H(\alpha) = 0$ *if and only if* α *is trivial*.
4. $I(\alpha; \beta|\gamma) = H(\alpha|\gamma)$ *for* $\alpha \prec \beta$.
5. $H(\alpha|\gamma) = I(\alpha; \beta|\gamma) + H(\alpha|\beta \wedge \gamma)$.
6. $I(\alpha; \beta_1|\gamma) \le I(\alpha; \beta_2|\gamma)$ *if* $\beta_1 \prec \beta_2$.
7. $H(\alpha|\beta_1) \ge H(\alpha|\beta_2)$ *if* $\beta_1 \prec \beta_2$.
8. $I(\alpha; \beta|\gamma) \le H(\alpha|\gamma) \le H(\alpha) \le H_0(\alpha)$.
9. $H(\alpha \wedge \beta|\gamma) = H(\alpha|\gamma) + H(\beta|\alpha \wedge \gamma)$.
10. $I(\alpha; \beta) = H(\gamma) + I(\alpha; \beta|\gamma)$ *if* $\gamma \prec \alpha$ *and* $\gamma \prec \beta$.

Proof:

1. If $\alpha \wedge \beta \perp\!\!\!\perp \gamma$ then $0 = I(\alpha \wedge \beta; \gamma) = I(\alpha; \gamma) + I(\beta; \gamma|\alpha) = I(\beta; \gamma) + I(\alpha; \gamma|\beta)$. Hence, $I(\alpha; \gamma) = I(\alpha; \gamma|\beta) = 0$. Using this fact, we obtain $I(\alpha; \beta|\gamma) = I(\alpha; \gamma) + I(\alpha; \beta|\gamma) = I(\alpha; \beta \wedge \gamma) = I(\alpha; \beta) + I(\alpha; \gamma|\beta) = I(\alpha; \beta)$.
2. We have $0 = H(\alpha|\beta) = I(\alpha; \alpha|\beta)$ if and only if $\alpha \perp\!\!\!\perp \alpha|\beta$, which is equivalent to $\alpha \precsim \beta$.
3. We have $0 = H(\alpha) = I(\alpha; \alpha|\{\Omega\})$ if and only if $\alpha \precsim \{\Omega\}$, which is equivalent to α being trivial.
4. By $\alpha \perp\!\!\!\perp \beta|\alpha \wedge \gamma$, $I(\alpha; \beta|\alpha \wedge \gamma) = 0$. The sequel is easy: $H(\alpha|\gamma) = I(\alpha; \alpha|\gamma) = I(\alpha; \alpha|\gamma) + I(\alpha; \beta|\alpha \wedge \gamma) = I(\alpha; \alpha \wedge \beta|\gamma) = I(\alpha; \beta|\gamma)$

5. By property (4), we have $H(\alpha|\gamma) = I(\alpha; \alpha \wedge \beta|\gamma) = I(\alpha; \beta|\gamma) + I(\alpha; \alpha|\beta \wedge \gamma) = I(\alpha; \beta|\gamma) + H(\alpha|\beta \wedge \gamma)$.
6. If $\beta_1 \prec \beta_2$, then $\beta_2 = \beta_1 \wedge \beta_2$. Hence $I(\alpha; \beta_2|\gamma) = I(\alpha; \beta_1 \wedge \beta_2|\gamma) = I(\alpha; \beta_1|\gamma) + I(\alpha; \beta_1|\beta_2 \wedge \gamma) \geq I(\alpha; \beta_1|\gamma)$.
7. By property (4) we have $H(\alpha|\gamma) = I(\alpha; \alpha \wedge \beta|\gamma)$. Hence for $\beta_1 \prec \beta_2$, being equivalent to $\beta_2 = \beta_1 \wedge \beta_2$, we obtain $H(\alpha|\beta_2) = H(\alpha|\beta_1 \wedge \beta_2) = I(\alpha; \alpha|\beta_1 \wedge \beta_2) \leq I(\alpha; \alpha|\beta_1 \wedge \beta_2) + I(\alpha; \beta_2|\beta_1) = I(\alpha; \alpha \wedge \beta_1|\beta_2) = H(\alpha|\beta_1)$.
8. Follows from properties (5) and (7).
9. We have $H(\alpha \wedge \beta|\gamma) = I(\alpha \wedge \beta; \alpha \wedge \beta|\gamma) = I(\alpha \wedge \beta; \alpha|\gamma) + I(\alpha \wedge \beta; \beta|\alpha \wedge \gamma) = H(\alpha|\gamma) + H(\beta|\alpha \wedge \gamma)$ by property (4).
10. By property (4) we have $H(\gamma) = I(\alpha \wedge \gamma; \gamma)$. Hence $I(\alpha; \beta) = I(\alpha; \beta \wedge \gamma) = I(\alpha; \gamma) + I(\alpha; \beta|\gamma) = I(\alpha \wedge \gamma; \gamma) + I(\alpha; \beta|\gamma) = H(\gamma) + I(\alpha; \beta|\gamma)$. $\square$

In particular, claim (6) of Theorem 5.8 is called the data-processing inequality. It asserts that the amount of information as measured by entropy and mutual information does not decrease when we refine partitions. As a result, the amount of information contained in a random variable does not change when we apply a reversible operation to it. Claim (10) of Theorem 5.8 will be used to prove the ergodic decomposition of excess entropy in Section 5.4.

Last but not least, we note this fact concerning equality of conditional mutual information for partitions that are almost surely equal.

Theorem 5.9: *We have $I(\alpha; \beta|\gamma) = I(\alpha'; \beta'|\gamma')$ if $\alpha \sim \alpha', \beta \sim \beta', \gamma \sim \gamma'$.*

Proof: For events with a strictly positive probability, there is a bijection between events $A_i \cap B_j \cap C_k \in \alpha \wedge \beta \wedge \gamma$ and $A_i' \cap B_j' \cap C_k' \in \alpha' \wedge \beta' \wedge \gamma'$ such that $P((A_i \cap B_j \cap C_k) \triangle (A_i' \cap B_j' \cap C_k')) = 0$. Hence, $I(\alpha; \beta|\gamma)$ and $I(\alpha'; \beta'|\gamma')$ are sums over equal expressions. $\square$

The above-discussed algebraic identities are called the polymatroid identities. It turns out that Shannon information measures also satisfy certain inequalities which cannot be reduced to the polymatroid inequalities. This topic will not be pursued here. An interested reader is referred to Yeung (2002).

5.2 Block Entropy and Its Limits

In this section, we will consider Shannon information measures for discrete stationary processes, as half-formally sketched in Section 1.7, where we have introduced the block entropy, the entropy rate, and the excess entropy. The terminology of this section and the results are largely borrowed from Crutchfield

and Feldman (2003). Primarily, we will be interested in the entropy of sequences
of consecutive variables, called blocks. The blocks of random variables will be
written as

$$X_j^k := (X_j, X_{j+1}, \dots, X_k) \tag{5.13}$$

by an analogy to the notation for strings $x_j^k := (x_j, x_{j+1}, \dots, x_k)$.

In the following, we will assume that $(X_i)_{i \in \mathbb{Z}}$ is a stationary process on a proba-
bility space $(\mathbb{X}^{\mathbb{Z}}, \mathcal{X}^{\mathbb{Z}}, P)$, where $X_j : \mathbb{X}^{\mathbb{Z}} \ni (x_i)_{i \in \mathbb{Z}} \mapsto x_j \in \mathbb{X}$ and the alphabet $\mathbb{X}$ is
countable. We observe that the entropy of a block drawn from a stationary process
depends only on the block length.

Definition 5.7 (*Block entropy*): The entropy of a block of variables drawn from
the discrete stationary process $(X_i)_{i \in \mathbb{Z}}$ will be denoted as

$$H(n) := H_P(n) := H(X_1^n) = H(X_{i+1}^{i+n}) \tag{5.14}$$

For convenience, we also put $H(0) := H_P(0) := 0$.

As we have mentioned in Section 1.9, Hilberg (1990) supposed that the block
entropy for texts in natural language grows like a power law, say,

$$H(n) = Cn^\beta, \quad C > 0, \quad 0 < \beta < 1 \tag{5.15}$$

We will treat this hypothesis as a sufficient motivation to investigate the general
behavior of the block entropy for stationary processes. In this section, we will not
construct any particular process that satisfies a condition resembling hypothesis
(5.15), but we will only investigate general properties of function $H(n)$.

Let us observe that $H(n) < \infty$ if and only if $H(1) < \infty$, since $H(1) \le H(n) \le
nH(1)$ by the polymatroid identities. In this case, the process will be called a finite
entropy process. As we will see, by Theorem 5.3, some differences of the block
entropy of a finite entropy process equal to the conditional entropy and mutual
information. Let Δ be the difference operator,

$$\Delta F(n) := F(n) - F(n-1) \tag{5.16}$$

Using this operator, we can state a few interesting propositions. The first result
states that the first difference of $H(n)$ is equal to conditional entropy, whereas the
second difference of $H(n)$ is equal to conditional mutual information.

Theorem 5.10: *The block entropy of a finite entropy process satisfies*

$$\Delta H(n) = H(X_n | X_1^{n-1}) \tag{5.17}$$
$$\Delta^2 H(n) = -I(X_1; X_n | X_2^{n-1}) \tag{5.18}$$

where $H(X_1 | X_1^0) := H(X_1)$ *and* $I(X_1; X_2 | X_2^1) := I(X_1; X_2)$.

Proof: By Theorem 5.3, we have

$$H(X_n|X_1^{n-1}) = H(X_1^n) - H(X_1^{n-1})$$
$$= H(n) - H(n-1) = \Delta H(n) \tag{5.19}$$
$$-I(X_1; X_n|X_2^{n-1}) = H(X_1^n) - H(X_1^{n-1}) - H(X_2^n) + H(X_2^{n-1})$$
$$= H(n) - 2H(n-1) + H(n-2) = \Delta^2 H(n) \tag{5.20}$$

□

Hence, for a stationary process, block entropy $H(n)$ is positive, $H(n) \geq 0$, growing, $\Delta H(n) \geq 0$, and concave, $\Delta^2 H(n) \leq 0$, as a function of n.

Now, we will investigate the behavior of block entropy for large arguments. First of all, we note that block entropy $H(n)$ is subadditive, i.e. $H(n) + H(m) - H(n + m) = I(X_1^n; X_{n+1}^{n+m}) \geq 0$. The following proposition for subadditive and superadditive sequences is due to Fekete (1923).

Theorem 5.11 *(Fekete lemma):* *Let a sequence of real numbers $(a_n)_{n\in\mathbb{N}}$ be superadditive, i.e. let it satisfy $a_{n+m} \geq a_n + a_m$. Then there exists limit*

$$\lim_{n\to\infty} \frac{a_n}{n} = \sup_{n\in\mathbb{N}} \frac{a_n}{n} \tag{5.21}$$

Conversely, if sequence $(a_n)_{n\in\mathbb{N}}$ is subadditive, i.e. it satisfies $a_{n+m} \leq a_n + a_m$, then there exists limit

$$\lim_{n\to\infty} \frac{a_n}{n} = \inf_{n\in\mathbb{N}} \frac{a_n}{n} \tag{5.22}$$

Proof: Sequence $(a_n)_{n\in\mathbb{N}}$ is subadditive if and only if $(-a_n)_{n\in\mathbb{N}}$ is superadditive. Without loss of generality, let us assume that $(a_n)_{n\in\mathbb{N}}$ is superadditive. Fix a number $l < \sup_{n\in\mathbb{N}} a_n/n$. Choose another number n such that $a_n \geq nl$. For $m \geq n$, we may write $m = qn + r$, where $q \geq 1$ and $0 \leq r < n$ are some integers. By the superadditivity, we have $a_m \geq qa_n + a_r$. Hence,

$$\frac{a_m}{m} \geq \frac{qa_n + a_r}{qn + r} \geq \frac{qnl + a_r}{qn + r} \xrightarrow[m\to\infty]{} l \tag{5.23}$$

Since l was chosen arbitrarily, we hence obtain

$$\liminf_{n\to\infty} \frac{a_m}{m} \geq \sup_{n\in\mathbb{N}} \frac{a_n}{n} \tag{5.24}$$

which is equivalent to the claim. □

For sequences that are concave, the Fekete lemma can be strengthened.

Theorem 5.12: *Let a sequence of real numbers* $(a_n)_{n \in \{0\} \cup \mathbb{N}}$, *where* $a_0 = 0$, *be concave, i.e. let it satisfy* $\Delta^2 a_n \leq 0$. *Then sequence* $(a_n)_{n \in \mathbb{N}}$ *is subadditive and*

$$\lim_{n \to \infty} \frac{a_n}{n} = \inf_{n \in \mathbb{N}} \frac{a_n}{n} = \lim_{n \to \infty} \Delta a_n = a_1 + \sum_{n=2}^{\infty} \Delta^2 a_n \qquad (5.25)$$

Proof: By the concavity, the sequence of differences $(\Delta a_n)_{n \in \{0\} \cup \mathbb{N}}$ is decreasing. Hence,

$$a_{n+m} = a_n + \sum_{k=n+1}^{n+m} \Delta a_k \leq a_n + a_0 + \sum_{k=1}^{m} \Delta a_k = a_n + a_m \qquad (5.26)$$

so $(a_n)_{n \in \mathbb{N}}$ is subadditive. Subsequently, it is sufficient to show

$$b := \lim_{n \to \infty} \frac{a_n}{n} = b' := \lim_{n \to \infty} \Delta a_n \qquad (5.27)$$

where limit b exists by the Fekete lemma (Theorem 5.11) and limit b' exists since sequence $(\Delta a_n)_{n \in \{0\} \cup \mathbb{N}}$ is decreasing. By the first equality in (5.26), we also obtain

$$a_n + m\Delta a_n \leq a_{n+m} \leq a_n + m\Delta a_{m+n} \qquad (5.28)$$

Putting $n = m$, dividing by n, and letting $n \to \infty$ yields

$$b + b' \leq 2b \leq b + b' \qquad (5.29)$$

Hence, $b = b'$. □

Since the block entropy satisfies the assumptions of the above Proposition, then we may define the following limiting quantity, called the entropy rate.

Definition 5.8 *(Entropy rate):* The entropy rate of a finite entropy stationary process $(X_i)_{i \in \mathbb{Z}}$ is defined as

$$h := h_P := \lim_{n \to \infty} \frac{H(n)}{n} = \inf_{n \in \mathbb{N}} \frac{H(n)}{n} = \lim_{n \to \infty} H(X_1 | X_{-n}^0) = H(1) + \sum_{n=2}^{\infty} \Delta^2 H(n) \qquad (5.30)$$

We have $0 \leq h \leq H(1)$. By the third equivalent definition in Eq. (5.30), the entropy rate measures how much unpredictable a single observation X_1 is if we know the infinite past $X_{-\infty}^0$. Thus, the entropy rate is a certain measure of asymptotic indeterminism of the process. In particular, we have $h = 0$ if condition (5.15) is satisfied. The case of a stationary process with a vanishing entropy rate will be given a name.

Definition 5.9 *(Asymptotically deterministic process):* A finite entropy stationary process $(X_i)_{i \in \mathbb{Z}}$ is called asymptotically deterministic if $h = 0$.

For a concave sequence $(a_n)_{n \in \{0\} \cup \mathbb{N}}$, it also makes sense to investigate the limit of differences $2a_n - a_{2n}$. There are four equivalent expressions for this quantity.

Theorem 5.13: *Let a sequence of real numbers $(a_n)_{n \in \{0\} \cup \mathbb{N}}$, where $a_0 = 0$, be concave, i.e. let it satisfy $\Delta^2 a_n \leq 0$. Define*

$$e'_n := 2a_n - a_{2n}, \quad e''_n := a_n - n\Delta a_n, \quad e'''_n := a_n - nb \qquad (5.31)$$

where $b := \lim_{n \to \infty} \Delta a_n$. We have

1. $0 \leq e''_n \leq e'_n \leq e''_{2n}$ and $e''_n \leq e'''_n$.
2. e'_n, e''_n, and e'''_n are increasing.
3. $\lim_{n \to \infty} e'_n = \lim_{n \to \infty} e''_n = \lim_{n \to \infty} e'''_n = -\sum_{n=2}^{\infty} (n-1)\Delta^2 a_n$.

Proof: Observe the chain rule $a_n = \sum_{i=1}^{n} \Delta a_i$. Hence, we obtain

$$e'_n = \sum_{i=1}^{n} [\Delta a_i - \Delta a_{i+n}] = -\sum_{i=1}^{n} \sum_{j=1}^{n} \Delta^2 a_{i+j}$$

$$= -\sum_{k=2}^{n} (k-1)\Delta^2 a_k - \sum_{k=n+1}^{2n} (2n-k+1)\Delta^2 a_k \qquad (5.32)$$

$$e''_n = \sum_{i=1}^{n} [\Delta a_i - \Delta a_n] = -\sum_{i=1}^{n} \sum_{j=i+1}^{n} \Delta^2 a_j$$

$$= -\sum_{k=2}^{n} (k-1)\Delta^2 a_k \qquad (5.33)$$

$$e'''_n = e''_n + n[\Delta a_n - b]$$

$$= -\sum_{k=2}^{n} (k-1)\Delta^2 a_k - \sum_{k=n+1}^{\infty} n\Delta^2 a_k \qquad (5.34)$$

Since terms $-\Delta^2 a_k$ are all positive, hence, the claims follow. $\qquad \square$

Since the block entropy satisfies the assumptions of the above proposition, then we may define another limiting quantity, called the excess entropy.

Definition 5.10 *(Excess entropy):* The excess entropy of a finite entropy stationary process $(X_i)_{i \in \mathbb{Z}}$ is defined as

$$E := E_P := \lim_{n \to \infty} I(X_{-n+1}^0; X_1^n) = \lim_{n \to \infty} [H(n) - n\Delta H(n)]$$

$$= \lim_{n \to \infty} [H(n) - nh] = -\sum_{n=2}^{\infty} (n-1)\Delta^2 H(n) \qquad (5.35)$$

We have $0 \leq E \leq \infty$. By the first equivalent definition in Eq. (5.35), the excess entropy measures how much dependence there is between the infinite past $X_{-\infty}^0$

and the infinite future X_1^∞. Thus, excess entropy is a certain global measure of dependence in the process. In particular, we have $E = \infty$ if condition (5.15) is satisfied. The case of a stationary process with an infinite excess entropy will be also given a name.

Definition 5.11 *(Finitary and infinitary processes)*: A finite entropy stationary process $(X_i)_{i\in\mathbb{Z}}$ is called finitary if $E < \infty$ and infinitary if $E = \infty$.

Now, let us briefly discuss the entropy rate and the excess entropy for the three classes of processes introduced in Section 2.4. As we will see, in typical cases, none of these processes is asymptotically deterministic or infinitary. The following propositions without proofs are left as easy exercises.

Theorem 5.14: *The entropy rate and excess entropy of an IID process $(X_i)_{i\in\mathbb{Z}}$ are*

$$h = H(X_i) \in [0, \log \#\mathbb{X}] \tag{5.36}$$
$$E = 0 \tag{5.37}$$

Theorem 5.15: *The entropy rate and excess entropy of a stationary Markov process $(X_i)_{i\in\mathbb{Z}}$ are*

$$h = H(X_i|X_{i-1}) \in [0, \log \#\mathbb{X}] \tag{5.38}$$
$$E = I(X_i; X_{i-1}) \in [0, k\log \#\mathbb{X}] \tag{5.39}$$

Theorem 5.16: *The block entropy, the entropy rate, and the excess entropy of a stationary kth order Markov process $(X_i)_{i\in\mathbb{Z}}$ are*

$$H(n) = (n-k)H(k+1) - (n-k-1)H(k) \text{ for } n > k \tag{5.40}$$
$$h = H(k+1) - H(k) \in [0, \log \#\mathbb{X}] \tag{5.41}$$
$$E = (k+1)H(k) - kH(k+1) = H(k) - kh \in [0, k\log \#\mathbb{X}] \tag{5.42}$$

In particular, we can see that a kth order Markov process can be infinitary only when the alphabet is infinite. If the alphabet is finite, then a kth order Markov process is asymptotically deterministic if and only if the distribution of the process is a periodic measure.

In general, there are no closed expressions for the entropy rate and excess entropy of a hidden Markov process (Birch, 1962; Jacquet et al., 2008; Travers and Crutchfield, 2014). We can give, however, simple bounds. In particular, the excess entropy of a finite-state process is finite, as the following theorem implies.

Theorem 5.17: *The entropy rate and the excess entropy of a stationary hidden Markov process $(X_i)_{i \in \mathbb{Z}}$ with the underlying Markov process $(Y_i)_{i \in \mathbb{Z}}$ satisfy inequalities*

$$h \geq H(X_i|Y_{i-1}) \in [0, \log \#\mathbb{X}] \tag{5.43}$$

$$E \leq I(Y_i; Y_{i-1}) \in [0, \log \#\mathbb{Y}] \tag{5.44}$$

Proof: Since random variables X_i^{i+n-1} and X_{i-n}^{i-1} are conditionally probabilistically independent given random variables Y_i^{i+n-1} and Y_{i-n}^{i-1}, we have

$$h = \inf_{n \in \mathbb{N}} H(X_i|X_{i-n}^{i-1}) \geq \inf_{n \in \mathbb{N}} H(X_i|Y_{i-n}^{i-1}) = H(X_i|Y_{i-1}) \tag{5.45}$$

$$E = \sup_{n \in \mathbb{N}} I(X_i^{i+n-1}; X_{i-n}^{i-1}) \leq \sup_{n \in \mathbb{N}} I(Y_i^{i+n-1}; Y_{i-n}^{i-1}) = I(Y_i; Y_{i-1}) \tag{5.46}$$

$\square$

Concluding this section, we can see that a stationary hidden Markov process $(X_i)_{i \in \mathbb{Z}}$ is not infinitary or asymptotically deterministic, respectively, if the underlying Markov process $(Y_i)_{i \in \mathbb{Z}}$ is not infinitary or asymptotically deterministic. Consequently, hidden Markov processes are finitary when the alphabet of the underlying Markov process is finite, i.e. if the process is finite-state. Some infinitary hidden Markov processes where the alphabets of the underlying Markov processes are infinite were discussed by Travers and Crutchfield (2014) and Dębowski (2014). We consider these examples kind of artificial. In Section 8.3 and Chapter 11, we will construct some different examples of infinitary and asymptotically deterministic processes, which are more motivated by natural language phenomena.

5.3 Shannon Measures for Fields

An interesting question arises whether the entropy rate and the excess entropy of a stationary process can be interpreted as the conditional entropy and the mutual information for infinite sequences of random variables. In other words, we ask whether we can define $H(X_1|X_{-\infty}^0)$ and $I(X_{-\infty}^0; X_1^\infty)$ in such a way that

$$h = \lim_{n \to \infty} H(X_1|X_{-n}^0) = H(X_1|X_{-\infty}^0) \tag{5.47}$$

$$E = \lim_{n \to \infty} I(X_{-n+1}^0; X_1^n) = I(X_{-\infty}^0; X_1^\infty) \tag{5.48}$$

Such a hypothetical representation presumes that we generalize the Shannon information measures to classes of events larger than partitions, such as arbitrary fields.

To make a historical remark, we note that generalization of Shannon information measures to arbitrary fields is a classical topic in information theory dating back to Gelfand et al. (1956). The original approach proposed by Gelfand et al. (1956) and developed later by Dobrushin (1959) and Pinsker (1964), goes through an integral over the Radon–Nikodym derivative of a conditional product measure with respect to a joint measure. This approach is doomed to fail if the regular conditional probability does not exist, since conditional product measure may fail to exist in this case either (Sazonov, 1964; Swart, 1996). A simpler and fully general approach, to be presented in this section, was essentially proposed by Wyner (1978) and Dębowski (2009) independently, with some later corrections by Dębowski (2020) concerning the proof of Theorem 2 by Dębowski (2009).

The definition proposed by Wyner (1978) and Dębowski (2009) reads:

Definition 5.12 *(Shannon information measures)*: Let $(\Omega, \mathcal{J}, P)$ be a probability space. For a class $\mathcal{C} \subseteq \mathcal{J}$ and an event $E \in \mathcal{J}$, let $P(E|\mathcal{C}) := P(E|\sigma_p(\mathcal{C}))$ be the conditional probability of E with respect to the smallest complete σ-field containing $\mathcal{C}$. Subsequently, for an arbitrary class $\mathcal{C}$ and countable partitions α and β, we define the pointwise conditional entropy and mutual information

$$H(\alpha||\mathcal{C}) := H_{P(\cdot|\mathcal{C})}(\alpha), \quad I(\alpha;\beta||\mathcal{C}) := I_{P(\cdot|\mathcal{C})}(\alpha;\beta) \tag{5.49}$$

Finally, for arbitrary fields $\mathcal{A}$ and $\mathcal{B}$ and a class $\mathcal{C}$, we define the average conditional entropy and mutual information

$$H(\mathcal{A}|\mathcal{C}) := \sup_{\alpha \subseteq \mathcal{A}} \mathbf{E}\, H(\alpha||\mathcal{C}), \quad I(\mathcal{A};\mathcal{B}|\mathcal{C}) := \sup_{\alpha \subseteq \mathcal{A}, \beta \subseteq \mathcal{B}} \mathbf{E}\, I(\alpha;\beta||\mathcal{C}) \tag{5.50}$$

where the suprema are taken over all finite partitions of $\mathcal{A}$ and $\mathcal{B}$.

Let us note that if $\mathcal{A}$, $\mathcal{B}$, and $\mathcal{C}$ are finite fields, then the above definitions reduce to respective Shannon information measures for finite partitions:

Theorem 5.18: *If $\mathcal{A}_i = \sigma(\alpha_i)$ are generated by finite partitions then:*

1. $H(\mathcal{A}_1) = H(\alpha_1)$.
2. $H(\mathcal{A}_1|\mathcal{A}_3) = H(\alpha_1|\alpha_3)$.
3. $I(\mathcal{A}_1;\mathcal{A}_2) = I(\alpha_1;\alpha_2)$.
4. $I(\mathcal{A}_1;\mathcal{A}_2|\mathcal{A}_3) = I(\alpha_1;\alpha_2|\alpha_3)$.

Proof: The claims follow by monotonicity of Shannon information measures, i.e. claim (6) of Theorem 5.8. □

We also stress that Definition 5.12, in contrast to the earlier expositions by Dobrushin (1959) and Pinsker (1964), is simpler and does not require regular conditional probability. In fact, the expressions on the right-hand sides of Eqs. (5.50) are well-defined for all $\mathcal{A}$, $\mathcal{B}$, and $\mathcal{C}$ since conditional probabilities $P(\ \cdot\ |\mathcal{C})$ are $\mathcal{J}$-measurable. No problems arise either when the conditional probability is not regular since the conditional distribution $(P(A|\mathcal{C}))_{A\in\alpha}$ restricted to a finite partition α is almost surely a probability distribution.

Similarly as in the discrete case, for arbitrary random variables Y_i with image spaces $(\mathbb{Y}_i, \mathcal{Y}_i)$ and preimages $\mathcal{A}_i = Y_i^{-1}(\mathcal{Y}_i)$, we will also define the Shannon information measures

$$I(Y_1;Y_2|Y_3) := I(\mathcal{A}_1;\mathcal{A}_2|\mathcal{A}_3) \tag{5.51}$$

$$I(Y_1;Y_2) := I(\mathcal{A}_1;\mathcal{A}_2) \tag{5.52}$$

$$H(Y_1|Y_2) := H(\mathcal{A}_1|\mathcal{A}_2) \tag{5.53}$$

$$H(Y_1) := H(\mathcal{A}_1) \tag{5.54}$$

As previously declared, sometimes we will apply a notation where we mix random variables and classes of events as arguments of these functions.

Shannon information measures as generalized above turn out to satisfy all generalized polymatroid identities as well as some continuity. To begin, we notice the following expression of them in terms of conditional mutual information.

Theorem 5.19: *Let $\mathcal{A}_1$ and $\mathcal{A}_2$ be fields. We have*

1. $I(\mathcal{A}_1;\mathcal{A}_2) = I(\mathcal{A}_1;\mathcal{A}_2|\{\emptyset,\Omega\})$.
2. $H(\mathcal{A}_1|\mathcal{A}_2) = I(\mathcal{A}_1;\mathcal{A}_1|\mathcal{A}_2)$.
3. $H(\mathcal{A}_1) = I(\mathcal{A}_1;\mathcal{A}_1|\{\emptyset,\Omega\})$.

Proof: The claims follow since $P(A|\{\emptyset,\Omega\}) = P(A)$ almost surely and $I(\alpha_1;\alpha_1') \le I(\alpha_1 \wedge \alpha_1';\alpha_1 \wedge \alpha_1') = H(\alpha_1 \wedge \alpha_1')$. □

Before we derive other polymatroid identities for the generalized Shannon information measures, it is necessary to discuss quite a few technical facts concerning continuity of these measures. Although conditional mutual information has usually been discussed for σ-fields, Definition 5.12 makes sense also for fields. This point of view is convenient to prove continuity.

We will write $\mathcal{B}_n \uparrow \mathcal{B}$ for a sequence $(\mathcal{B}_n)_{n\in\mathbb{N}}$ of fields such that $\mathcal{B}_1 \subseteq \mathcal{B}_2 \subseteq ... \subseteq \mathcal{B} = \bigcup_{n\in\mathbb{N}} \mathcal{B}_n$. (Field $\mathcal{B}$ need not be a σ-field.) By an analogy to this notation, for a sequence $(a_n)_{n\in\mathbb{N}}$ of real numbers, we will also write $a_n \uparrow a$ if $a_i \le a_{i+1}$ and $\lim_{n\to\infty} a_n = a$.

Theorem 5.20 *(Continuity of mutual information):* For any fields $\mathcal{A}$, $\mathcal{B}_n \uparrow \mathcal{B}$, and $\mathcal{C}$, we have $I(\mathcal{A};\mathcal{B}_n|\mathcal{C}) \uparrow I(\mathcal{A};\mathcal{B}|\mathcal{C})$.

Proof: The claim holds by claim (6) of Theorem 5.8 since every partition of $\mathcal{B} = \bigcup_{n\in\mathbb{N}}\mathcal{B}_n$ is a partition of $\mathcal{B}_m$ for all but finitely many m. □

Theorem 5.21: *For finite fields $\mathcal{A}$ and $\mathcal{B}$ and any fields $\mathcal{C}_n \uparrow \mathcal{C}$, we have* $\lim_{n\to\infty} I(\mathcal{A};\mathcal{B}|\mathcal{C}_n) = I(\mathcal{A};\mathcal{B}|\mathcal{C})$.

Proof: Finite fields are generated by finite partitions. Let $\mathcal{A} = \sigma(\alpha)$ and $\mathcal{B} = \sigma(\beta)$, where $\alpha = \{A_1,\dots,A_k\}$. By monotonicity of Shannon information measures, i.e. claim (6) of Theorem 5.8, we have $I(\mathcal{A};\mathcal{B}|\mathcal{C}_n) = \mathbf{E}\, I_{P(\cdot|\mathcal{C}_n)}(\alpha;\beta)$, where $I_{P(\cdot|\mathcal{C}_n)}(\alpha;\beta) \in [0, \log k]$ by claim (8) of Theorem 5.8. Hence, the claim follows by the Levy law (3.68) and the dominated convergence (Theorem 3.10). □

Recall the symmetric difference (2.32) and its properties discussed in Problem 2.4. We observe the following bound:

Theorem 5.22 *(Continuity of entropy):* Fix an $\epsilon \in (0, e^{-1}]$ and a field $\mathcal{C}$. For finite partitions $\alpha = \{A_i\}_{i=1}^I$ and $\alpha' = \{A_i'\}_{i=1}^I$ such that $P(A_i \triangle A_i') \le \epsilon$ for all $i \in \{1,\dots,I\}$, we have

$$|H(\alpha|\mathcal{C}) - H(\alpha'|\mathcal{C})| \le I\sqrt{\epsilon}\log\frac{I}{\sqrt{\epsilon}} \tag{5.55}$$

Proof: We have the expectation $\int P(A_i \triangle A_i'|\mathcal{C})dP = P(A_i \triangle A_i') \le \epsilon$. Hence, by the Markov inequality (Theorem 3.11) we obtain

$$P(P(A_i \triangle A_i'|\mathcal{C}) \ge \sqrt{\epsilon}) \le \sqrt{\epsilon} \tag{5.56}$$

Denote

$$B = (P(A_i \triangle A_i'|\mathcal{C}) < \sqrt{\epsilon} \text{ for all } i \in \{1,\dots,I\}) \tag{5.57}$$

From the left Bonferroni inequality (2.31), we obtain $P(B^c) \le I\sqrt{\epsilon}$. Subsequently, we observe that $|H(\alpha||\mathcal{C}) - H(\alpha'||\mathcal{C})| \le \log I$ holds almost surely since the Shannon entropy of a partition is uniformly bounded by the logarithm of the partition's cardinality. Hence,

$$|H(\alpha|\mathcal{C}) - H(\alpha'|\mathcal{C})| = \left|\int [H(\alpha|\mathcal{C}) - H(\alpha'|\mathcal{C})]dP\right|$$
$$\le P(B^c)\log I + \int_B |H(\alpha||\mathcal{C}) - H(\alpha'||\mathcal{C})|dP$$
$$\le I\sqrt{\epsilon}\log I + \int_B |H(\alpha||\mathcal{C}) - H(\alpha'||\mathcal{C})|dP \tag{5.58}$$

Function $-x\log x$ is subadditive and increasing for $x \in (0, e^{-1}]$. In particular, we have $|(x+y)\log(x+y) - x\log x| \le -y\log y$ for $x, y \ge 0$. Thus, on the event B, we obtain

$$
\begin{aligned}
|H(\alpha||\mathcal{C}) - H(\alpha'||\mathcal{C})| &= \left| \sum_{i=1}^{I} P(A_i'|\mathcal{C}) \log P(A_i'|\mathcal{C}) - \sum_{i=1}^{I} P(A_i|\mathcal{C}) \log P(A_i|\mathcal{C}) \right| \\
&\leq - \sum_{i=1}^{I} |P(A_i|\mathcal{C}) - P(A_i'|\mathcal{C})| \log |P(A_i|\mathcal{C}) - P(A_i'|\mathcal{C})| \\
&\leq - \sum_{i=1}^{I} P(A_i \triangle A_i'|\mathcal{C}) \log P(A_i \triangle A_i'|\mathcal{C}) \\
&\leq -I\sqrt{\epsilon} \log \sqrt{\epsilon}
\end{aligned}
\tag{5.59}
$$

Plugging (5.59) into (5.58) yields the claim. $\qquad\square$

Now we can prove the invariance of conditional mutual information with respect to extending fields to complete σ-fields. For this aim, we will use Theorem 3.28.

Theorem 5.23 (*Invariance of completion*): *Let $\mathcal{A}$, $\mathcal{B}$, $\mathcal{C}$, and $\mathcal{D}$ be subfields of $\mathcal{J}$. We have*

$$
I(\mathcal{A}; \mathcal{B}|\mathcal{C}) = I(\mathcal{A}; \sigma_P(\mathcal{B})|\mathcal{C}) = I(\mathcal{A}; \mathcal{B}|\sigma_P(\mathcal{C}))
\tag{5.60}
$$

Proof: Equality $I(\mathcal{A}; \mathcal{B}|\mathcal{C}) = I(\mathcal{A}; \mathcal{B}|\sigma_P(\mathcal{C}))$ holds by Definition 5.12. It remains to prove equality $I(\mathcal{A}; \mathcal{B}|\mathcal{C}) = I(\mathcal{A}; \sigma_P(\mathcal{B})|\mathcal{C})$. For this goal, it suffices to show that for any $\epsilon > 0$ and any finite partitions $\alpha \subseteq \mathcal{A}$ and $\beta' \subseteq \sigma_P(\mathcal{B})$ there exists a finite partition $\beta \subseteq \mathcal{B}$ such that

$$
|I(\alpha; \beta|\mathcal{C}) - I(\alpha; \beta'|\mathcal{C})| < \epsilon
\tag{5.61}
$$

Fix then some $\epsilon > 0$ and finite partitions $\alpha := \{A_i\}_{i=1}^{I} \subseteq \mathcal{A}$ and $\beta' := \{B_j'\}_{j=1}^{J} \subseteq \sigma(\mathcal{B})$. Invoking Theorem 3.28, we know that for each $\eta > 0$ there exists a class of sets $\{C_j\}_{j=1}^{J} \subseteq \mathcal{B}$, which need not be a partition, such that

$$
P(C_j \triangle B_j') \leq \eta
\tag{5.62}
$$

for all $j \in \{1, \dots, J\}$. Let us put $B_{J+1}' := \emptyset$ and let us construct sets $D_0 := \emptyset$ and $D_j := \bigcup_{k=1}^{j} C_k$ for $j \in \{1, \dots, J\}$. Subsequently, we put $B_j := C_j \backslash D_{j-1}$ for $j \in \{1, \dots, J\}$ and $B_{J+1} := \Omega \backslash D_J$. In this way, we obtain a partition $\beta := \{B_j\}_{j=1}^{J+1} \subseteq \mathcal{B}$.

The next step of the proof is showing an analogue of bound (5.62) for partitions β and β'. To begin, for $j \in \{1, \dots, J\}$, we have

$$
P(B_j \triangle B_j') = P((C_j \backslash D_{j-1}) \triangle B_j') \leq P(C_j \triangle B_j') + P(D_{j-1} \cap B_j')
$$

$$\leq \eta + \sum_{k=1}^{j-1} P(C_k \cap B'_j)$$

$$\leq \eta + \sum_{k=1}^{j-1} [P(B'_k \cap B'_j) + P((C_k \cap B'_j) \triangle (B'_k \cap B'_j))]$$

$$\leq \eta + \sum_{k=1}^{j-1} [0 + P(C_k \triangle B'_k)] \leq j\eta \qquad (5.63)$$

Now, we observe for $j, k \in \{1, \dots, J\}$ and $j \neq k$ that

$$P(C_j) \geq P(B'_j) - P(C_j \triangle B'_j) \geq P(B'_j) - \eta \qquad (5.64)$$

$$P(C_j \cap C_k) \leq P(B'_j \cap B'_k) + P((C_j \cap C_k) \triangle (B'_j \cap B'_k))$$

$$\leq 0 + P(C_j \triangle B'_j) + P(C_k \triangle B'_k) \leq 2\eta \qquad (5.65)$$

Hence, using the right Bonferroni inequality (2.31), we obtain

$$P(B_{J+1} \triangle B'_{J+1}) = P((\Omega \backslash D_J) \triangle \emptyset) = P(\Omega \backslash D_J) = 1 - P(D_J)$$

$$\leq 1 - \sum_{1 \leq j \leq J} P(C_j) + \sum_{1 \leq j < k \leq J} P(C_j \cap C_k)$$

$$\leq 1 - \sum_{1 \leq j \leq J} P(B'_j) + J\eta + \sum_{1 \leq j < k \leq J} 2\eta = J^2 \eta \qquad (5.66)$$

Resuming our bounds, we obtain

$$P((A_i \cap B_j) \triangle (A_i \cap B'_j)) \leq P(B_j \triangle B'_j) \leq J^2 \eta \qquad (5.67)$$

for all $i \in \{1, \dots, I\}$ and $j \in \{1, \dots, J+1\}$. Then invoking Theorem 5.22 yields

$$|I(\alpha; \beta | \mathcal{C}) - I(\alpha; \beta' | \mathcal{C})| \leq |H(\alpha \wedge \beta | \mathcal{C}) - H(\alpha \wedge \beta' | \mathcal{C})| + |H(\beta | \mathcal{C}) - H(\beta' | \mathcal{C})|$$

$$\leq I(J+1)\sqrt{J^2 \eta} \log \frac{I(J+1)}{\sqrt{J^2 \eta}} + (J+1)\sqrt{J^2 \eta} \log \frac{J+1}{\sqrt{J^2 \eta}} \qquad (5.68)$$

Taking η sufficiently small, we obtain (5.61), which is the desired claim. $\qquad \square$

Let us denote the join of fields $\mathcal{A}$ and $\mathcal{B}$ as $\mathcal{A} \wedge \mathcal{B} := \sigma_0(\mathcal{A} \cup \mathcal{B})$. Some consequence of Theorem 5.23 is this approximation result proved by Dobrushin (1959) and Pinsker (1964) and used by Wyner (1978) to demonstrate a generalized chain rule. Applying the invariance of completion, we supply a different proof than Dobrushin (1959) and Pinsker (1964).

Theorem 5.24 (*Split of join*): *Let $\mathcal{A}$, $\mathcal{B}$, $\mathcal{C}$, and $\mathcal{D}$ be fields. We have*

$$I(\mathcal{A}; \mathcal{B} \wedge \mathcal{C} | \mathcal{D}) = \sup_{\alpha \subseteq \mathcal{A}, \beta \subseteq \mathcal{B}, \gamma \subseteq \mathcal{C}} \mathbf{E} \, I(\alpha; \beta \wedge \gamma || \mathcal{D}) \qquad (5.69)$$

where the supremum is taken over all finite subpartitions.

Proof: Define class

$$\mathcal{E} := \bigcup_{\beta \subseteq \mathcal{B}, \gamma \subseteq \mathcal{C}} \sigma(\beta \wedge \gamma) \tag{5.70}$$

It can be easily verified that $\mathcal{E}$ is a field such that $\sigma(\mathcal{E}) = \sigma(\mathcal{B} \wedge \mathcal{C})$. By definition of $\mathcal{E}$, for all finite partitions $\beta \subseteq \mathcal{B}$ and $\gamma \subseteq \mathcal{C}$, we have $\beta \wedge \gamma \subseteq \mathcal{E}$. Moreover, for each finite partition $\varepsilon \subseteq \mathcal{E}$, there exist finite partitions $\beta \subseteq \mathcal{B}$ and $\gamma \subseteq \mathcal{C}$ such that partition $\beta \wedge \gamma$ is finer than ε. Hence, by Theorem 5.20, we obtain in this case

$$\mathbf{E}\, I(\alpha; \varepsilon || \mathcal{D}) \le \mathbf{E}\, I(\alpha; \beta \wedge \gamma || \mathcal{D}) \le I(\alpha; \mathcal{E} | \mathcal{D}) \tag{5.71}$$

In consequence, by Theorem 5.23, we obtain the claim

$$I(\mathcal{A}; \mathcal{B} \wedge \mathcal{C} | \mathcal{D}) = I(\mathcal{A}; \mathcal{E} | \mathcal{D}) = \sup_{\alpha \subseteq \mathcal{A}, \varepsilon \subseteq \mathcal{E}} \mathbf{E}\, I(\alpha; \varepsilon || \mathcal{D})$$

$$= \sup_{\alpha \subseteq \mathcal{A}, \beta \subseteq \mathcal{B}, \gamma \subseteq \mathcal{C}} \mathbf{E}\, I(\alpha; \beta \wedge \gamma || \mathcal{D}) \tag{5.72}$$

□

The last auxiliary technical statement we need is as follows:

Theorem 5.25 *(Convergence of conditioning):* *Let $\alpha = \{A_i\}_{i=1}^{I}$ be a finite partition and let $\mathcal{C}$ be a field. For each $\epsilon > 0$, there exists a finite partition $\gamma' \subseteq \sigma_P(\mathcal{C})$ such that for any partition $\gamma \subseteq \sigma_P(\mathcal{C})$ finer than γ' we have*

$$|H(\alpha | \mathcal{C}) - H(\alpha | \gamma)| \le \epsilon \tag{5.73}$$

Proof: Fix an $\epsilon > 0$. For each $n \in \mathbb{N}$ and $A \in \mathcal{J}$, partition

$$\gamma_A := \{((k-1)/n < P(A | \mathcal{C}) \le k/n) : k \in \{0, 1, \dots, n\}\} \tag{5.74}$$

is finite and belongs to $\sigma_P(\mathcal{C})$. If we consider partition $\gamma' := \wedge_{i=1}^{I} \gamma_{A_i}$, it remains finite and still satisfies $\gamma' \subseteq \sigma_P(\mathcal{C})$. Let a partition $\gamma \subseteq \sigma_P(\mathcal{C})$ be finer than γ'. Then

$$|P(A_i | \mathcal{C}) - P(A_i | \gamma)| \le 1/n \tag{5.75}$$

almost surely for all $i \in \{1, \dots, I\}$. We also have

$$|H(\alpha | \mathcal{C}) - H(\alpha | \gamma)| \le \int |H(\alpha || \mathcal{C}) - H(\alpha || \gamma)| dP \tag{5.76}$$

From the proof of Theorem 5.22, we recall that function $-x \log x$ is subadditive and increasing for $x \in (0, e^{-1}]$. Applying this observation likewise, for $n \ge e$ we obtain almost surely

$$|H(\alpha || \mathcal{C}) - H(\alpha || \gamma)| = \left| \sum_{i=1}^{I} P(A_i | \mathcal{C}) \log P(A_i | \mathcal{C}) - \sum_{i=1}^{I} P(A_i | \gamma) \log P(A_i | \gamma) \right|$$

$$\leq -\sum_{i=1}^{I}|P(A_i|\mathcal{C}) - P(A_i|\gamma)|\log|P(A_i|\mathcal{C}) - P(A_i|\gamma)|$$

$$\leq \frac{I\log n}{n} \tag{5.77}$$

Taking n so large that $n^{-1}I\log n \leq \epsilon$ yields the claim. $\qquad\square$

Once we have established continuity of generalized Shannon information measures and the series of technical results, we may inspect whether the polymatroid identities for partitions can be generalized for arbitrary fields. That is, we are going to establish some generalizations of Theorems 5.4–5.7. It would be highly convenient to establish maximally general statements. As we will see, this can be done in full generality, without assuming regular conditional probability.

In the first step of our constructions, let us generalize relations $\prec$, $\precsim$, and $\sim$ defined for partitions. We use the operation of completion σ_P introduced in Section 3.7.

1. We will say that field $\mathcal{B}$ is finer than field $\mathcal{A}$, written $\mathcal{A} \prec \mathcal{B}$, if $\mathcal{A} \subseteq \mathcal{B}$. For finite partitions α and β, we have $\sigma(\alpha) \prec \sigma(\beta)$ if and only if $\alpha \prec \beta$.
2. We will say that field $\mathcal{B}$ is almost surely finer than field $\mathcal{A}$, written $\mathcal{A} \precsim \mathcal{B}$, if $\mathcal{A} \subseteq \sigma_P(\mathcal{B})$. Equivalently, we have $\mathcal{A} \precsim \mathcal{B}$ if and only if for each event $A \in \mathcal{A}$, there exists an event $B \in \mathcal{B}$ such that $P(A \triangle B) = 0$. For finite partitions α and β, we have $\sigma(\alpha) \precsim \sigma(\beta)$ if and only if $\alpha \precsim \beta$.
3. We will say that fields $\mathcal{A}$ and $\mathcal{B}$ are almost surely equal, written $\mathcal{A} \sim \mathcal{B}$, if $\sigma_P(\mathcal{A}) = \sigma_P(\mathcal{B})$. For finite partitions α and β, we have $\sigma(\alpha) \sim \sigma(\beta)$ if and only if $\alpha \sim \beta$.

Similarly, we may generalize the concepts of independence and conditional independence.

1. We will say that fields $\mathcal{A}$ and $\mathcal{B}$ are independent, written $\mathcal{A} \perp\!\!\!\perp \mathcal{B}$, if for all $A \in \mathcal{A}$ and $B \in \mathcal{B}$, we have $P(A \cap B) = P(A)P(B)$. For partitions α and β, we have $\sigma(\alpha) \perp\!\!\!\perp \sigma(\beta)$ if and only if $\alpha \perp\!\!\!\perp \beta$.
2. We will say that fields $\mathcal{A}$ and $\mathcal{B}$ are conditionally independent given a field $\mathcal{C}$, written $\mathcal{A} \perp\!\!\!\perp \mathcal{B}|\mathcal{C}$, if for all $A \in \mathcal{A}$ and $B \in \mathcal{B}$, we have $P(A \cap B|\mathcal{C}) = P(A|\mathcal{C})P(B|\mathcal{C})$ almost surely. We have $\mathcal{A} \perp\!\!\!\perp \mathcal{B}|\{\emptyset, \Omega\}$ if and only if $\mathcal{A} \perp\!\!\!\perp \mathcal{B}$. For finite partitions α, β, and γ, we have $\sigma(\alpha) \perp\!\!\!\perp \sigma(\beta)|\sigma(\gamma)$ if and only if $\alpha \perp\!\!\!\perp \beta|\gamma$.

Extending these concepts, we will say that arbitrary random variables Y_i with image spaces $(\mathbb{Y}_i, \mathcal{Y}_i)$ are (conditionally) independent if the respective preimages $Y_i^{-1}(\mathcal{Y}_i)$ are (conditionally) independent. Moreover, we will say that a stochastic process $(Y_i)_{i \in \mathbb{T}}$ is a collection of (conditionally) independent random variables (given Z) if for all $\mathbb{A} \subseteq \mathbb{T}$, variables $(Y_i)_{i \in \mathbb{A}}$ and $(Y_i)_{i \in \mathbb{T} \setminus \mathbb{A}}$ are (conditionally) independent (given Z).

Now, we prove the analogues of polymatroid identities, i.e. Theorems 5.4–5.7. First, Theorem 5.4 can be generalized without a problem.

Theorem 5.26: *Let $\mathcal{A}$ and $\mathcal{B}$ be fields. We have $\mathcal{A} \prec \mathcal{B} \iff \mathcal{B} = \mathcal{A} \wedge \mathcal{B} \implies \mathcal{A} \precsim \mathcal{B} \iff \mathcal{B} \sim \mathcal{A} \wedge \mathcal{B}$.*

Proof: We have

1. $\mathcal{A} \prec \mathcal{B} \iff \mathcal{A} \subseteq \mathcal{B} \subseteq \sigma_P(\mathcal{B}) \implies \mathcal{A} \precsim \mathcal{B}$.
2. $\mathcal{B} = \mathcal{A} \wedge \mathcal{B} \iff \mathcal{A} \wedge \mathcal{B} \subseteq \mathcal{B} \iff \mathcal{A} \subseteq \mathcal{B} \iff \mathcal{A} \prec \mathcal{B}$.
3. $\mathcal{B} \sim \mathcal{A} \wedge \mathcal{B} \iff \mathcal{A} \wedge \mathcal{B} \subseteq \sigma_P(\mathcal{B}) \iff \mathcal{A} \subseteq \sigma_P(\mathcal{B}) \iff \mathcal{A} \precsim \mathcal{B}$. $\quad\square$

Theorem 5.5 can be also generalized.

Theorem 5.27: *Let $\mathcal{A}$, $\mathcal{B}$, and $\mathcal{C}$ be fields. We have*

1. *$I(\mathcal{A}; \mathcal{B}|\mathcal{C}) = I(\mathcal{B}; \mathcal{A}|\mathcal{C})$.*
2. *$I(\mathcal{A}; \mathcal{B}|\mathcal{C}) \geq 0$ with equality if and only if $\mathcal{A} \perp\!\!\!\perp \mathcal{B}|\mathcal{C}$.*

Proof:
1. Trivial.
2. We have $I(\mathcal{A}; \mathcal{B}|\mathcal{C}) = 0$ if and only if $I(\alpha; \beta||\mathcal{C}) = 0$ almost surely for all finite partitions $\alpha \subseteq \mathcal{A}$ and $\beta \subseteq \mathcal{B}$. The latter holds if and only if $\mathcal{A} \perp\!\!\!\perp \mathcal{B}|\mathcal{C}$. $\quad\square$

The generalization of Theorem 5.6 is as follows:

Theorem 5.28: *Let $\mathcal{A}$, $\mathcal{B}$, and $\mathcal{C}$ be fields. We have*

$$\mathcal{A} \precsim \mathcal{B} \iff \mathcal{A} \perp\!\!\!\perp \mathcal{A}|\mathcal{B} \implies \mathcal{A} \perp\!\!\!\perp \mathcal{C}|\mathcal{B} \tag{5.78}$$

Proof: We have $P(A|\mathcal{B})P(A'|\mathcal{B}) = P(A \cap A'|\mathcal{B})$ for all $A, A' \in \mathcal{A}$ almost surely if and only if $P(A|\mathcal{B}) \in \{0, 1\}$ for all $A \in \mathcal{A}$ almost surely. The last condition is equivalent to $\mathcal{A} \subseteq \sigma_P(\mathcal{B})$ by the following reasoning. First, for an $A \in \mathcal{A}$, let $\tilde{A} = (P(A|\mathcal{B}) = 1)$. If $P(A|\mathcal{B}) \in \{0, 1\}$ almost surely, we have

$$P(\tilde{A}) = \int_{\tilde{A}} dP = \int_{\tilde{A}} P(A|\mathcal{B})dP \tag{5.79}$$

$$P(A) = \int P(A|\mathcal{B})dP = \int_{\tilde{A}} P(A|\mathcal{B})dP \tag{5.80}$$

$$P(A \cap \tilde{A}) = \int_{\tilde{A}} P(A|\mathcal{B})dP \tag{5.81}$$

Hence, $P(A \triangle \tilde{A}) = P(A) + P(\tilde{A}) - 2P(A \cap \tilde{A}) = 0$, so $\mathcal{A} \subseteq \sigma_P(\mathcal{B})$. To prove the converse, we notice that $\mathcal{A} \subseteq \sigma_P(\mathcal{B})$ implies $P(A|\mathcal{B}) = I_A \in \{0, 1\}$ almost surely. Moreover, if we have $P(A|\mathcal{B}) \in \{0,1\}$ for all $A \in \mathcal{A}$, almost surely then $P(A|\mathcal{B})P(C|\mathcal{B}) = P(A \cap C|\mathcal{B})$ for all $A \in \mathcal{A}$ and $C \in \mathcal{C}$ almost surely. Thus, we have proved (5.78). □

Theorem 5.7, i.e. the chain rule, can be also restated in full generality.

Theorem 5.29 *(Chain rule):* *Let $\mathcal{A}$, $\mathcal{B}$, $\mathcal{C}$, and $\mathcal{D}$ be fields. We have*

$$I(\mathcal{A}; \mathcal{B} \wedge \mathcal{C}|\mathcal{D}) = I(\mathcal{A}; \mathcal{B}|\mathcal{D}) + I(\mathcal{A}; \mathcal{C}|\mathcal{B} \wedge \mathcal{D}) \tag{5.82}$$

Proof: Let $\mathcal{A}$, $\mathcal{B}$, $\mathcal{C}$, and $\mathcal{D}$ be arbitrary fields, and let α, β, γ, and δ be finite partitions. The point of our departure is the chain rule for finite partitions

$$I(\alpha; \beta \wedge \gamma) = I(\alpha; \beta) + I(\alpha; \gamma|\beta) \tag{5.83}$$

demonstrated in Theorem 5.7. By Definition 5.12 and Theorems 5.23–5.25, conditional mutual information $I(\mathcal{A}; \mathcal{B}|\mathcal{C})$ can be approximated by $I(\alpha; \beta|\gamma)$, where we take appropriate limits of refined finite partitions with a certain care.

In particular, by Theorems 5.23–5.25, taking sufficiently fine finite partitions of arbitrary fields $\mathcal{B}$ and $\mathcal{C}$, the chain rule (5.83) for finite partitions implies

$$I(\alpha; \mathcal{B} \wedge \mathcal{C}) = I(\alpha; \mathcal{B}) + I(\alpha; \mathcal{C}|\mathcal{B}) \tag{5.84}$$

where all expressions are finite. Hence, we also obtain

$$\begin{aligned}
0 &= [I(\alpha; \mathcal{B} \wedge \mathcal{C} \wedge \mathcal{D}) - I(\alpha; \mathcal{D}) - I(\alpha; \mathcal{B} \wedge \mathcal{C}|\mathcal{D})] \\
&\quad - [I(\alpha; \mathcal{B} \wedge \mathcal{D}) - I(\alpha; \mathcal{D}) - I(\alpha; \mathcal{B}|\mathcal{D})] \\
&\quad - [I(\alpha; \mathcal{B} \wedge \mathcal{C} \wedge \mathcal{D}) - I(\alpha; \mathcal{B} \wedge \mathcal{D}) - I(\alpha; \mathcal{C}|\mathcal{B} \wedge \mathcal{D})] \\
&= I(\alpha; \mathcal{B}|\mathcal{D}) + I(\alpha; \mathcal{C}|\mathcal{B} \wedge \mathcal{D}) - I(\alpha; \mathcal{B} \wedge \mathcal{C}|\mathcal{D}) \tag{5.85}
\end{aligned}$$

where all expressions are finite. Having established the above claim for finite α, we generalize it to

$$I(\mathcal{A}; \mathcal{B} \wedge \mathcal{C}|\mathcal{D}) = I(\mathcal{A}; \mathcal{B}|\mathcal{D}) + I(\mathcal{A}; \mathcal{C}|\mathcal{B} \wedge \mathcal{D}) \tag{5.86}$$

for an arbitrary field $\mathcal{A}$, taking its appropriately fine finite partitions. □

Now, we will produce some corollaries of the generalized polymatroid identities. We say that a σ-field $\mathcal{A}$ is trivial if $P(A) \in \{0,1\}$ for all $A \in \mathcal{A}$. We have this corollary of Theorems 5.19 and 5.26–5.29, which is an analogue of Theorem 5.8.

Theorem 5.30: *For any fields, we have*

1. $I(\mathcal{A}; \mathcal{B}|\mathcal{C}) = I(\mathcal{A}; \mathcal{B})$ *if* $\mathcal{A} \wedge \mathcal{B} \perp\!\!\!\perp \mathcal{C}$.
2. $H(\mathcal{A}|\mathcal{B}) = 0$ *if and only if* $\mathcal{A} \precsim \mathcal{B}$.

3. $H(\mathcal{A}) = 0$ *if and only if* $\mathcal{A}$ *is trivial.*
4. $I(\mathcal{A}; \mathcal{B}|\mathcal{C}) = H(\mathcal{A}|\mathcal{C})$ *for* $\mathcal{A} \prec \mathcal{B}$.
5. $H(\mathcal{A}|\mathcal{C}) = I(\mathcal{A}; \mathcal{B}|\mathcal{C}) + H(\mathcal{A}|\mathcal{B} \wedge \mathcal{C})$.
6. $I(\mathcal{A}; \mathcal{B}_1|\mathcal{C}) \le I(\mathcal{A}; \mathcal{B}_2|\mathcal{C})$ *if* $\mathcal{B}_1 \prec \mathcal{B}_2$.
7. $H(\mathcal{A}|\mathcal{B}_1) \ge H(\mathcal{A}|\mathcal{B}_2)$ *if* $\mathcal{B}_1 \prec \mathcal{B}_2$.
8. $I(\mathcal{A}; \mathcal{B}|\mathcal{C}) \le H(\mathcal{A}|\mathcal{C}) \le H(\mathcal{A})$.
9. $H(\mathcal{A} \wedge \mathcal{B}|\mathcal{C}) = H(\mathcal{A}|\mathcal{C}) + H(\mathcal{B}|\mathcal{A} \wedge \mathcal{C})$.
10. $I(\mathcal{A}; \mathcal{B}) = H(\mathcal{C}) + I(\mathcal{A}; \mathcal{B}|\mathcal{C})$ *if* $\mathcal{C} \prec \mathcal{A}$ *and* $\mathcal{C} \prec \mathcal{B}$.

Proof: The proof is analogous to the proof of Theorem 5.8. □

As another exercise, we leave proving the following expressions of Shannon information measures in terms of joint entropies, which generalizes Theorem 5.3.

Theorem 5.31: *Let* $\mathcal{A}_i$ *be fields. If the negative terms are finite, we have*

1. $I(\mathcal{A}_1; \mathcal{A}_2|\mathcal{A}_3) = H(\mathcal{A}_1|\mathcal{A}_3) + H(\mathcal{A}_2|\mathcal{A}_3) - H(\mathcal{A}_1 \wedge \mathcal{A}_2|\mathcal{A}_3)$.
2. $I(\mathcal{A}_1; \mathcal{A}_2) = H(\mathcal{A}_1) - H(\mathcal{A}_1|\mathcal{A}_2) = H(\mathcal{A}_1) + H(\mathcal{A}_2) - H(\mathcal{A}_1 \wedge \mathcal{A}_2)$.
3. $H(\mathcal{A}_1|\mathcal{A}_2) = H(\mathcal{A}_1 \wedge \mathcal{A}_2) - H(\mathcal{A}_2)$.

Finally, we state equality of conditional mutual information for fields that are almost surely equal.

Theorem 5.32: *We have* $I(\mathcal{A}; \mathcal{B}|\mathcal{C}) = I(\mathcal{A}'; \mathcal{B}'|\mathcal{C}')$ *if* $\mathcal{A} \sim \mathcal{A}', \mathcal{B} \sim \mathcal{B}',$ *and* $\mathcal{C} \sim \mathcal{C}'$.

Proof: For $\mathcal{C} \sim \mathcal{C}'$, we have $P(E|\mathcal{C}) = P(E|\mathcal{C}')$ almost surely. Assume next $\mathcal{A} \sim \mathcal{A}'$ and $\mathcal{B} \sim \mathcal{B}'$. Then for any partitions $\alpha \subseteq \mathcal{A}$ and $\beta \subseteq \mathcal{B}$, there exist partitions $\alpha' \subseteq \mathcal{A}'$ and $\beta' \subseteq \mathcal{B}'$ such that $\alpha \sim \alpha'$ and $\beta \sim \beta'$. Hence, by Theorem 5.9, we obtain $I(\alpha; \beta||\mathcal{C}) = I(\alpha'; \beta'||\mathcal{C}')$ almost surely. Taking expectations and suprema, we obtain the claim. □

5.4 Block Entropy Limits Revisited

Let us come back to stationary processes. In view of the developments in the previous section, we can indeed interpret entropy rate and excess entropy as conditional entropy and mutual information, respectively.

Theorem 5.33: *Consider a finite entropy stationary process* $(X_i)_{i \in \mathbb{Z}}$, *where* $X_i : \Omega \to \mathbb{X}$. *We have*

$$h = \lim_{n \to \infty} H(X_1|X_{-n}^0) = H(X_1|X_{-\infty}^0) \text{ if } \mathbb{X} \text{ is finite} \tag{5.87}$$

$$E = \lim_{n \to \infty} I(X_{-n+1}^0; X_1^n) = I(X_{-\infty}^0; X_1^\infty) \tag{5.88}$$

Proof: The claims follow by Theorems 5.20, 5.21, and 5.23. □

The above result matters not only for aesthetic reasons. Theorem 5.33 is the first step to prove the ergodic decomposition of the entropy rate and excess entropy. This topic requires some preparation. First, we need to denote some special σ-fields. The first definition is just a convenient notation.

Definition 5.13 *(Preimage σ-field)*: Let X be a random variable with an image space $(\mathbb{X}, \mathcal{X})$ on a measurable space $(\Omega, \mathcal{J})$. The preimage σ-field of X will be denoted as

$$\sigma(X) := X^{-1}(\mathcal{X}) \tag{5.89}$$

Next, we define the σ-fields of blocks, semi-infinite past and future, and tail σ-fields.

Definition 5.14 *(Some σ-fields of a process)*: Consider a stochastic process $(X_i)_{i\in\mathbb{Z}}$ on a probability space $(\mathbb{X}^{\mathbb{Z}}, \mathcal{X}^{\mathbb{Z}}, P)$, where $X_j : \mathbb{X}^{\mathbb{Z}} \ni (x_i)_{i\in\mathbb{Z}} \mapsto x_j \in \mathbb{X}$. We define

- the σ-fields of blocks

$$\mathcal{G}_j := \sigma(X_j), \quad \mathcal{G}_j^k := \sigma(X_j^k) \tag{5.90}$$

- the σ-fields of semi-infinite past and future

$$\mathcal{G}_{-\infty}^j := \sigma\left(\bigcup_{i\leq j}\mathcal{G}_i\right), \quad \mathcal{G}_j^\infty := \sigma\left(\bigcup_{i\geq j}\mathcal{G}_i\right) \tag{5.91}$$

- the tail σ-fields of past and future

$$\mathcal{G}_{-\infty} := \bigcap_{i\in\mathbb{Z}}\mathcal{G}_{-\infty}^i, \quad \mathcal{G}_\infty := \bigcap_{i\in\mathbb{Z}}\mathcal{G}_i^\infty \tag{5.92}$$

Another simple definition we need is a concept of the random ergodic measure of a stationary process. This is just the conditional distribution of the process with respect to the invariant σ-field.

Definition 5.15 *(Random ergodic measure)*: For the process as in Definition 5.14, let $\mathcal{J}$ be the invariant σ-field for the shift operation (4.1). If σ-field $\mathcal{X}$ is countably generated, we define the random ergodic measure of the process as the conditional probability measure

$$F(A) := P(A|\mathcal{J}) \tag{5.93}$$

Moreover, we define the σ-field of this random variable as

$$\mathcal{F} := \sigma\left(\bigcup_{A \in \mathcal{X}^{\mathbb{Z}}} \sigma(F(A)) \right) \tag{5.94}$$

Now, we will state some auxiliary proposition concerning the above defined σ-fields. The proposition states that the invariant σ-field is contained in the intersection of completed tail σ-fields. For the sake of this statement, let $\mathbb{S}$ be the class of stationary measures on the measurable space $(\mathcal{X}^{\mathbb{Z}}, \mathcal{X}^{\mathbb{Z}})$ with the shift operation (4.1). Let $\sigma_{\mathbb{S}}(\mathcal{A})$ be the completion of a class of events $\mathcal{A}$ with respect to the class of probability measures $\mathbb{S}$, defined in (3.85).

Theorem 5.34: *If the σ-field $\mathcal{X}$ is countably generated, then*

$$\sigma_{\mathbb{S}}(\mathcal{I}) = \sigma_{\mathbb{S}}(\mathcal{F}) \subseteq \sigma_{\mathbb{S}}(\mathcal{G}_{-\infty}) \cap \sigma_{\mathbb{S}}(\mathcal{G}_{\infty}) \tag{5.95}$$

Proof: Let σ-fields $\mathcal{G}_i = \sigma(\mathcal{A}_i)$ be generated by countable classes $\mathcal{A}_i = (A_{i1}, A_{i2}, ...)$. Let $B_1, B_2, ...$ be some enumeration of sets A_{ij}. We have $\mathcal{X}^{\mathbb{Z}} = \mathcal{G}_{-\infty}^{\infty} = \sigma(\mathcal{K})$, where $\mathcal{K} = \sigma_0(\{B_1, B_2, ...\})$ is a countable field. Consider a stationary measure $P \in \mathbb{S}$. By Theorem 3.29, we obtain $\sigma_P(\mathcal{F}) = \sigma_P(\mathcal{I}) = \sigma_P(\mathcal{B})$, where

$$\mathcal{B} = \sigma\left(\bigcup_{K \in \mathcal{K}} \sigma(F(K)) \right) \tag{5.96}$$

On the other hand, by the Birkhoff ergodic theorem (Theorem 4.6), we have $F(K) = \Phi(K)$ P-almost surely for each $K \in \mathcal{K}$. Since each $K \in \mathcal{K}$ belongs to some $\mathcal{G}_j^k$, then $\Phi(K)$ is measurable with respect to $\mathcal{G}_i^{\infty}$ for each $i \in \mathbb{Z}$. Hence, $\Phi(K)$ is $\mathcal{G}_{\infty}$-measurable. In this way, we obtain $\sigma_P(\mathcal{B}) \subseteq \sigma_P(\mathcal{G}_{\infty})$. Applying the Birkhoff ergodic theorem to operation T^{-1}, we obtain the complementary claim $\sigma_P(\mathcal{B}) \subseteq \sigma_P(\mathcal{G}_{-\infty})$. In this way, we derive

$$\sigma_P(\mathcal{I}) = \sigma_P(\mathcal{F}) \subseteq \sigma_P(\mathcal{G}_{-\infty}) \cap \sigma_P(\mathcal{G}_{\infty}) \tag{5.97}$$

Applying the intersection over all $P \in \mathbb{S}$ to all terms of (5.97) yields the claim. $\square$

Subsequently, we may state the main result of this section, namely, the ergodic decomposition of entropy rate and excess entropy. Continuing our earlier notation, by h_F and E_F we will denote the entropy rate and the excess entropy of the random ergodic measure, i.e. h_F and E_F are the random variables which take the values of entropy rate and excess entropy of some processes distributed according to the values of the random measure F.

Theorem 5.35 (*Ergodic decomposition*): *Consider a finite entropy stationary process $(X_i)_{i \in \mathbb{Z}}$, where $X_i : \Omega \to \mathbb{X}$ and $\mathbb{X}$ is countable. We have*

$$h = \mathbf{E}\, h_F \quad \text{if } \mathbb{X} \text{ is finite} \tag{5.98}$$

$$E = H(\mathcal{I}) + \mathbf{E}\, E_F \tag{5.99}$$

Proof: Random variables h_F and E_F are measurable since they are limits of measurable random variables by the definition of entropy rate and excess entropy. In particular, by Theorem 5.33, we have

$$h_F = \lim_{n\to\infty} H_F(X_1|X^0_{-n}) = H_F(X_1|X^0_{-\infty}) \text{ if } \mathbb{X} \text{ is finite} \tag{5.100}$$

$$E_F = \lim_{n\to\infty} I_F(X^0_{-n+1};X^n_1) = I_F(X^0_{-\infty};X^\infty_1) \tag{5.101}$$

Since $H_F(X^k_j) = H(X^k_j||\mathcal{I})$, we hence obtain

$$\mathbf{E}\,h_F = \mathbf{E}\lim_{n\to\infty} H_F(X_1|X^0_{-n}) = \lim_{n\to\infty}\mathbf{E}\,H_F(X_1|X^0_{-n})$$
$$= \lim_{n\to\infty} H(X_1|X^0_{-n}\wedge\mathcal{I}) = H(X_1|X^0_{-\infty}\wedge\mathcal{I}) \text{ if } \mathbb{X} \text{ is finite} \tag{5.102}$$

$$\mathbf{E}\,E_F = \mathbf{E}\lim_{n\to\infty} I_F(X^0_{-n+1};X^n_1) = \lim_{n\to\infty}\mathbf{E}\,I_F(X^0_{-n+1};X^n_1)$$
$$= \lim_{n\to\infty} I(X^0_{-n+1};X^n_1|\mathcal{I}) = I(X^0_{-\infty};X^\infty_1|\mathcal{I}) \tag{5.103}$$

by the dominated and monotone convergence, respectively (Theorems 3.10 and 3.8) and further by Theorems 5.20, 5.21, and 5.23. Consequently, the application of Theorems 5.34 and 5.32 and claim (10) of Theorem 5.30 yields

$$h = H(X_1|X^0_{-\infty}) = H(X_1|X^0_{-\infty}\wedge\mathcal{I}) = \mathbf{E}\,h_F \text{ if } \mathbb{X} \text{ is finite} \tag{5.104}$$

$$E = I(X^0_{-\infty};X^\infty_1) = H(\mathcal{I}) + I(X^0_{-\infty};X^\infty_1|\mathcal{I}) = H(\mathcal{I}) + \mathbf{E}\,E_F \tag{5.105}$$

□

By formula (5.98), a stationary process is asymptotically deterministic, i.e. it has the zero entropy rate if and only if its random ergodic measure is asymptotically deterministic almost surely. In contrast, by (5.99), a stationary process is infinitary, i.e. it has an infinite excess entropy not only if its random ergodic measure is infinitary with a strictly positive probability but also when the invariant σ-field has the infinite entropy. Decomposition (5.98) is a result by Gray and Davisson (1974a, b). Decomposition (5.99) is a result by Dębowski (2009). Some generalization of formula (5.99) was derived by Löhr (2009) for statistical complexity, i.e. the entropy of prediction process, or so-called ε-machine – see Problem 5.8.

In the next two sections, we will see how Theorem 5.35 combined with certain related results yields a characterization of some asymptotically deterministic and infinitary processes. These results shed some light onto a possible reason why texts in natural language should be modeled by an infinitary process, exhibiting the relaxed Hilberg hypothesis (1.71). The plausible reason can be the phenomenon of strong nonergodicity, which was introduced half-formally in Section 1.11. We will develop these ideas further in Chapter 8.

5.5   Convergence of Entropy

To characterize some asymptotically deterministic processes, let us link the convergence of simple random variables with inclusion of σ-fields, via conditional entropy. The following two results provide another perspective of the Riesz theorem (Theorem 3.27) and its applications, such as Theorem 3.29.

We begin with two simple results, which we will call the Fano inequalities. The first inequality is actually due to Fano (1961), the other is a simple converse.

Theorem 5.36   *(Fano inequalities):   Let function $\mathcal{H}(p)$ be given by*

$$\mathcal{H}(p) := -p \log p - (1-p)\log(1-p), \quad p \in (0,1) \tag{5.106}$$

and $\mathcal{H}(0) := \mathcal{H}(1) := 0$ to assure continuity. Let $X : \Omega \to \mathbb{X}$ be a simple random variable, where $\#\mathbb{X} = N$. We have

1. *For any $\mathcal{A}$-measurable random variable $\tilde{X}$,*

$$H(X|\mathcal{A}) \le \mathcal{H}(P(X = \tilde{X})) + [1 - P(X = \tilde{X})]\log(N-1) \tag{5.107}$$

2. *Let $\tilde{X}$ be an $\mathcal{A}$-measurable random variable such that $\tilde{X} = x$ if $P(X = x|\mathcal{A}) \ge P(X = x'|\mathcal{A})$ for all $x' \in \mathbb{X}$. Then*

$$H(X|\mathcal{A}) \ge \frac{\mathcal{H}(1/N)}{1 - 1/N} \times [1 - P(X = \tilde{X})] \tag{5.108}$$

Proof: Let $Y = \mathbf{1}\{X = \tilde{X}\}$. As for statement (1), we obtain

$$
\begin{aligned}
H(X|\mathcal{A}) &\le H(X|\tilde{X}) = H(X, Y|\tilde{X}) = H(Y|\tilde{X}) + H(X|Y, \tilde{X}) \\
&\le H(Y) + H(X|Y, \tilde{X}) \\
&\le \mathcal{H}(P(X = \tilde{X})) + [1 - P(X = \tilde{X})]\log(N-1) \tag{5.109}
\end{aligned}
$$

As for statement (2), by concavity of $\mathcal{H}(p)$, we have

$$\mathcal{H}(p) \ge \mathcal{H}(q)\frac{1-p}{1-q} + \mathcal{H}(1)\frac{p-q}{1-q} = \mathcal{H}(q)\frac{1-p}{1-q} \tag{5.110}$$

for $p \in [q, 1]$. In particular,

$$
\begin{aligned}
H(X|\mathcal{A}) \ge H(Y|\mathcal{A}) &\ge \mathbf{E}\left[\mathcal{H}(P(X = \tilde{X}|\mathcal{A}))\right] \\
&\ge \frac{\mathcal{H}(1/N)}{1 - 1/N} \times \mathbf{E}\left[1 - P(X = \tilde{X}|\mathcal{A})\right] \\
&= \frac{\mathcal{H}(1/N)}{1 - 1/N} \times [1 - P(X = \tilde{X})] \tag{5.111}
\end{aligned}
$$

$\square$

Now, we can show the main theorem of this section.

Theorem 5.37:   *Let X be a simple random variable. Consider σ-fields $\mathcal{A}_n \uparrow \mathcal{A}$. The following statements are equivalent:*

1. $\lim_{n\to\infty} P(X = X_n) = 1$ for some $\mathcal{A}_n$-measurable simple variables X_n.
2. $\lim_{n\to\infty} H(X|\mathcal{A}_n) = 0$.
3. $H(X|\mathcal{A}) = 0$.
4. X is $\sigma_p(\mathcal{A})$-measurable.

Proof: Statements (1) and (2) are equivalent by the Fano inequalities (Theorem 5.36). Statements (2) and (3) are equivalent by claim (3) of Theorem 5.23, whereas statements (3) and (4) are equivalent by claim (3) of Theorem 5.30.   □

It is an important assumption in the above theorem that random variable X assumes only a finite number of values. Consider a random variable X that takes values in natural numbers and has an infinite entropy $H(X) = \infty$. Let $Z_k = 1$ for $X \geq k$ and $Z_k = 0$ else. We have $H(X|Z_1^k) = \infty$ since $H(X) = H(X|Z_1^k) + H(Z_1^k)$ and $H(Z_1^k) \leq k$. But $H(X|(Z_k)_{k\in\mathbb{N}}) = H(X|X) = 0$.

Theorem 5.37 gives rise to the following characterization of asymptotically deterministic processes in the finite alphabet case. Namely, a process is asymptotically deterministic if and only if each random variable is a measurable function of its infinite past.

Theorem 5.38:   *Consider a stationary process $(X_i)_{i\in\mathbb{Z}}$, where $X_i : \Omega \to \mathbb{X}$ and $\mathbb{X}$ is finite. We have $h = 0$ if and only if there exists a measurable function $f : \mathbb{X}^{\mathbb{N}} \to \mathbb{X}$ such that $X_i = f(X_{-\infty}^{i-1})$ almost surely for all $i \in \mathbb{Z}$.*

Proof: By Theorems 5.37 and 5.33, $h = 0$ if and only if each X_i is measurable with respect to the completion of the σ-field generated by the infinite past $X_{-\infty}^{i-1}$. This is equivalent to the existence of a measurable function f such that $X_i = f(X_{-\infty}^{i-1})$ almost surely.   □

The results mentioned in this section come from Dębowski (2009). Theorem 5.38 suggests that asymptotic determinism – or the zero entropy rate – should not be satisfied by natural language. It seems unlikely that we could perfectly predict the next symbol of any text given an infinite amount of previously generated texts. There should exist some hidden variables involved in the text generation which would make such a perfect prediction impossible.

5.6   Entropy as Self-Information

The ergodic decomposition of excess entropy (5.99) illustrates that the concept of entropy as self-information $H(\mathcal{A}) := I(\mathcal{A}; \mathcal{A})$ arises naturally when the chain rule for conditional mutual information is considered. It should be noted, however, that self-information $H(\mathcal{A})$ is infinite if the σ-field $\mathcal{A}$ is nonatomic.

Theorem 5.39: $H(\mathcal{A}) = \infty$ if $\mathcal{A}$ contains a nonatomic sub-σ-field.

Proof: Let $(Z_k)_{k \in \mathbb{N}}$ be an $\mathcal{A}$-measurable fair-coin process. Such a process exists by Theorem 3.5. For any natural number k, we have $H(\mathcal{A}) \geq H(Z_1^k) = \sum_{i=1}^{k} H(Z_k) = k$. Hence, the claim follows: □

Thus, for a real random variable Y, the self-information entropy $H(Y)$ should not be confused with the differential entropy defined as

$$h(Y) := -\int p(y) \log p(y) dy = -\int p \log p \, d\lambda \qquad (5.112)$$

where λ is the Lebesgue measure and $p = dP(Y \in \cdot)/d\lambda$ is the probability density. Although the appropriate difference of differential entropies for two real random variables equals the mutual information, usually we have $h(Y) \neq H(Y)$.

Now let us specialize the above result to the ergodic decomposition of excess entropy (5.99). Necessarily, when the invariant σ-field has infinite entropy, then the process is nonergodic. This observation motivates this definition.

Definition 5.16 *(Strongly nonergodic process):* A process $(X_i)_{i \in \mathbb{Z}}$ will be called strongly nonergodic if $\sigma_P(\mathcal{J})$ contains a nonatomic sub-σ-field.

In the half-formal Section 1.11, we have introduced a different definition of strongly nonergodic processes, which is more complicated but motivated by modeling semantics of texts in natural language. In Section 8.3, we will prove formally that these two distinct definitions are equivalent.

By the ergodic decomposition of excess entropy, strongly nonergodic processes are infinitary. This gives a partial characterization of infinitary processes.

Theorem 5.40: *Strongly nonergodic stationary processes satisfy $E = \infty$.*

Proof: By definition of a strongly nonergodic process, $\sigma_P(\mathcal{J})$ contains a nonatomic sub-σ-field. Hence, $E = H(\mathcal{J}) = \infty$ by Theorems 5.39 and 5.35. □

Theorem 5.40 was noticed by Dębowski (2009). Less formalized intuitions about links between nonergodicity and infinite excess entropy were presented independently later by Crutchfield and Marzen (2015).

Now, we will present a simple example of a strongly nonergodic process. It begins with a familiar example, the Bernoulli processes.

Definition 5.17 *(Bernoulli processes)*: A process $(X_i)_{i \in \mathbb{Z}}$, where $X_i : \Omega \to \{0, 1\}$, is called a Bernoulli($\theta$) process, where $\theta \in [0, 1]$, if

$$P(X_1^n = x_1^n) = \prod_{i=1}^{n} \theta^{x_i}(1 - \theta)^{1-x_i} \tag{5.113}$$

A process $(X_i)_{i \in \mathbb{Z}}$, where $X_i : \Omega \to \{0, 1\}$, is called a mixture Bernoulli process if

$$P(X_1^n = x_1^n) = \int_0^1 \left(\prod_{i=1}^{n} \theta^{x_i}(1 - \theta)^{1-x_i} \right) d\theta$$

$$= \frac{(\sum_{i=1}^{n} x_i)!(n - \sum_{i=1}^{n} x_i)!}{(n + 1)!} \tag{5.114}$$

The plain Bernoulli processes are ergodic.

Theorem 5.41: *The Bernoulli(θ) processes are ergodic.*

Proof: Bernoulli(θ) processes are IID processes. They are mixing by Theorem 4.14 and hence, they are ergodic. □

In contrast, the mixture Bernoulli process satisfies the desired property of strong nonergodicity.

Theorem 5.42: *The mixture Bernoulli process is strongly nonergodic.*

Proof: The mixture Bernoulli process is a convex combination of uncountably many different stationary ergodic measures and hence, by the second ergodic decomposition theorem (Theorem 4.20), it is nonergodic. Now, we will show it is strongly nonergodic. Let $(X_i)_{i \in \mathbb{Z}}$ be a Bernoulli(θ) process with respect to measure P_θ, where $\theta \in [0, 1]$, and a mixture Bernoulli process with respect to measure P. Then by the uniqueness of the ergodic decomposition we have

$$P(A) = \int_0^1 P_\theta(A) d\theta \tag{5.115}$$

Now define random variable

$$Y := \lim_{n \to \infty} \frac{1}{n} \sum_{i=1}^{n} X_i \tag{5.116}$$

Random variable Y is $\mathcal{J}$-measurable. By the ergodic theorem, equality $Y = \int X_i dP_\theta = \theta$ holds P_θ-almost surely. Thus,

$$P(Y \le r) = \int_0^1 1\{\theta \le r\} d\theta = r \tag{5.117}$$

Hence, the distribution of Y with respect to measure P is the Lebesgue measure. Since the Lebesgue measure is nonatomic by Theorem 3.4, then σ-field $\mathcal{J}$ contains a nonatomic sub-σ-field. □

By Theorems 5.40 and 5.42, the mixture Bernoulli process satisfies $E = \lim_{n\to\infty} I(X_1^n; X_{n+1}^{2n}) = \infty$. In the concluding result of this section, we will show that the block mutual information for this process grows much slower than implied by Hilberg's hypothesis (5.15).

Theorem 5.43: *For the mixture Bernoulli process, $I(X_1^n; X_{n+1}^{2n}) \leq \log(n+1)$.*

Proof: Let $S_n := \sum_{i=1}^{n} X_i$. Variable S_n is a function of X_1^n and it can be shown that X_1^n and X_{n+1}^{2n} are conditionally independent given S_n. Hence,

$$I(X_1^n; X_{n+1}^{2n}) = I(S_n; X_{n+1}^{2n}) + I(X_1^n; X_{n+1}^{2n} | S_n)$$
$$= I(S_n; X_{n+1}^{2n}) \leq H(S_n) \leq \log(n+1) \tag{5.118}$$

since random variable S_n assumes $n+1$ distinct values. □

In general, a function $Z = f(X)$ of a random variable X is called a sufficient statistic for predicting a random variable Y if X and Y are conditionally independent given Z. Like in (5.118), it can be shown that inequality $I(X;Y) \leq H(Z)$ holds in this case. For the mixture Bernoulli process, we have shown that the mutual information $I(X_1^n; X_{n+1}^{2n})$ grows quite slowly since we can construct a sufficient statistic with a relatively small entropy. In contrast, in Section 8.3 and Chapter 11, we will formally discuss the Santa Fe processes, half-formally mentioned in Section 1.11. Santa Fe processes are some simple strongly nonergodic processes, more motivated by the semantics of natural texts, for which the block mutual information grows at a power-law rate, as implied for instance by the relaxed Hilberg hypothesis (1.71). The sufficient statistic for these processes takes a form which is quite different from the statistic $S_n = \sum_{i=1}^{n} X_i$ for the Bernoulli process.

Problems

5.1 Consider a random variable $N : \Omega \to \mathbb{N} \cup \{0\}$ and a number $q \in (0,1)$. Show that
$$H(N) \leq -\log(1-q) - \mathbf{E}\, N \log q \tag{5.119}$$

5.2 Show that a stationary process $(X_i)_{i\in\mathbb{N}}$ over a finite alphabet is mixing if and only if for any $k \in \mathbb{N}$ we have
$$\lim_{n\to\infty} I(X_1^k; X_{n+1}^{n+k}) = 0 \tag{5.120}$$

5.3 Show that for any stochastic process $(X_i)_{i\in\mathbb{N}}$ over a finite alphabet and any $k \in \mathbb{N}$, we have

$$\lim_{n\to\infty} I(X_1^k; X_{n+1}^{n+k} | X_{k+1}^n) = 0 \qquad (5.121)$$

5.4 Consider a random variable $N : \Omega \to \mathbb{N}\backslash\{1\}$ whose distribution is

$$P(N_i = n) = \frac{C(\gamma)}{n(\log n)^\gamma} \qquad (5.122)$$

where $\gamma > 1$ and $C(\gamma) := \left[\sum_{n=2}^{\infty} n^{-1}(\log n)^{-\gamma}\right]^{-1}$. Show that its entropy satisfies $H(N) = \infty$ for $1 < \gamma \le 2$ and $H(N) < \infty$ for $\gamma > 2$.

5.5 Let $(N_i)_{i\in\mathbb{Z}}$ be a stationary Markov process where variables $N_i : \Omega \to \mathbb{N}\backslash\{1\}$ take values in natural numbers but 1 and have the marginal distribution of N in (5.122). Let $(X_i)_{i\in\mathbb{Z}}$ with $X_i = f(N_i)$ for a function $f : \mathbb{N}\backslash\{1\} \to \mathbb{X}$ with $\mathbb{X}$ finite be a stationary hidden Markov process. Show that there exist constants $K(\gamma) < \infty$ such that the block mutual information of process $(X_i)_{i\in\mathbb{Z}}$ satisfies

$$I(X_{-n+1}^0; X_1^n) \le \begin{cases} K(\gamma)\, n^{2-\gamma}, & 1 < \gamma < 2 \\ K(\gamma)\log n, & \gamma = 2 \\ K(\gamma), & \gamma > 2 \end{cases} \qquad (5.123)$$

This result was shown by Dębowski (2014).

5.6 Let X and Y be discrete random variables. An important problem in the interpretation of mutual information between X and Y is whether this quantity is close to the Shannon entropy of a certain random variable W depending on X and Y. Such entropy $H(W)$ is usually called common information between X and Y. There are two simple choices of common information. Respectively, the Gács–Körner common information and the modified Wyner common information are defined as

$$C_{\mathrm{GK}}(X;Y) := \sup_{W:W=f(X)=g(Y)} H(W) \qquad (5.124)$$

$$C_{\mathrm{MW}}(X;Y) := \inf_{W:X\perp\!\!\!\perp Y|W} H(W) \qquad (5.125)$$

Show that

$$0 \le C_{\mathrm{GK}}(X;Y) \le I(X;Y) \le C_{\mathrm{MW}}(X;Y) \le \min\{H(X), H(Y)\} \quad (5.126)$$

Can these inequalities be strict? The original Wyner common information is defined as the infimum of mutual information $I(X, Y; W)$ rather than entropy $H(W)$. The above results were shown by Gács and Körner (1973) and Wyner (1975).

5.7 Suppose that we add some side information to the definition of the Gács–Körner common information. Formally, such side information can be represented as a random variable N. Show that

$$I(X;Y) \geq H(W) - 2H(N) \text{ if } W = f(X,N) = g(Y,N) \qquad (5.127)$$

5.8 Consider a stochastic process $(X_i)_{i \in \mathbb{Z}}$ on a probability space $(\mathbb{X}^{\mathbb{Z}}, \mathcal{X}^{\mathbb{Z}}, P)$, where $X_j : \mathbb{X}^{\mathbb{Z}} \ni (x_i)_{i \in \mathbb{Z}} \mapsto x_j \in \mathbb{X}$ and the alphabet $\mathbb{X}$ is finite. The prediction process $(S_i)_{i \in \mathbb{Z}}$, introduced by Knight (1975) and called the ϵ-machine by Shalizi and Crutchfield (2001), is a sequence of random measures

$$S_i : \mathbb{X}^{\mathbb{N}} \in A \mapsto S_i(A) := P(X_i^\infty \in A | X_{-\infty}^{i-1}) \qquad (5.128)$$

Show that
(a) $(S_i)_{i \in \mathbb{Z}}$ is a stationary process.
(b) $(S_i)_{i \in \mathbb{Z}}$ is a nondiscrete Markov process, i.e. $S_{-\infty}^{t-1} \perp\!\!\!\perp S_{t+1}^{-\infty} | S_t$.
(c) $(X_i)_{i \in \mathbb{Z}}$ is a nondiscrete hidden Markov process with the underlying Markov process $(S_i)_{i \in \mathbb{Z}}$, i.e. $X_{t+1} \perp\!\!\!\perp X_{t+2} \perp\!\!\!\perp \cdots \perp\!\!\!\perp X_{t+k} | S_{t+1}^{t+k+1}$.
(d) $(S_i)_{i \in \mathbb{Z}}$ is unifilar given $(X_i)_{i \in \mathbb{Z}}$, i.e. $H(S_{t+1} | S_t, X_t) = 0$.
(e) $H(X_i | S_i) = h$, where h is the entropy rate of $(X_i)_{i \in \mathbb{Z}}$.
(f) $C \geq E$, where $C := H(S_i)$ is called the statistical complexity of $(X_i)_{i \in \mathbb{Z}}$ and E is the excess entropy of $(X_i)_{i \in \mathbb{Z}}$.
(g) $C = H(\mathcal{J}) + \mathbf{E} C_F$, where $\mathcal{J}$ is the invariant σ-field, F is the random ergodic measure of $(X_i)_{i \in \mathbb{Z}}$, and C_F is the statistical complexity of F.

Remark: These results were formally demonstrated by Löhr (2009), based on half-formal considerations by Shalizi (2001) and Shalizi and Crutchfield (2001). The idea of the prediction process was introduced by Knight (1975) in the more general setting of continuous time. Unifilar processes are the stochastic analogue of deterministic finite-state automata, see Hopcroft and Ullman (1979). Recent computational experiments by Braverman et al. (2019) and Hahn and Futrell (2019) seem to show that both excess entropy and statistical complexity of natural language are large.

6

Equipartition and Universality

When we deal with empirical data, such as texts in natural language, and we wish to predict them, often we do not know their exact probability distribution in advance. Consequently, we may wish to apply some statistical procedure to infer this distribution incrementally from the data while making predictions. It is not guaranteed in general, however, whether the loss that we incur in this case is similar to the loss that we would incur if we used the true distribution of the data-generating process. It turns out that for stationary processes, there exist indeed effective procedures, called universal distributions, which are capable of approximating any unknown process with the minimal achievable loss – equal to the entropy rate of the respective ergodic component of the process. This fact is a consequence of the Birkhoff ergodic theorem.

The existence of universal distributions is of a great practical importance since, as it will become clear from Section 7.1, these distributions can be used as general purpose data-compressors. From this perspective, computationally simpler universal procedures such as the Lempel–Ziv code (Ziv and Lempel, 1977) and irreducible grammar-based codes (Kieffer and Yang, 2000) attract more attention in practice. While studying natural language, however, we would like to support a different point of view into universal distributions, namely, whether they admit some kind of a combinatorial interpretation resembling the theorems about facts and words, mentioned in Section 1.12 and formally to be developed in Section 8.4. Such universal distributions can shed light onto emergence of contiguous words in the stream of human speech in particular and can help to answer how to count the number words in languages where there are no spaces between words, like Chinese or Japanese, see de Marcken (1996).

From the latter perspective, universal distributions such as Prediction by Partial Matching (PPM) (Ryabko, 1984; Cleary and Witten, 1984) and minimal grammar-based codes (Charikar et al., 2005; Dębowski, 2011) appear especially interesting. Namely, it turns out that the pointwise mutual information of these distributions is upper bounded by the number of distinct word-like strings

Information Theory Meets Power Laws: Stochastic Processes and Language Models, First Edition.
Łukasz Dębowski.
© 2021 John Wiley & Sons, Inc. Published 2021 by John Wiley & Sons, Inc.

detected in the compressed text multiplied by a slowly growing term. Here, the advantage of the PPM probability over minimal grammar-based codes lies in that the former is computable in a reasonable time, whereas the latter are not. Moreover, the slowly growing term for the PPM probability can be uniformly bounded in contrast to minimal grammar-based codes. Consequently, in this chapter, we will limit ourselves to studying the PPM probability, whereas for the analogous analysis of minimal grammar-based codes, we relegate the reader to Problem 7.4.

Thus, in this chapter, we will study the abstract phenomenon of universal distributions and the related asymptotic equipartition property. In its basic form, the asymptotic equipartition property, called also the Shannon–McMillan–Breiman (SMB) theorem or the ergodic theorem of information theory, equates the almost sure rate of pointwise entropy with the expected entropy rate of the stationary ergodic process over a finite alphabet. This version of the SMB theorem will be proved in Section 6.1 using the Birkhoff ergodic theorem, the Barron lemma, and the Levy laws. There exist, however, a few generalizations of this fact, which apply different techniques and which will be discussed later.

To discuss some of these generalizations, it is convenient to consider universal distributions. In Section 6.2, we will formally define them as such that the rate of their pointwise entropy is less than the entropy rate for any stationary ergodic process almost surely. This condition states that the loss of a universal distribution is minimal. It is not yet guaranteed, however, that universal distributions exist.

In Section 6.3, we will exhibit an example of a universal distribution called PPM probability. The PPM probability has been already introduced half-formally in Section 1.8. The PPM probability is a convex combination of incrementally built Markov models of growing orders for the predicted data. It is effectively computable and its pointwise entropy is roughly equal to the conditional entropy of the empirical distribution, as we show doing lengthy calculations. More precisely, the absolute difference between the PPM pointwise entropy and the empirical conditional entropy is upper bounded by the empirical vocabulary size multiplied by the logarithm of the data length. For this reason, it can be easily shown that PPM probability is universal.

Knowing that universal distributions exist, in Section 6.4, we will revisit the SMB theorem, and we will generalize it to stationary nonergodic processes over a finite alphabet. In this generalization, the rates of various pointwise entropies are equal to the expected entropy rate of the random ergodic measure of the process, which can be different for different ergodic components of the process.

In Section 6.5, we will come back to analyzing PPM probabilities. We will introduce the PPM order of empirical data, which is the order of the incrementally built Markov model which contributes the largest probability mass to the total PPM probability. We will show that the total PPM probability is approximately the probability of the Markov model of the PPM order. Asymptotically, the PPM order constitutes an upper bound for the true Markov order of the data-generating process. In the second turn, we will investigate the PPM-based pointwise mutual information. Based on the bounds developed in Section 6.3, we will show that the PPM-based pointwise mutual information is upper bounded by the empirical vocabulary size of the PPM order multiplied by the logarithm of the data length. This result is a preparation for the theorem about facts and words, which will be discussed in Section 8.4.

6.1 SMB Theorem

For two random variables X and Y on a probability space $(\Omega, \mathcal{J}, P)$, let us define random variables being random probabilities

$$P(X)(\omega) := P(X = X(\omega)) \tag{6.1}$$

$$P(X|Y)(\omega) := P(X = X(\omega)|Y)(\omega) \tag{6.2}$$

where $P(A|Y) = P(A|\mathcal{Y})$ is the conditional probability of event A with respect to the σ-field $\mathcal{Y}$ generated by random variable Y (see Section 3.5). In particular, we can express the Shannon entropy of X and the conditional Shannon entropy of X given Y as

$$H(X) = \mathbf{E}[-\log P(X)] \tag{6.3}$$

$$H(X|Y) = \mathbf{E}[-\log P(X|Y)] \tag{6.4}$$

Moreover, if $(X_i)_{i \in \mathbb{Z}}$ is a stationary process, where $X_i : \Omega \to \mathbb{X}$ and the alphabet $\mathbb{X}$ is finite, then by Theorem 5.33 the entropy rate h of process $(X_i)_{i \in \mathbb{Z}}$ can be equivalently expressed as

$$h = \lim_{k \to \infty} H(X_1|X_{-k}^0) = H(X_1|X_{-\infty}^0) \tag{6.5}$$

The following proposition, called the SMB theorem or the ergodic theorem of information theory or the asymptotic equipartition property, states that the limiting rate of the pointwise entropy $-\log P(X_1^n)$ is equal to the entropy rate h of process $(X_i)_{i \in \mathbb{Z}}$ almost surely and in expectation for stationary ergodic processes. As we will see, this fact is a consequence of the Gibbs and Barron inequalities, the ergodic theorem, identity (6.5), and the chain rule for conditional probability. The proof of the SMB theorem which we present here is due to Algoet and Cover (1988). We note that an alternative proof of equality (6.6) was established in Theorem 5.12.

Let us recall that $\sum_{k=1}^{\infty} k^{-2} = \pi^2/6$. Hence, $\pi(n) := 6\pi^{-2}n^{-2}$ is a probability distribution on natural numbers, $n \in \mathbb{N}$.

Theorem 6.1 *(SMB theorem):* *Let $(X_i)_{i \in \mathbb{Z}}$ be a stationary ergodic process on $(\Omega, \mathcal{J}, P)$, where $X_i : \Omega \to \mathbb{X}$ and the alphabet $\mathbb{X}$ is finite. Let h be the entropy rate of $(X_i)_{i \in \mathbb{Z}}$. We have*

$$\lim_{n \to \infty} \frac{1}{n}\mathbf{E}[-\log P(X_1^n)] = h \tag{6.6}$$

$$\lim_{n \to \infty} \frac{1}{n}[-\log P(X_1^n)] = h \text{ almost surely} \tag{6.7}$$

Proof: First let us prove equality (6.7). Define

$$P_k(X_1^n) := P(X_1^k)\prod_{i=1}^{n-k} P(X_{i+k}|X_i^{i+k-1}) \tag{6.8}$$

and observe that

$$P(X_1^n|X_{-\infty}^0) = \prod_{i=1}^{n} P(X_i|X_{-\infty}^{i-1}) \tag{6.9}$$

Now by the Barron lemma (Theorem 3.20), when $\pi(n)P_k(X_1^n)$ with $\pi(n) := 6\pi^{-2}n^{-2}$ is a semidistribution and $P(X_1^n)$ is the probability distribution, we obtain that

$$-\log P_k(X_1^n) \geq -\log P(X_1^n) - \log \pi(n) \tag{6.10}$$

for all but finitely many n holds P-almost surely. Next, consider conditional measure $S(A) = P(A|X_{-\infty}^0)$. When $\pi(n)P(X_1^n)$ with $\pi(n) := 6\pi^{-2}n^{-2}$ is a semidistribution and $P(X_1^n|X_{-\infty}^0)$ is the probability distribution, the Barron lemma yields that

$$-\log P(X_1^n) \geq -\log P(X_1^n|X_{-\infty}^0) - \log \pi(n) \tag{6.11}$$

for all but finitely many n holds S-almost surely for all values of S. Since $P(A) = \int S(A)dP$, the above inequality holds for all but finitely many n also P-almost surely. Hence,

$$\limsup_{n \to \infty} \frac{1}{n}[-\log P(X_1^n)] \leq \limsup_{n \to \infty} \frac{1}{n}[-\log P_k(X_1^n)] \tag{6.12}$$

$$\liminf_{n \to \infty} \frac{1}{n}[-\log P(X_1^n)] \geq \liminf_{n \to \infty} \frac{1}{n}[-\log P(X_1^n|X_{-\infty}^0)] \tag{6.13}$$

holds P-almost surely. But by the Birkhoff ergodic theorem (Theorem 4.6),

$$\lim_{n \to \infty} \frac{1}{n}[-\log P_k(X_1^n)] = H(X_1|X_{-k+1}^0) \tag{6.14}$$

$$\lim_{n \to \infty} \frac{1}{n}[-\log P(X_1^n|X_{-\infty}^0)] = H(X_1|X_{-\infty}^0) \tag{6.15}$$

holds P-almost surely. Thus, using formula (6.5), we obtain (6.7).

Now, we will demonstrate (6.6). Observe that, since $\mathbf{E}_P Y = \mathbf{E}_P \mathbf{E}_S Y$, then by the Gibbs inequality (Theorem 3.18) we have

$$\mathbf{E}_P[-\log P_k(X_1^n)] \geq \mathbf{E}_P[-\log P(X_1^n)] \geq \mathbf{E}_P[-\log P(X_1^n | X_{-\infty}^0)] \tag{6.16}$$

Since

$$\lim_{n \to \infty} \frac{1}{n} \mathbf{E}_P[-\log P_k(X_1^n)] = H(X_1 | X_{-k+1}^0) \tag{6.17}$$

$$\lim_{n \to \infty} \frac{1}{n} \mathbf{E}_P[-\log P(X_1^n | X_{-\infty}^0)] = H(X_1 | X_{-\infty}^0) \tag{6.18}$$

we obtain (6.6). □

There are a few generalizations of the above SMB theorem: for stationary nonergodic processes (see Theorem 6.9), for two-sided asymptotically mean stationary (AMS) processes (see Theorem 10.14), for one-sided AMS processes (Gray and Kieffer, 1980), for stationary processes over a countably infinite alphabet (Chung, 1961) and for probability densities (Barron, 1985a). In Section 8.2, we will also present a power-law analogue of the SMB theorem for the pointwise and expected block mutual information (Theorem 8.9), derived by Dębowski (2015a). In contrast, in this chapter, we would like to show that a simple path to some generalizations of the SMB theorem leads through existence of universal semidistributions, which will be formally defined in Section 6.2.

6.2 Universal Semidistributions

Let us recall the definition of a semidistribution (Definition 3.13), which is an incomplete discrete probability distribution. When we have a discrete stochastic process, which is an infinite sequence of discrete random variables, it is natural to consider an infinite sequence of semidistributions for blocks of increasing length.

Definition 6.1 *(Semidistribution sequence):* A sequence of semidistributions $q_k : \mathbb{X}^k \to \mathbb{R}$, where $k \in \mathbb{N}$, will be written as $Q = (q_k)_{k \in \mathbb{N}} : \mathbb{X}^* \to \mathbb{R}$. In this case we will write $Q(x_1^k) := q_k(x_1^k)$.

As a consequence of previous results, we have the following fact about sequences of semidistributions as applied to stationary ergodic processes.

Theorem 6.2: *Let $(X_i)_{i \in \mathbb{Z}}$ be a stationary ergodic process on $(\Omega, \mathcal{J}, P)$, where $X_i : \Omega \to \mathbb{X}$ and the alphabet $\mathbb{X}$ is finite. Let h be the entropy rate of $(X_i)_{i \in \mathbb{Z}}$ and let*

$Q : \mathbb{X}^* \to \mathbb{R}$ *be a sequence of semidistributions. We have*

$$\frac{1}{n}\mathbf{E}[-\log Q(X_1^n)] \geq h \tag{6.19}$$

$$\liminf_{n \to \infty} \frac{1}{n}[-\log Q(X_1^n)] \geq h \text{ almost surely} \tag{6.20}$$

Proof: The claim follows by the Gibbs inequality (Theorem 3.18), the Barron lemma (Theorem 3.20), and the SMB theorem (Theorem 6.1). □

It turns out that a stronger property can be satisfied. Here we will state a technical definition which is easy to check for particular instances of semidistributions.

Definition 6.2 (*Universal semidistribution sequence*): A sequence of semidistributions $Q : \mathbb{X}^* \to \mathbb{R}$, where alphabet $\mathbb{X}$ is finite, is called universal if

$$-\log Q(x_1^n) \leq cn \tag{6.21}$$

for a $c > 0$ and for every stationary ergodic process $(X_i)_{i \in \mathbb{Z}}$, where $X_i : \Omega \to \mathbb{X}$, we have

$$\limsup_{n \to \infty} \frac{1}{n}[-\log Q(X_1^n)] \leq h \text{ almost surely} \tag{6.22}$$

In Section 6.3, we will construct a simple example of a universal semidistribution, called the PPM probability. Right after that we will generalize Theorem 6.2 and the SMB theorem to arbitrary stationary processes over a finite alphabet.

6.3 PPM Probability

The PPM probability is an important simple example of a universal semidistribution. As we may remember from the half-formal exposition in Section 1.8, the PPM probability works by approximating the distribution of the unknown process by adaptive Markov approximations of the process, where the joint probabilities of blocks are estimated by the empirical frequencies of these blocks incremented by one.

The exact definition of the PPM probability distribution is as follows.

Definition 6.3 (*PPM probability*): Consider an alphabet $\mathbb{X} = \{a_1, a_2, ..., a_D\}$, where $D \geq 2$. We define the frequency of a string $w_1^k \in \mathbb{X}^*$ in a string $x_1^n \in \mathbb{X}^*$ where $1 \leq k \leq n$ as

$$N(w_1^k | x_1^n) := \sum_{i=1}^{n-k+1} \mathbf{1}\{x_i^{i+k-1} = w_1^k\} \tag{6.23}$$

Subsequently, we define the total PPM probability

$$PPM(x_1^n) := \frac{6}{\pi^2} \left[D^{-n} + \sum_{k=0}^{\infty} \frac{PPM_k(x_1^n)}{(k+2)^2} \right] \tag{6.24}$$

where

$$PPM_k(x_1^n) := \prod_{i=1}^{n} PPM_k(x_i | x_1^{i-1}) \tag{6.25}$$

$$PPM_k(x_i | x_1^{i-1}) := \begin{cases} \dfrac{1}{D} & k > i - 2 \\[2ex] \dfrac{N(x_{i-k}^i | x_1^{i-1}) + 1}{N(x_{i-k}^{i-1} | x_1^{i-2}) + D} & \text{else} \end{cases} \tag{6.26}$$

As we can see, term $PPM(x_1^n)$ is a convex combination of terms $PPM_k(x_1^n)$. Each term $PPM_k(x_1^n)$ is an estimator of the probability of block x_1^n based on the Markov model of order k, where the Markov model of order -1 is the uniform distribution. We should note that these Markov estimators are adaptive, i.e. the respective transition probabilities $PPM_k(x_i | x_1^{i-1})$ are re-estimated given each new observation x_i. Moreover, we notice that term $PPM_k(x_i | x_1^{i-1})$ is a conditional probability distribution and thus both $PPM_k(x_1^n)$ and $PPM(x_1^n)$ are probability distributions,

$$\sum_{x_i \in \mathbb{X}} PPM_k(x_i | x_1^{i-1}) = \sum_{x_1^n \in \mathbb{X}^n} PPM_k(x_1^n) = \sum_{x_1^n \in \mathbb{X}^n} PPM(x_1^n) = 1 \tag{6.27}$$

since $\sum_{k=1}^{\infty} k^{-2} = \pi^2/6$.

The total PPM probability is effectively computable. To show this, we observe that sum (6.24) contains only finitely many distinct elements.

Theorem 6.3:   *We have*

$$PPM(x_1^n) = D^{-n} + \frac{6}{\pi^2} \sum_{k=0}^{n-2} \frac{PPM_k(x_1^n) - D^{-n}}{(k+2)^2} \tag{6.28}$$

Proof: By the definition of the PPM probability, we have $PPM_k(x_1^n) = D^{-n}$ for $k > n - 2$. Hence,

$$\begin{aligned} PPM(x_1^n) &= \frac{6}{\pi^2} \left[D^{-n} + \sum_{k=0}^{\infty} \frac{PPM_k(x_1^n)}{(k+2)^2} \right] \\ &= D^{-n} + \frac{6}{\pi^2} \sum_{k=0}^{\infty} \frac{PPM_k(x_1^n) - D^{-n}}{(k+2)^2} \\ &= D^{-n} + \frac{6}{\pi^2} \sum_{k=0}^{n-2} \frac{PPM_k(x_1^n) - D^{-n}}{(k+2)^2} \end{aligned} \tag{6.29}$$

$\square$

Summation (6.28) can be further truncated, see the Problem 6.6.

Now, we will do some algebra in order to prove a few other properties of the PPM probability. Let us introduce a convenient shorthand

$$\langle q \rangle := -\log q \qquad (6.30)$$

which will be mostly used to denote pointwise entropies of PPM distributions.

Theorem 6.4: *The pointwise entropy of the distribution* PPM$_k$ *is*

$$\langle \mathrm{PPM}_k(x_1^n) \rangle = \begin{cases} n \log D, & k > n - 2 \\ k \log D - \displaystyle\sum_{u \in \mathbb{X}^k} \log \frac{(D-1)! \prod_{a \in \mathbb{X}} N(ua|x_1^n)!}{(N(u|x_1^{n-1}) + D - 1)!}, & else \end{cases} \qquad (6.31)$$

Proof: Let $k > n - 2$. Then $\mathrm{PPM}_k(x_1^n) = D^{-n}$. Else, by (6.26), we obtain

$$\mathrm{PPM}_k(x_1^n) = D^{-k-1} \prod_{i=k+2}^{n} \frac{N(x_{i-k}^i|x_1^{i-1}) + 1}{N(x_{i-k}^{i-1}|x_1^{i-2}) + D}$$

$$= D^{-k} \prod_{i=k+1}^{n} \frac{N(x_{i-k}^i|x_1^i)}{N(x_{i-k}^{i-1}|x_1^{i-1}) + D - 1}$$

$$= D^{-k} \prod_{u \in \mathbb{X}^k} \frac{(D-1)! \prod_{a \in \mathbb{X}} N(ua|x_1^n)!}{(N(u|x_1^{n-1}) + D - 1)!} \qquad (6.32)$$

In view of this we obtain the claim. □

In the following, we will recall the Stirling approximation.

Theorem 6.5 (*Stirling approximation*): *For any natural number $n \geq 1$, we have*

$$n \log n - n \log e \leq \log n! \leq (n+1) \log(n+1) - n \log e \qquad (6.33)$$

Proof: We have

$$\ln n! = \sum_{j=1}^{n} \ln j! \in \left(\int_0^n \ln x \, dx, \int_1^{n+1} \ln x \, dx \right) \qquad (6.34)$$

whereas

$$\int_a^b \ln x \, dx = [x \ln x - x]_a^b \qquad (6.35)$$

Hence, we obtain (6.33). □

Let us adopt the convention that $0 \log 0 = 0$. Then formula (6.33) remains valid for $n = 0$. For $n > k$, $u \in \mathbb{X}^k$, and a block of random variables x_1^n, let us denote the smoothed empirical distribution

$$N_n(a, u) := \begin{cases} N(ua|X_1^n) & \text{if } a \in \mathbb{X} \\ D - 1 & \text{if } a = \lambda \end{cases} \tag{6.36}$$

$$N_n(u) := \sum_{a \in F} N_n(a, u) = N(u|X_1^{n-1}) + D - 1 \tag{6.37}$$

$$N_n := \sum_{u \in C_n, a \in F} N_n(a, u) = n - k + D^k(D - 1) \tag{6.38}$$

where $C_n := \{u \in \mathbb{X}^k : X_1^{n-1} = vuz \text{ for some } v, z \in \mathbb{X}^*\}$ and $F := \mathbb{X} \cup \{\lambda\}$. Moreover, if we define random variables A_n and U_n distributed according to the smoothed empirical distribution

$$P(A_n = a, U_n = u) = \frac{N_n(a, u)}{N_n} \tag{6.39}$$

then we observe the identity

$$N_n H(A_n | U_n) = \sum_{u \in C_n} \left[N_n(u) \log N_n(u) - \sum_{a \in F} N_n(a, u) \log N_n(a, u) \right] \tag{6.40}$$

In contrast, in view of Theorems 6.4 and 6.5, we obtain the bounds

$$k \log D + \sum_{u \in C_n} \left[N_n(u) \log N_n(u) - \sum_{a \in F} (N_n(a, u) + 1) \log(N_n(a, u) + 1) \right]$$

$$\leq \langle \mathrm{PPM}_k(X_1^n) \rangle \leq$$

$$k \log D + \sum_{u \in C_n} \left[(N_n(u) + 1) \log(N_n(u) + 1) - \sum_{a \in F} N_n(a, u) \log N_n(a, u) \right] \tag{6.41}$$

Let us observe that $(x + 1) \log(x + 1) \leq x \log x + \log[e(x + 1)]$ since $\log x \leq (x - 1) \log e$. Hence, we obtain an important bound

$$D \# C_n \log[e(n - k + D)]$$
$$\leq \langle \mathrm{PPM}_k(X_1^n) \rangle - k \log D - N_n H(A_n | U_n) \leq$$
$$\#C_n \log[e(n - k + D)] \tag{6.42}$$

since $N_n(u), N_n(a, u) \leq n - k + D - 1$.

In view of bound (6.42), we obtain the following result.

Theorem 6.6:   *Consider a stationary ergodic process* $(X_i)_{i \in \mathbb{Z}}$. *For* $k \geq 0$, *we have*

$$\lim_{n \to \infty} \frac{1}{n} \langle \mathrm{PPM}_k(X_1^n) \rangle = H(X_{k+1}|X_1^k) \; P \text{ -almost surely} \tag{6.43}$$

Proof: Consider a string $u \in \mathbb{X}^k$. By the Birkhoff ergodic theorem (Theorem 4.6), the smoothed empirical distribution converges to the true distribution,

$$\lim_{n \to \infty} \frac{N_n(a, u)}{N_n} = P(X_1^{k+1} = ua) \; P\text{-almost surely} \tag{6.44}$$

Hence,

$$\lim_{n \to \infty} \frac{N_n H(A_n | U_n)}{n} = H(X_{k+1}|X_1^k) \; P\text{-almost surely} \tag{6.45}$$

Concluding, in view of bound (6.42) and inequality $\#C_n \leq D^k$, we obtain (6.43). $\qquad\qquad\square$

Now we can demonstrate universality of the PPM probability.

Theorem 6.7:   *The PPM probability is a universal semidistribution.*

Proof: Consider a stationary ergodic process $(X_i)_{i \in \mathbb{Z}}$. Since

$$\langle \mathrm{PPM}(x_1^n) \rangle \leq \langle \mathrm{PPM}_k(x_1^n) \rangle + \log \frac{\pi^2}{6} + 2\log(k+2) \tag{6.46}$$

hence by Theorem 6.6 we obtain

$$\limsup_{n \to \infty} \frac{1}{n} \langle \mathrm{PPM}(X_1^n) \rangle \leq \inf_{k \in \mathbb{N}} H(X_{k+1}|X_1^k) = h \; P\text{-almost surely} \tag{6.47}$$

Moreover, we have

$$\langle \mathrm{PPM}(x_1^n) \rangle \leq n \log D + \log \frac{\pi^2}{6} \tag{6.48}$$

In this way, we have proved the claim. $\qquad\qquad\square$

Let $n > k$ and $w \in \mathbb{X}^k$. In the following, we would like to derive a somewhat more symmetric bound for the pointwise entropy of the PPM distribution in terms of the plain empirical distribution

$$M_n(b, w) := N(wb|X_1^n) \text{ for } b \in \mathbb{X} \tag{6.49}$$

$$M_n(w) := \sum_{b \in \mathbb{X}} M_n(b, w) = N(w|X_1^{n-1}) \tag{6.50}$$

$$M_n := \sum_{w \in C_n} \sum_{b \in \mathbb{X}} M_n(b, w) = n - k \tag{6.51}$$

If we define random variables B_n and W_n distributed according to the plain empirical distribution

$$P(B_n = b, W_n = w) = \frac{M_n(b, w)}{M_n} \tag{6.52}$$

and we define random variable $E_n = \mathbf{1}\{A_n = \lambda\}$, then we observe the identity

$$H(A_n|U_n) = H(A_n, E_n|U_n) = H(E_n|U_n) + H(A_n|E_n, U_n)$$
$$= H(E_n|U_n) + \frac{M_n}{N_n} H(B_n|W_n) \tag{6.53}$$

Now, since $\log x \le (x-1) \log e$ and $M_n(u) \le n - k$, we observe

$$0 \le N_n H(E_n|U_n)$$
$$= \sum_{u \in C_n} \left[M_n(u) \log \frac{M_n(u) + D - 1}{M_n(u)} + (D-1) \log \frac{M_n(u) + D - 1}{D - 1} \right]$$
$$\le \sum_{u \in C_n} \left[(D-1) \log e + (D-1) \log \frac{M_n(u) + D - 1}{D - 1} \right]$$
$$\le (D-1) \# C_n \log[e(n - k + D)] \tag{6.54}$$

Plugging the above two results to bound (6.42), we obtain the desired bound

$$|\langle \mathrm{PPM}_k(X_1^n) \rangle - k \log D - (n-k) H(B_n|W_n)| \le D \# C_n \log[e(n - k + D)] \tag{6.55}$$

Some beautification of the above formula will be performed now. We will introduce two important concepts. Denoting the empirical entropy, i.e. the kth order conditional entropy of the empirical distribution

$$h_k(X_1^n) := H(B_n|W_n) \tag{6.56}$$

and the empirical vocabulary, i.e. the set of distinct substrings of length k as

$$V_k(X_1^n) := \{u \in \mathbb{X}^k : X_1^n = vuz \text{ for some } v, z \in \mathbb{X}^*\} \tag{6.57}$$

formula (6.55) can be rewritten as

$$|\langle \mathrm{PPM}_k(X_1^n) \rangle - k \log D - (n-k) h_k(X_1^n)| \le D \# V_k(X_1^n) \log[e(n - k + D)] \tag{6.58}$$

We will use this bound in Section 6.5.

The PPM probability can be turned into an effective algorithm for estimation of the entropy rate. The general idea of universal coding dates back to Kolmogorov (1965). The PPM probability with minor differences to Definition 6.3 was successively developed by Krichevsky and Trofimov (1981), Ryabko (1984), and Cleary and Witten (1984), where the latest authors proposed the acronym "PPM," being the most popular name now. Applying the Shannon–Fano coding, see Chapter 7, the PPM probability can be turned also into a very good procedure for data compression, as elaborated by Cleary and Witten (1984). The PPM probability is but a single example of universal semidistributions. Other important examples of universal semidistributions, also called universal codes, are the Lempel–Ziv code (Ziv and Lempel, 1977) and grammar-based codes (Kieffer and Yang, 2000; Charikar et al., 2005; Dębowski, 2011; Ochoa and Navarro, 2019). These universal codes work according to different principles, applying directly the ideas of data compression to be developed in Section 7.1, see also Problem 7.4.

6.4 SMB Theorem Revisited

The existence of universal semidistributions has profound consequences for generalizing the SMB theorem to nonergodic stationary processes. First of all, Theorem 6.2 can be strengthened for a universal semidistribution to a full analogue of the SMB theorem for ergodic processes (Theorem 6.1).

Theorem 6.8: *Let $(X_i)_{i\in\mathbb{Z}}$ be a stationary ergodic process on $(\Omega, \mathcal{J}, P)$, where $X_i : \Omega \to \mathbb{X}$ and the alphabet $\mathbb{X}$ is finite. For a universal semidistribution $Q : \mathbb{X}^* \to \mathbb{R}$, we have*

$$\lim_{n\to\infty} \frac{1}{n}\mathbf{E}[-\log Q(X_1^n)] = h \qquad (6.59)$$

$$\lim_{n\to\infty} \frac{1}{n}[-\log Q(X_1^n)] = h \text{ almost surely} \qquad (6.60)$$

Proof: By (6.20) and (6.22), we obtain (6.60). Since we have (6.21), then from (6.60) we obtain (6.59) by the dominated convergence (Theorem 3.10). □

Thus, universal semidistributions allow us to estimate the entropy rate of an unknown stationary ergodic process. For this aim, it is sufficient to exhibit just a single instance of a universal semidistribution, such as the PPM probability, and it will be good for all stationary ergodic processes.

Moreover, since universal semidistribution exists, then we easily obtain a joint generalization of Theorem 6.8 and the SMB theorem for stationary nonergodic processes. The idea of the presented proof seems new.

Theorem 6.9 *(SMB theorem): Let $(X_i)_{i\in\mathbb{Z}}$ be a stationary process on $(\Omega, \mathcal{J}, P)$, where $X_i : \Omega \to \mathbb{X}$ and the alphabet $\mathbb{X}$ is finite. Let $\mathcal{J}$ be the invariant σ-field of the process $(X_i)_{i\in\mathbb{Z}}$, let $F(A) = P(A|\mathcal{J})$ be the random ergodic measure of the process,* *and let h_F be the entropy rate of $(X_i)_{i \in \mathbb{Z}}$ with respect to measure F. For a universal semidistribution $Q : \mathbb{X}^* \rightarrow \mathbb{R}$ we have*

$$\lim_{n \to \infty} \frac{1}{n} \mathbf{E}[-\log Q(X_1^n)] = \lim_{n \to \infty} \frac{1}{n} \mathbf{E}[-\log P(X_1^n)]$$

$$= \lim_{n \to \infty} \frac{1}{n} \mathbf{E}[-\log F(X_1^n)] = h = \mathbf{E} h_F \qquad (6.61)$$

$$\lim_{n \to \infty} \frac{1}{n}[-\log Q(X_1^n)] = \lim_{n \to \infty} \frac{1}{n}[-\log P(X_1^n)]$$

$$= \lim_{n \to \infty} \frac{1}{n}[-\log F(X_1^n)] = h_F \ \ almost \ surely \qquad (6.62)$$

Proof: Notice that $P(A) = \int F(A)dP$. Thus, a given statement holds *P*-almost surely if it holds *F*-almost surely for all values of *F*. Hence, by Theorem 6.8, we have

$$\lim_{n \to \infty} \frac{1}{n}[-\log Q(X_1^n)] = \lim_{n \to \infty} \frac{1}{n}[-\log F(X_1^n)] = h_F \ \ P\text{-almost surely} \qquad (6.63)$$

Now, by the Barron lemma (Theorem 3.20), when $\pi(n)Q(X_1^n)$ with $\pi(n) := 6\pi^{-2}n^{-2}$ is a semidistribution and $P(X_1^n)$ is the probability distribution, we obtain that

$$-\log Q(X_1^n) \geq -\log P(X_1^n) - 2\log n \qquad (6.64)$$

for all but finitely many n holds *P*-almost surely. In contrast, when $\pi(n)P(X_1^n)$ with $\pi(n) := 6\pi^{-2}n^{-2}$ is a semidistribution and $F(X_1^n)$ is the probability distribution, we obtain that

$$-\log P(X_1^n) \geq -\log F(X_1^n) - 2\log n \qquad (6.65)$$

for all but finitely many n holds *F*-almost surely. Hence, we obtain (6.62).

Next we will prove (6.61). Observe that, since $\mathbf{E}_P Y = \mathbf{E}_P \mathbf{E}_F Y$, then by the Gibbs inequality (Theorem 3.18), we have

$$\mathbf{E}_P[-\log Q(X_1^n)] \geq \mathbf{E}_P[-\log P(X_1^n)] \geq \mathbf{E}_P[-\log F(X_1^n)] \qquad (6.66)$$

Since we have (6.21) and $\mathbf{E}_F[-\log F(X_1^n)] \leq n \log \#\mathbb{X}$ holds *P*-almost surely then by the dominated convergence (Theorem 3.10), we obtain from (6.6) and (6.59) that

$$\lim_{n \to \infty} \frac{1}{n} \mathbf{E}_P[-\log Q(X_1^n)] = \lim_{n \to \infty} \frac{1}{n} \mathbf{E}_P[-\log F(X_1^n)] = \mathbf{E}_P h_F \qquad (6.67)$$

Now we notice that $\mathbf{E}_P h_F = h_P$ by Theorem 5.35. From this, we deduce (6.61). □

We can see now that the existence of universal semidistributions is important not only from a practical point of view of entropy rate estimation and data compression – to be developed in Section 7.1, but also from a theoretical point of view of establishing some properties of stochastic processes. As we will observe later, in Theorem 10.14, the SMB theorem can be further generalized to some nonstationary processes using a similar technique.

6.5 PPM-based Statistics

In this section, we will analyze some statistics of individual strings defined via the PPM probability. Our primary motivation is to provide some auxiliary results for the theorems about facts and words (Theorems 8.22 and 8.23), some linguistically motivated results concerning stationary processes which we have mentioned in the half-formal Section 1.12. On our way, in this section, we will demonstrate some results of partly an independent interest, concerning the PPM order of a string and superadditivity of empirical entropy.

First of all, the PPM probability can be imagined as a family of competing distributions PPM_k, out of which the one is chosen effectively which yields the largest probability for a given sequence x_1^n. As we will see now, this interpretation is valid. For this goal, we will introduce the concept of the PPM order of a string.

Definition 6.4 *(PPM order)*: Let us define the largest of all probabilities $PPM_k(x_1^n)$ as

$$PPM_G(x_1^n) := \max\{PPM_k(x_1^n) : k \geq -1\} \tag{6.68}$$

where we put $PPM_{-1}(x_1^n) := D^{-n}$. The PPM order of sequence x_1^n is

$$G(x_1^n) := \min\{k \geq -1 : PPM_G(x_1^n) = PPM_k(x_1^n)\} \tag{6.69}$$

The PPM order is bounded by the string length.

Theorem 6.10: *We have $G(x_1^n) \leq n - 2$.*

Proof: By the definition of the PPM probability, we have $PPM_k(x_1^n) = D^{-n} = PPM_{-1}(x_1^n)$ for $k > n - 2$. Hence, the claim follows: □

As we have announced, we will show now that the pointwise entropy of the PPM probability for a sequence is approximately equal to the pointwise entropy of the distribution PPM_k which contributes the largest probability mass.

Theorem 6.11: *We have*

$$0 \leq \langle PPM(x_1^n) \rangle - \langle PPM_G(x_1^n) \rangle \leq \log \frac{\pi^2}{6} + 2\log n \tag{6.70}$$

Proof: The claim follows by inequality (6.46) and Theorem 6.10. □

Moreover, the PPM order may be interpreted as an almost sure upper bound of the Markov order of a stationary ergodic process. Some simple formal definition of the Markov order makes use of conditional entropies.

Definition 6.5 *(Conditional entropies):* For a stationary ergodic process $(X_i)_{i \in \mathbb{Z}}$ over a finite alphabet $\mathbb{X} = \{a_1, a_2, \ldots, a_D\}$, let us write the conditional entropies

$$h_k := \begin{cases} H(X_1 | X^0_{-k+1}) & k \geq 0 \\ \log D & k = -1 \end{cases} \qquad (6.71)$$

Now we can define the Markov order of a stationary ergodic process.

Definition 6.6 *(Markov order):* For a stationary ergodic process $(X_i)_{i \in \mathbb{Z}}$, we define the Markov order

$$M := \inf\{k \geq -1 : h_k = h\} \qquad (6.72)$$

where the infimum of the empty set equals infinity, $\inf \emptyset := \infty$.

We note that equality $M = k$ holds for a finite k if and only if the process is a kth order Markov process and is not an lth order Markov process for $l < k$, whereas non-Markovian processes have the Markov order $M = \infty$.

As we have announced, we will demonstrate now that the PPM order is a semi-consistent estimator of the Markov order. That is, the PPM order is an almost sure upper bound of the Markov order.

Theorem 6.12: *For a stationary ergodic process $(X_i)_{i \in \mathbb{Z}}$, we have*

$$\liminf_{n \to \infty} G(X^n_1) \geq M \text{ almost surely} \qquad (6.73)$$

Proof: Suppose by contradiction that $\liminf_{n \to \infty} G(X^n_1) < M$. Then there exists an integer number $k \geq -1$ such that $h_k > h$ and $G(X^n_1) \leq k$ for infinitely many n. Then by Theorem 6.6, we obtain

$$\limsup_{n \to \infty} \frac{1}{n} \langle PPM_G(X^n_1) \rangle \geq h_k > h \qquad (6.74)$$

But $\langle PPM_G(X^n_1) \rangle \leq \langle PPM(X^n_1) \rangle$ and the PPM probability is universal by Theorem 6.7. We obtained a contradiction. Hence, we have (6.73). □

It was conjectured by Csiszar and Shields (2000) that the PPM order is a consistent estimator of the Markov order indeed, i.e. for any stationary ergodic process, we have

$$\lim_{n \to \infty} G(X^n_1) = M \text{ almost surely} \qquad (6.75)$$

Besides this conjecture, Csiszar and Shields (2000) as well as Morvai and Weiss (2005) demonstrated consistency of a few other estimators of the Markov order, where the estimator by Morvai and Weiss (2005) works also for processes over a countably infinite alphabet. Still, the consistency of the PPM order is an interesting open problem.

In the following, we would like to show several results concerning some statistical analogues of Shannon mutual information, leading us toward the theorems about facts and words in Section 8.4. First, we will prove that the empirical entropy is superadditive, which upon a closer inspection corresponds to positivity of certain conditional Shannon mutual information.

Theorem 6.13: *For* $0 \le k < n, m - n < m$, *we have*

$$(n - k)h_k(X_1^n) + (m - n - k)h_k(X_{n+1}^m) - (m - k)h_k(X_1^m) \le 0 \qquad (6.76)$$

Proof: For $w \in \mathbb{X}^k$ and $b \in \mathbb{X}$, we have

$$N(wb|X_1^m) = N(wb|X_1^n) + N(wb|X_{n-k}^{n+k}) + N(wb|X_{n+1}^m) \qquad (6.77)$$

Let us define random variable Z distributed as

$$P(Z = i) = \begin{cases} \dfrac{n - k}{m - k} & \text{if } i = 1 \\[2mm] \dfrac{k}{m - k} & \text{if } i = 2 \\[2mm] \dfrac{m - n - k}{m - k} & \text{if } i = 3 \end{cases} \qquad (6.78)$$

and random variables B and W distributed according to the conditional distribution

$$P(B = b, W = w|Z = i) = \begin{cases} \dfrac{N(wb|X_1^n)}{n - k} & \text{if } i = 1 \\[2mm] \dfrac{N(wb|X_{n-k}^{n+k})}{k} & \text{if } i = 2 \\[2mm] \dfrac{N(wb|X_{n+1}^m)}{m - k} & \text{if } i = 3 \end{cases} \qquad (6.79)$$

Then we obtain

$$0 \le I(B; Z|W) = H(B|W) - H(B|W, Z)$$
$$= h_k(X_1^m) - \frac{n - k}{m - k}h_k(X_1^n) - \frac{k}{m - k}h_k(X_{n-k}^{n+k}) - \frac{m - n - k}{m - k}h_k(X_{n+1}^m)$$
$$\le h_k(X_1^m) - \frac{n - k}{m - k}h_k(X_1^n) - \frac{m - n - k}{m - k}h_k(X_{n+1}^m) \qquad (6.80)$$

which yields the claim after regrouping. $\qquad\square$

We will see two applications of the above theorem. The first one will concern the speed of convergence of the encoding rate, whereas the second one will be linked with the theorems about facts and words. To begin, Theorem 6.13 can be used to prove that the empirical entropy is smaller in expectation than the conditional entropy.

Theorem 6.14: *For a stationary ergodic process $(X_i)_{i \in \mathbb{Z}}$, we have*

$$(n - k)\mathbf{E}h_k(X_1^n) \leq \frac{(2^m n - k)}{2^m}\mathbf{E}h_k(X_1^{2^m n}) \leq nh_k \tag{6.81}$$

Proof: By Theorem 6.6 and formula (6.58), $\lim_{n \to \infty} h_k(X_1^n) = h_k$. Hence, by the dominated convergence (Theorem 3.10), $\lim_{n \to \infty} \mathbf{E}h_k(X_1^n) = h_k$. Consequently, the claim follows by stationarity and Theorem 6.13. □

As a result we obtain an upper bound for the expectation $\mathbf{E}\langle\mathrm{PPM}(X_1^n)\rangle$ in terms of conditional entropies h_k of a stationary ergodic process.

Theorem 6.15: *For a stationary ergodic process $(X_i)_{i \in \mathbb{Z}}$, we have*

$$\mathbf{E}\langle\mathrm{PPM}(X_1^n)\rangle \leq \min_{D \leq k \leq n-2}[nh_k + D^{k+3}\log(en)] \tag{6.82}$$

Proof: By inequalities (6.46) and (6.58) and Theorems 6.10 and 6.14, we obtain

$$\mathbf{E}\langle\mathrm{PPM}(X_1^n)\rangle$$

$$\leq \min_{-1 \leq k \leq n-2}\left[nh_k + D^{k+1}\log[e(n - k + D)] + k\log D + 2\log(k + 2) + \log\frac{\pi^2}{6}\right] \tag{6.83}$$

Subsequently, we simplify this expression, noting that $\frac{\pi^2}{6} < 2 \leq D$ and using inequality $(x - 1)\log e \geq \log x$ for $x = D^{k+1}$ in particular. □

In particular, if we know the rate of convergence of the conditional entropy h_k to the entropy rate h, then by Theorem 6.15 we may bound the rate of convergence of the encoding rate $\mathbf{E}\langle\mathrm{PPM}(X_1^n)\rangle/n$ to the entropy rate h. The following comment to the above result is due. Shields (1993) showed that for each semidistribution Q and any function $f : \mathbb{N} \to \mathbb{R}$ satisfying condition $\lim_{n \to \infty} f(n)/n = 0$, there exist stationary ergodic processes $(X_i)_{i \in \mathbb{Z}}$ such that the inequality

$$\mathbf{E}[-\log Q(X_1^n)] - H(X_1^n) \geq f(n) \tag{6.84}$$

holds for infinitely many n. In spite of this nonexistence of a nontrivial universal upper bound for redundancy $\mathbf{E}[-\log Q(X_1^n)] - H(X_1^n)$, we have shown above that knowing the asymptotics of conditional entropy h_k implies a nontrivial upper bound for redundancy $\mathbf{E}\langle \mathrm{PPM}(X_1^n)\rangle - H(X_1^n)$ of the PPM probability. For some particular calculations, see Problem 6.7.

Now, let us move on to proving some auxiliary results for the theorems about facts and words to be discussed in Section 8.4. For this aim, the substrings of string x_1^n of length $G(x_1^n)$ will be called the PPM words of string x_1^n.

Definition 6.7 *(PPM vocabulary)*: The PPM vocabulary of a string x_1^n is defined as

$$V_G(x_1^n) := V_{G(x_1^n)}(x_1^n) \tag{6.85}$$

where we put $V_{-1}(x_1^n) := \emptyset$, and we have $V_0(x_1^n) = \{\lambda\}$.

We will show that the PPM order is a sort of the optimal length of a word-like substring in sequence x_1^n. That is, we will demonstrate that the PPM mutual information between two parts of a string is roughly less than the number of PPM words in the string multiplied by the logarithm of the string length. This will be the second application of Theorem 6.13 and the first use of inequality (6.58).

Theorem 6.16 *(PPM mutual information and words)*: For $G(X_1^m) < n, m - n < m$, we have

$$\langle \mathrm{PPM}(X_1^n)\rangle + \langle \mathrm{PPM}(X_{n+1}^m)\rangle - \langle \mathrm{PPM}(X_1^m)\rangle$$

$$\leq (G(X_1^m) + 3)\log D + 3D[\#V_G(X_1^m) + 2]\log[e(m + D)] \tag{6.86}$$

Proof: Consider $0 \leq k < n, m - n < m$. By formula (6.58), we have

$$|\langle \mathrm{PPM}_k(X_1^n)\rangle - k\log D - (n - k)h_k(X_1^n)| \leq D\#V_k(X_1^n)\log[e(n + D)] \tag{6.87}$$

whereas by Theorem 6.13, we have

$$(n - k)h_k(X_1^n) + (m - n - k)h_k(X_{n+1}^m) - (m - k)h_k(X_1^m) \leq 0 \tag{6.88}$$

Hence,

$$\langle \mathrm{PPM}_k(X_1^n)\rangle + \langle \mathrm{PPM}_k(X_{n+1}^m)\rangle - \langle \mathrm{PPM}_k(X_1^m)\rangle$$

$$\leq k\log D + D\#V_k(X_1^n)\log[e(n + D)]$$

$$+ D\#V_k(X_{n+1}^m)\log[e(m - n + D)] + D\#V_k(X_1^m)\log[e(m + D)]$$

$$\leq k\log D + 3D\#V_k(X_1^m)\log[e(m + D)] \tag{6.89}$$

Similarly, we obtain

$$\langle \mathrm{PPM}_{-1}(X_1^n) \rangle + \langle \mathrm{PPM}_{-1}(X_{n+1}^m) \rangle - \langle \mathrm{PPM}_{-1}(X_1^m) \rangle$$

$$= 0 = 0 \log D + 3D \# V(-1|X_1^n) \log[e(n+D)] \tag{6.90}$$

Now, let $g = G(X_1^m)$. Since

$$\langle \mathrm{PPM}(X_1^m) \rangle \geq \langle \mathrm{PPM}_g(X_1^m) \rangle \tag{6.91}$$

and

$$\langle \mathrm{PPM}(X_j^l) \rangle \leq \langle \mathrm{PPM}_g(X_j^l) \rangle + \log \frac{\pi^2}{6} + 2 \log(g+2) \tag{6.92}$$

we obtain

$$\langle \mathrm{PPM}(X_1^n) \rangle + \langle \mathrm{PPM}(X_{n+1}^m) \rangle - \langle \mathrm{PPM}(X_1^m) \rangle$$

$$\leq \langle \mathrm{PPM}_g(X_1^n) \rangle + \langle \mathrm{PPM}_g(X_{n+1}^m) \rangle - \langle \mathrm{PPM}_g(X_1^m) \rangle$$

$$+ 2 \log \frac{\pi^2}{6} + 4 \log(g+2) \tag{6.93}$$

Plugging the above to the earlier obtained bounds for $g < n, m - n < m$ yields the claim

$$\langle \mathrm{PPM}(X_1^n) \rangle + \langle \mathrm{PPM}(X_{n+1}^m) \rangle - \langle \mathrm{PPM}(X_1^m) \rangle$$

$$\leq (g+1) \log D + 3D \# V_g(X_1^m) \log[e(m+D)] + 2 \log \frac{\pi^2}{6} + 4 \log(g+2)$$

$$\leq (G(X_1^m) + 3) \log D + 3D[\# V_G(X_1^m) + 2] \log[e(m+D)] \tag{6.94}$$

where the last transition is due to $g \leq m - 2$, $\frac{\pi^2}{6} < 2 \leq D$, and $V_g(X_1^m) = V_G(x_1^m)$.
□

A similar bound holds in expectation with no restriction on n.

Theorem 6.17 *(Expected PPM mutual information and words):* *For any $n \geq 1$, we have*

$$\mathbf{E}[\langle \mathrm{PPM}(X_1^n) \rangle + \langle \mathrm{PPM}(X_{n+1}^{2n}) \rangle - \langle \mathrm{PPM}(X_1^{2n}) \rangle]$$

$$\leq (3\mathbf{E}G(X_1^{2n}) + 5) \left[\log D + \frac{1}{n} \right] + 3D \left[\mathbf{E} \# V_G(X_1^{2n}) + 2 \right] \log[e(2n+D)] \tag{6.95}$$

Proof: We observe a uniform bound

$$\langle \mathrm{PPM}(X_1^n) \rangle + \langle \mathrm{PPM}(X_{n+1}^{2n}) \rangle - \langle \mathrm{PPM}(X_1^{2n}) \rangle \leq 2 \left[n \log D + \log \frac{\pi^2}{6} \right] \tag{6.96}$$

Subsequently, by Theorem 6.16 and the Markov inequality (Theorem 3.11), we obtain

$$\mathbf{E}[\langle \text{PPM}(X_1^n) \rangle + \langle \text{PPM}(X_{n+1}^{2n}) \rangle - \langle \text{PPM}(X_1^{2n}) \rangle]$$

$$\leq (EG(X_1^{2n}) + 3) \log D + 3D \, [\mathbf{E}\# V_G(X_1^{2n}) + 2] \log[e(2n + D)]$$

$$+ 2 \left[n \log D + \log \frac{\pi^2}{6} \right] P(G(X_1^{2n}) + 1 \geq n + 1)$$

$$\leq (EG(X_1^{2n}) + 3) \log D + 3D \, [\mathbf{E}\# V_G(X_1^{2n}) + 2] \log[e(2n + D)]$$

$$+ 2[n \log D + 1] \frac{\mathbf{E}(G(X_1^{2n}) + 1)}{n + 1}$$

$$\leq (3EG(X_1^{2n}) + 5) \left[\log D + \frac{1}{n} \right] + 3D \, [\mathbf{E}\# V_G(X_1^{2n}) + 2] \log[e(2n + D)]$$

$$\tag{6.97}$$

□

Inequality (6.95) will be used to prove the theorems about facts and words (Theorems 8.22 and 8.23), some important linguistically motivated results concerning general stationary processes over a finite alphabet. Inequality (6.86) and the respective theorems about facts and words were derived by Dębowski (2018a). Earlier, an analogue of inequality (6.86) was demonstrated for minimal grammar-based codes by Dębowski (2011) in the context of proving a theorem about facts and words for these codes, see Problem 7.4 and the longer remark that follows it.

Problems

6.1 Let $X : \Omega \to \mathbb{X}$ and $Y : \Omega \to \mathbb{Y}$ be discrete random variables.
(a) Show that

$$P(-\log P(X) \geq \log|\mathbb{X}| + m) \leq 2^{-m} \tag{6.98}$$

(b) Show that

$$P\left(\log \frac{P(X, Y)}{P(X)P(Y)} \geq \log \min\{|\mathbb{X}|, |\mathbb{Y}|\} + m \right) \leq 2^{-m} \tag{6.99}$$

$$P\left(\log \frac{P(X, Y)}{P(X)P(Y)} \leq -m \right) \leq 2^{-m} \tag{6.100}$$

(c) Show that there is a real number $C < \infty$ such that

$$0 \leq \mathbf{E} \left| \log \frac{P(X,Y)}{P(X)P(Y)} \right| - I(X,Y) \leq C \tag{6.101}$$

holds for any X and Y, where $I(X,Y)$ is the Shannon mutual information.

6.2 Let $\mathcal{H}(p)$ be the entropy of a two-point distribution defined in (5.111) and

$$\binom{n}{k} := \frac{n!}{k!(n-k)!} \tag{6.102}$$

be the binomial coefficient. Show that for $p \in [0,1]$ and $\lim_{n\to\infty} k_n/n = p$, we have

$$\lim_{n\to\infty} \frac{1}{n} \log \binom{n}{k_n} = \mathcal{H}(p) \tag{6.103}$$

6.3 Show that for a stationary process $(X_i)_{i\in\mathbb{Z}}$ over a finite alphabet and positive growing sequences $(k_n)_{n\in\mathbb{N}}$ and $(l_n)_{n\in\mathbb{N}}$ of natural numbers such that $\lim_{n\to\infty}(k_n + l_n) = \infty$, we have

$$\lim_{n\to\infty} \frac{1}{k_n + l_n + 1} [-\log P(X_{-k_n}^{l_n})] = h_F \text{ almost surely} \tag{6.104}$$

where h_F is the Shannon entropy rate of the random ergodic measure of the process.

6.4 Show that for a stationary process $(X_i)_{i\in\mathbb{Z}}$ over a countable alphabet, we have

$$(n-k)h_k(X_1^n) \leq - \sum_{i=k+1}^{n} \log P(X_i|X_{i-k}^{i-1}) \tag{6.105}$$

How can we use this inequality to prove universality of the PPM distribution? There is a similar inequality for the Lempel–Ziv code called the Ziv inequality (Cover and Thomas, 2006, lemma 13.5.5).

6.5 Show that $(n-k)h_k(x_1^n) \geq (n-k-1)h_{k+1}(x_1^n)$ for $k \geq 0$, whereas we have $h_k(x_1^n) = 0$ for $k \geq L(x_1^n)$, where $L(x_1^n)$ is the maximal repetition of string x_1^n defined in (1.112).

6.6 Consider the maximal repetition $L(x_1^n)$ of a string x_1^n. Obviously, we have $L(x_1^n) \leq n$. Show that

$$\text{PPM}_k(x_1^n) = D^{-n} \text{ for } k > L(x_1^n) \tag{6.106}$$

and in consequence

$$\mathrm{PPM}(x_1^n) = D^{-n} + \frac{6}{\pi^2} \sum_{k=0}^{L(x_1^n)} \frac{\mathrm{PPM}_k(x_1^n) - D^{-n}}{(k+2)^2} \tag{6.107}$$

$$G(x_1^n) \le L(x_1^n) \tag{6.108}$$

6.7 Show the following bounds for the encoding rate $\mathbf{E}\langle \mathrm{PPM}(X_1^n)\rangle/n$:

(a) If process $(X_i)_{i\in\mathbb{Z}}$ is a Markov process of order m, i.e. $h_m = h$, then

$$\frac{1}{n}\mathbf{E}\langle \mathrm{PPM}(X_1^n)\rangle - h \le \frac{C\log n}{n} \quad \text{for a } C > 0 \tag{6.109}$$

(b) If $h_k - h \le A^{-k}$ for an $A \in (0,1)$, then

$$\frac{1}{n}\mathbf{E}\langle \mathrm{PPM}(X_1^n)\rangle - h \le Cn^{-\alpha} \text{ for } C > 0 \quad \text{and} \quad \alpha = \frac{\log A}{\log A + \log D} \tag{6.110}$$

(c) If $h_k - h \le k^{-\alpha}$ for an $\alpha > 0$, then

$$\frac{1}{n}\mathbf{E}\langle \mathrm{PPM}(X_1^n)\rangle - h \le \frac{C}{(\log n)^\alpha} \text{ for } C > 0 \tag{6.111}$$

7

Coding and Computation

As we have mentioned in the introduction to Chapter 5, there are two standard approaches to the definition of the amount of information: one initiated by Shannon, which applies probability, and another proposed by Kolmogorov, which uses algorithms. It turns out that in both approaches, the amount of information contained in an object is roughly equal to the minimal number of binary digits that we need to describe the object. In fact, we have not touched this interpretation yet and we would like to discuss it in this chapter. In a single package, we will present the coding interpretation of the Shannon entropy and its close links with Kolmogorov's algorithmic complexity, which implements the idea of the minimal coding in a direct way. This chapter develops formally ideas sketched in the half-formal Sections 1.7 and 1.10.

The general idea of coding, an ingenious invention of the twentieth century, is to represent arbitrary objects from a countable set – such as letters, words, sentences, or whole finite texts – as unique finite sequences of binary digits. With the advent of personal computers and mobile devices, this idea seems as obvious as the idea of the alphabet, but it took some effort to discover profound implications of this idea for foundations of mathematics, computer science, and statistics. In particular, the idea of coding led to discovery of unprovable statements in mathematics by Gödel (1931), which was further strengthened by Chaitin (1975b) as unprovability of randomness.

Of course, the abstract idea of coding without computation would not be so important for the general public. As computers have become ubiquitous and indispensable for human civilization, scientists have started applying the metaphor of computation everywhere: to living organisms, to laws of physics, and to human brain. However, it is not obvious whether the most popular mathematical model of computation such as Turing machines is directly applicable and relevant to all these cases. The natural computation may differ to the mathematical computation in many important details and be more complicated. But simple mathematical

Information Theory Meets Power Laws: Stochastic Processes and Language Models, First Edition.
Łukasz Dębowski.
© 2021 John Wiley & Sons, Inc. Published 2021 by John Wiley & Sons, Inc.

models are always a good starting point to learn to imagine what else could be possible. For this reason, it may be good to get a primer in basic formal models of coding and computation.

The organization of this chapter is as follows. Section 7.1 reports basic ideas of coding. Codes are functions that represent countably many distinct objects as code words, i.e. finite sequences of fixed symbols such as binary digits. It is a desirable property that if we concatenate code words, then we can decipher the corresponding sequence of encoded objects. Codes that possess this property are called uniquely decodable and they satisfy a simple albeit important inequality, called the Kraft inequality. By the Kraft inequality, for a random variable assuming values in the objects to be encoded, its Shannon entropy is roughly equal to the minimal expected length of a uniquely decodable code. The minimum is roughly achieved by the Shannon–Fano code.

In Section 7.2, we make first steps toward algorithmic information theory. We introduce Turing machines, a traditional mathematical model of a general-purpose computer. Subsequently, we define the Kolmogorov complexity of an object as the length of the shortest binary program for a universal Turing machine that computes a binary representation of this object. We provide some simple bounds for Kolmogorov complexity but, as we also show, Kolmogorov complexity is uncomputable, i.e. there is no finite program that computes it for all arguments.

Section 7.3 is devoted to presenting parallels between Kolmogorov complexity and Shannon entropy. The respective results vary in their difficulty, but all of them rest on the idea of Shannon–Fano coding introduced in the first section. We show that Kolmogorov complexity not only satisfies the chain rule in a clear analogy to Shannon entropy but also its expectation is roughly equal to Shannon entropy if the compressed random variable has a computable probability distribution. The latter result is generalized to stochastic processes in a simple fashion.

Algorithmic information theory also sheds some light onto the limits of mathematics, the topic of Section 7.4. We discuss the Chaitin incompleteness theorem which complements the famous incompleteness theorem by Gödel. This proposition states that it is not possible to prove in any sound formal inference system that the Kolmogorov complexity of any particular object is substantially larger than the Kolmogorov complexity of the inference system itself. On the other hand, we know from the simple bounds discussed in Section 7.2 that all but finitely many objects have so large Kolmogorov complexity. Moreover, we can demonstrate that there exists a simply defined real number, called the halting probability Ω, whose binary approximations have the maximal Kolmogorov complexity. Number Ω is a source of many paradoxes. Not only its binary expansion cannot be computed but it also constitutes philosophers' stone, i.e. the validity of any mathematical statement can be simply computed given as many binary digits of number Ω as the statement length.

Halting probability Ω is a special case of a much more common object called algorithmically random sequences. In Section 7.5, we will study the definition and some properties thereof. An infinite sequence is called algorithmically random (in the Martin-Löf sense) with respect to a probability measure on sequences when the Kolmogorov complexity of its prefixes is close to the probability of the respective cylinder sets. This property holds for almost all sequences, so there are really uncountably many different algorithmically random sequences. There are a few other characterizations of algorithmically random sequences, which are equivalent by the Schnorr theorem, also discussed in Section 7.5.

7.1 Elements of Coding

The object of interest of coding theory are functions, called codes, that map elements of a countable set $\mathbb{X}$ into finite sequences over a countable set $\mathbb{Y}$. The set of these sequences is denoted as $\mathbb{Y}^* := \mathbb{Y}^0 \cup \mathbb{Y}^+$, where $\mathbb{Y}^+ := \bigcup_{n=1}^{\infty} \mathbb{Y}^n$, $\mathbb{Y}^0 := \{\lambda\}$, and λ is the empty sequence. Within the context of coding theory, sets $\mathbb{X}$ and $\mathbb{Y}$ are called alphabets, whereas finite sequences are preferably called strings. Of a special interest are binary codes, i.e. codes for which the output alphabet $\mathbb{Y}$ is the set $\{0, 1\}$. On the other hand, the input alphabet $\mathbb{X}$ consists typically of letters, digits, or even strings of symbols, such as words in natural language.

The first definition formalizes what we have said so far.

Definition 7.1 *(Code):* For two countable alphabets $\mathbb{X}$ and $\mathbb{Y}$, a function $C : \mathbb{X} \to \mathbb{Y}^*$ is called a code.

Strings $C(x)$ will be called code words. Since codes are mostly used to represent individual symbols of an input alphabet as distinct strings over the output alphabet, the following property is desired in the first step.

Definition 7.2 *(Nonsingular code):* A code $C : \mathbb{X} \to \mathbb{Y}^*$ is called nonsingular if for $x \neq x'$ and $x, x' \in \mathbb{X}$, we have $C(x) \neq C(x')$.

To recall, more generally, an injection $f : \mathbb{X} \to \mathbb{Y}$ is a mapping such that $f(x) \neq f(x')$ for any $x, x' \in \mathbb{X}$ such that $x \neq x'$.

Moreover, since we would like to use codes to represent sequences of input symbols as sequences of output symbols, it is convenient to define also this concept.

Definition 7.3 *(Code extensions):* Consider one-sided infinite sequences, written as $(x_i)_{i \in \mathbb{N}} = x_1 x_2 \cdots \in \mathbb{X}^{\mathbb{N}}$, and two-sided infinite sequences, written as $(x_i)_{i \in \mathbb{Z}} = \cdots x_{-1} x_0 . x_1 x_2 \cdots \in \mathbb{X}^{\mathbb{Z}}$ – mind the bold-face dot between the zeroth and the first symbol in the latter case. For a code $C : \mathbb{X} \to \mathbb{Y}^*$, we define its extensions $C^* : \mathbb{X}^* \to \mathbb{Y}^*$, $C^{\mathbb{N}} : \mathbb{X}^{\mathbb{N}} \to \mathbb{Y}^{\mathbb{N}}$, and $C^{\mathbb{Z}} : \mathbb{X}^{\mathbb{Z}} \to \mathbb{Y}^{\mathbb{Z}}$ as concatenations

$$C^*(x_1^n) := C(x_1)C(x_2)\cdots C(x_n) \tag{7.1}$$

$$C^{\mathbb{N}}((x_i)_{i\in\mathbb{N}}) := C(x_1)C(x_2)\cdots \tag{7.2}$$

$$C^{\mathbb{Z}}((x_i)_{i\in\mathbb{Z}}) := \cdots C(x_{-1})C(x_0).C(x_1)C(x_2)\cdots \tag{7.3}$$

where $x_i \in \mathbb{X}$.

As we have indicated, the main practical purpose of coding is to transmit some representations of strings written with symbols from an input alphabet through a communication channel which passes only strings consisting of symbols from a smaller output alphabet. Thus, the idea of a particularly good code is that we should be able to reconstruct coded symbols x_i from the concatenation of their code words $C(x_i)$. Formally speaking, the following property is desired.

Definition 7.4 (*Uniquely decodable code*): A code $C : \mathbb{X} \to \mathbb{Y}^*$ is called uniquely decodable if its finite extension $C^* : \mathbb{X}^* \to \mathbb{Y}^*$ is nonsingular.

The first class of examples of uniquely decodable codes is the class of comma-separated codes. The right metaphor for this construct is that there is a special symbol which like a comma separates the concatenated code words.

Definition 7.5 (*Comma-separated code*): A code $C : \mathbb{X} \to \mathbb{Y}^*$ is called a comma-separated code if $C(x) = G(x)c$, where $c \in \mathbb{Y}$ and $G : \mathbb{X} \to (\mathbb{Y}\backslash\{c\})^*$ is a nonsingular code.

Theorem 7.1: *Any comma-separated code is uniquely decodable.*

Proof: For a comma-separated code C, let us decompose $C(x) = G(x)c$. We first observe that $C(x_1)\cdots C(x_n) = C(x_1')\cdots C(x_m')$ holds only if $n = m$ (we have the same number of c's on both sides of the equality) and $G(x_i) = G(x_i')$ for $i = 1, \dots, n$. Since G is a nonsingular code, hence string $C(x_1)\cdots C(x_n)$ is the only image of sequence $(x_1, \dots, x_n)$ under the mapping C^*. Hence, code C is uniquely decodable. □

The second recipe for producing a uniquely decodable code is to fix the length of code words. In this way, we obtain fixed length codes.

Definition 7.6 (*Fixed-length code*): Let n be a natural number. A code $C : \mathbb{X} \to \mathbb{Y}^n$ is called a fixed-length code if it is nonsingular.

Theorem 7.2: *Any fixed-length code is uniquely decodable.*

Proof: Consider a fixed-length code C. We observe that

$$C(x_1) \cdots C(x_n) = C(x_1') \cdots C(x_m') \qquad (7.4)$$

holds only if $n = m$ (we have the same length of strings on both sides of the equality) and $C(x_i) = C(x_i')$ for $i = 1, \ldots, n$. Because code C is nonsingular, hence string $C(x_1) \cdots C(x_n)$ is the only image of $(x_1, \ldots, x_n)$ under the mapping C^*. Hence, code C is uniquely decodable. □

The third recipe for producing a uniquely decodable code makes use of the concept of a prefix-free set. A prefix-free set is such a set of strings that none of its elements is a prefix of any other element.

Definition 7.7 *(Prefix-free set):* A prefix-free set $A \subseteq \mathbb{Y}^*$ is a set of strings such that for any $a_1, a_2 \in A$ condition $a_1 \neq a_2$ implies $a_1 \neq a_2 u$ for any string $u \in \mathbb{Y}^+$.

A prefix-free code is a nonsingular code with code words forming a prefix-free set. It turns out that these codes are also uniquely decodable.

Definition 7.8 *(Prefix-free code):* A code $C : \mathbb{X} \to A$ is called a prefix-free code if the image A is a prefix-free set and C is nonsingular.

Theorem 7.3: *Any prefix-free code is uniquely decodable.*

Proof: Let C be a prefix-free code. We observe that $C(x)w = C(x')w'$ holds if and only if $x = x'$ and $w = w'$. Hence, equality $C(x_1) \cdots C(x_n) = C(x_1') \cdots C(x_m')$ holds if and only if $x_i = x_i'$ and $n = m$. Thus, code C is uniquely decodable. □

Now, we will present two useful codes for the set of natural numbers ($\mathbb{X} = \mathbb{N}$). First, we will consider the ordinary binary expansions $B(n)$ of natural numbers, an example given in Table 7.1. Code $B : \mathbb{N} \to \{0, 1\}^*$ is nonsingular but is not uniquely decodable. In particular, it is not comma-separated, fixed-length, or prefix-free.

There exists, however, a pretty good prefix-free, hence uniquely decodable, code for natural numbers called the Elias omega code (Elias, 1975). The algorithm for the Elias omega encoding is as follows:

(1) Put 0 at the end of the code.
(2) If the coded number is 1, stop. Otherwise, write the binary expansion $B(n)$, where n is the coded number, before the code.
(3) Repeat the previous step with the coded number equal to the number of digits written in the previous step minus 1.

Table 7.1 Examples of code words $B(n)$ and $E(n)$.

n	Binary expansion $B(n)$	Elias omega code word $E(n)$
1	1	0
2	10	10 0
3	11	11 0
4	100	10 100 0
5	101	10 101 0
6	110	10 110 0
7	111	10 111 0
8	1000	11 1000 0
...	...	...

In this way, we obtain the Elias code words $E(n)$, an example given also in Table 7.1. A shown by Elias (1975), the Elias omega code $E : \mathbb{N} \to \{0, 1\}^*$ is prefix-free.

Subsequently, let us estimate the lengths of codes $B(n)$ and $E(n)$. We have $|B(n)| = 1 + \lfloor \log n \rfloor$, whereas

$$|E(n)| = \begin{cases} |B(n)| + E(|B(n)| - 1), & n > 1 \\ 1, & n = 1 \end{cases} \tag{7.5}$$

We may define the iterated binary logarithm

$$\log^* n := \begin{cases} 1 + \lfloor \log n \rfloor + \log^* \lfloor \log n \rfloor, & n > 1 \\ 1, & n = 1 \end{cases} \tag{7.6}$$

Then $|E(n)| = \log^* n$. We also have inequality $|E(n)| < \log n + 2 \log \log n$ for all but finitely many natural numbers n.

It is of a practical importance to ask what is the shortest code on average to encode a given set of symbols, where the symbols appear with some given probabilities. Thus, for a probability distribution $p : \mathbb{X} \to \mathbb{R}$, we will be interested in the expected code length $\sum_{x \in \mathbb{X}} p(x) |C(x)|$. Specifically, we want to minimize this quantity for a given probability distribution. We note that both comma-separated codes and fixed-length codes have some advantages and drawbacks. If certain symbols appear more often than the other then comma-separated codes allow to code them as shorter strings and thus to spare space. On the other hand, if all symbols are equally probable then a fixed-length code without a comma occupies less space than the same code with a comma added. In contrast, as we will see, prefix-free codes allow us to encode a given random variable optimally.

There are two important results on our way, which we will state for binary codes to fix our attention. The first result says that each uniquely decodable code corresponds to some incomplete distribution – this correspondence is called the Kraft inequality. The subsequent proof of the first result is due to Brockway McMillan.

Theorem 7.4 (*Kraft inequality*): *For any uniquely decodable code* $C : \mathbb{X} \to \{0, 1\}^*$ *we have the inequality*

$$\sum_{x \in \mathbb{X}} 2^{-|C(x)|} \leq 1 \tag{7.7}$$

Proof: Consider an arbitrary $L \in \mathbb{N}$. Let $a(m, n, L)$ denote the number of sequences $(x_1, \ldots, x_n)$ such that $|C(x_i)| \leq L$ and the length of $C^*(x_1, \ldots, x_n)$ equals m. We have

$$\left(\sum_{x : |C(x)| \leq L} 2^{-|C(x)|} \right)^n = \sum_{m=1}^{nL} a(m, n, L) 2^{-m} \tag{7.8}$$

Because the code is uniquely decodable, we have $a(m, n, L) \leq 2^m$. Therefore

$$\sum_{x : |C(x)| \leq L} 2^{-|C(x)|} \leq (nL)^{1/n} \xrightarrow{n \to \infty} 1 \tag{7.9}$$

Letting $L \to \infty$, we obtain (7.7). □

The second result states that each incomplete distribution of form $\{2^{-l(i)}\}$ where $l(i)$ are natural numbers corresponds to some prefix-free code. The proof of the second result is due to Nicholas J. Pippenger.

Theorem 7.5 (*Inverse Kraft inequality*): *Let* $\mathbb{X} = \{1, 2, \ldots, N\}$ *or* $\mathbb{X} = \mathbb{N}$. *If function* $l : \mathbb{X} \to \mathbb{N}$ *satisfies inequality*

$$\sum_{i \in \mathbb{X}} 2^{-l(i)} \leq 1 \tag{7.10}$$

then we may construct a prefix-free code $C : \mathbb{X} \to \{0, 1\}^*$ *such that* $|C(i)| = l(i)$.

Proof: For a string $u \in \{0, 1\}^*$, let us denote the cylinder set

$$[u] := \{x_1^\infty \in \{0, 1\}^{\mathbb{N}} : x_1^n = u \text{ for } n = |u|\} \tag{7.11}$$

We observe that code C is prefix-free if and only if cylinder sets $[C(i)]$ and $[C(n)]$ are disjoint for $i \neq n$.

In the following we define code C by iteration as follows. First, we denote unions of cylinder sets $[C(i)]$ excluded before the nth iteration as $N(1) := \emptyset$ and $N(n) := \bigcup_{i=1}^{n-1} [C(i)]$ for $n > 1$. Next, we define $C(n)$ as the *first* element of set $\{0, 1\}^{l(n)}$ in the lexicographic order $\prec$ such that sets $[C(n)]$ and $N(n)$ are disjoint. It is obvious that code C defined in this way is prefix-free and satisfies $|C(n)| = l(n)$, as long as strings $C(n)$ with the requested property exist.

Now we will show that string $C(n)$ with the requested property exists if inequality (7.10) is satisfied. The proof of existence rests on this fact, which can be shown easily by induction on n:

(*) The difference $\{0, 1\}^{\mathbb{N}} \setminus N(n)$ is a finite union $\bigcup_{i=1}^{k} [u_i]$ of cylinder sets such that $[u_i] \cap [u_j] = \emptyset$, $u_i \prec u_j$, and $|u_i| > |u_j|$ for $i < j$.

Let $m = |u_k|$ for the above representation of $\{0, 1\}^{\mathbb{N}} \setminus N(n)$. By property (*), we obtain strict inequality

$$2^{-m+1} = 2^{-m} + 2^{-m-1} + \cdots > 1 - \sum_{i=1}^{n-1} 2^{-l(i)} \tag{7.12}$$

On the other hand, by the Kraft inequality (7.10), we have

$$1 - \sum_{i=1}^{n-1} 2^{-l(i)} \geq 2^{-l(n)} \tag{7.13}$$

Hence, we obtain that $2^{-l(n)} < 2^{-m+1}$, i.e. we have $2^{-l(n)} \leq 2^{-m}$. Thus, string $C(n)$ with the requested property exists. $\qquad \square$

By the above two results and the Gibbs inequality (Theorem 3.18), we may relate the problem of coding to Shannon entropy.

Theorem 7.6 *(Source coding inequality):* *For any uniquely decodable code C :* $\mathbb{X} \to \{0, 1\}^*$, *the expected length of the code satisfies inequality*

$$\sum_{x \in \mathbb{X}} p(x)|C(x)| \geq H(p) \tag{7.14}$$

where $H(p)$ is the Shannon entropy of the probability distribution $p : \mathbb{X} \to \mathbb{R}$.

Proof: Let us introduce function $q(x) = 2^{-|C(x)|}$. By the Kraft inequality (Theorem 7.4), function $q(x)$ is a semidistribution. Hence, by positivity of the Kullback–Leibler divergence (Theorem 3.18) we obtain the claim. $\qquad \square$

It turns out as well that the equality in the source coding inequality may be approximately satisfied. The respective code is called the Shannon–Fano code.

Definition 7.9 *(Shannon–Fano code):* A prefix-free code $C_p : \mathbb{X} \to \{0, 1\}^*$ is called a Shannon–Fano code for a probability distribution $p : \mathbb{X} \to \mathbb{R}$ if

$$|C_p(x)| = \lceil -\log p(x) \rceil \tag{7.15}$$

Theorem 7.7:   *The Shannon–Fano code exists for any discrete random variable and satisfies*

$$H(p) \leq \sum_{x \in \mathbb{X}} p(x)|C_p(x)| \leq H(p) + 1 \tag{7.16}$$

Proof: We have

$$\sum_{x \in \mathbb{X}} 2^{-\lceil -\log p(x) \rceil} \leq \sum_{x \in \mathbb{X}} 2^{\log p(x)} \leq 1 \tag{7.17}$$

Hence, Shannon–Fano codes exist by the inverse Kraft inequality (Theorem 7.5) – when we choose some one-to-one mapping $\mathbb{X} \to \{1, 2, \dots, N\}$ or $\mathbb{X} \to \mathbb{N}$. Consequently, inequality (7.16) follows by

$$-\log p(x) \leq |C_p(x)| \leq -\log p(x) + 1 \tag{7.18}$$

□

There is a code construction for a finite input alphabet, called the Huffman code, which is slightly better than the Shannon–Fano code. We will not discuss it here. An interested reader is referred to Csiszár and Körner (2011) and Cover and Thomas (2006). In Section 7.3, we will see an application of the above results to the prefix-free Kolmogorov complexity. However, first we have to define the latter concept.

7.2 Kolmogorov Complexity

In plain words, the Kolmogorov complexity of a string is defined as the length of the shortest program that computes the string. As such, the Kolmogorov complexity constitutes an algorithmic analogue of the Shannon entropy. The respective theory, called the algorithmic information theory, was founded by Solomonoff (1964) and Kolmogorov (1965) and later pursued among others by Zvonkin and Levin (1970), Gács (1974), and Chaitin (1975a, b). A modern introduction to the algorithmic information theory was written by Li and Vitányi (2008). Although the Kolmogorov complexity is not computable, it sets limits to what can be computed and gives rise to quite illuminating conceptual considerations.

Speaking slightly more formally, the Kolmogorov complexity of a string is usually defined as the length of the shortest program for a Turing machine such that the machine prints out the string and halts, after a finite number of computation steps. The Turing machine, which constitutes a standard mathematical model of the all-purpose computer, is defined as a deterministic finite-state automaton which moves along one or more infinite tapes filled with symbols from a fixed finite alphabet and which may read and write individual symbols. The finite-state automaton has a distinct start state, from which the computation begins, and a distinct halt state, at which the computation ends (Turing, 1936a, b).

The concrete value of the Kolmogorov complexity depends on the used Turing machine but many properties of the Kolmogorov complexity are universal. Some particular definition of the Kolmogorov complexity, called the prefix-free Kolmogorov complexity, is convenient to discuss links with the Shannon entropy. This sort of Kolmogorov complexity applies a particular class of Turing machines, called prefix-free Turing machines. The fundamental idea is to force that the accepted programs form a prefix-free set, so that they naturally form a prefix-free code for the described objects. This construction can be carried out in a few equivalent ways. Here we will use slightly nonstandard constructions motivated by our later convenience, see Chaitin (1975a).

The first definition is short but kind of opaque.

Definition 7.10 *(Prefix-free Turing machine):* A prefix-free Turing machine (with three binary tapes) is defined as a 4-tuple $S = (Q, s, h, \delta)$, where

(1) Q is a finite, nonempty set of states,
(2) $s \in Q$ is the start state,
(3) $h \in Q$ is the halt state,
    $\delta : Q\backslash\{h\} \times \{0, 1\} \times \{0, 1, B\} \times \{0, 1, B\} \to Q \times \{0, R\} \times \{L, 0, R\} \times \{0, 1, B\} \times \{L, 0, R\}$ is a function called the transition function.

The set of such prefix-free Turing machines is denoted $\mathcal{S}$.

The above formal description translates to the machine operation in the following way. The machine is given three infinite tapes:

(1) A unidirectional read-only tape $(X_k)_{k\in\mathbb{N}}$ filled with symbols 0 and 1.
(2) A bidirectional read-only tape $(Y_i)_{i\in\mathbb{Z}}$ filled with symbols 0, 1, and B.
(3) A bidirectional read-write tape $(Z_i)_{i\in\mathbb{Z}}$ filled with symbols 0, 1, and B.

The machine moves in discrete steps along these tapes in the way prescribed by the transition function. The initial machine state is the start state s and the machine reads symbols X_1, Y_1, and Z_1. The further machine operations are specified as follows: if the machine is in state a and reads symbol x from tape $(X_k)_{k\in\mathbb{N}}$, symbol y from tape $(Y_i)_{i\in\mathbb{Z}}$, and symbol z from tape $(Z_i)_{i\in\mathbb{Z}}$ then for

$$\delta(a, x, y, z) = (a', m_x, m_y, z', m_z) \tag{7.19}$$

the machine does these operations:

(1) The machine moves along tape $(X_k)_{k\in\mathbb{N}}$ one symbol to the right if $m_x = R$ or it does not move if $m_x = 0$.

(2) The machine moves along tape $(Y_i)_{i \in \mathbb{Z}}$ one symbol to the left if $m_y = L$ or one symbol to the right if $m_y = R$ or it does not move if $m_y = 0$.
(3) The machine writes symbol z' instead of symbol z on tape $(Z_i)_{i \in \mathbb{Z}}$.
(4) The machine moves along tape $(Z_i)_{i \in \mathbb{Z}}$ one symbol to the left if $m_z = L$ or one symbol to the right if $m_z = R$ or it does not move if $m_z = 0$.
(5) The machine assumes state a' in the next step.

The above steps are undertaken until the machine reaches the halt state h. Then the computation stops.

Let us write the set of both finite and infinite sequences as $\mathbb{Y}^\# := \mathbb{Y}^* \cup \mathbb{Y}^\mathbb{N}$. In the second definition, we make it formally precise how the machine computes a particular string and halts.

Definition 7.11 *(Halting):* We say that machine $S \in \mathcal{S}$ halts in a finite time $t \in \mathbb{N}$ on a program $z \in \{0,1\}^*$ with an oracle $q \in \{0,1\}^\#$ and returns a string $w \in \{0,1\}^*$ if:

(A) The machine in the start state reads symbols X_1, Y_1, and Z_1, and the initial state of the tapes is: $X_1^{|z|} = z, Y_i = B$ for $i < 0, Y_0^{|q|} = 0q$ and $Y_i = B$ for $i > |q|$ if q is a string or $Y_0^\infty = 1q$ if q is an infinite sequence, and $Z_i = B$ for all $i \in \mathbb{Z}$.
(B) The machine reaches the halt state after t steps and the machine in the halt state reads symbol $X_{|z|}$, whereas the final state of the tape $(Z_i)_{i \in \mathbb{Z}}$ is $Z_{j+1}^{j+|w|} = w$, with $Z_i = B$ for $i < j+1$ and $i > j+|w|$ for some j.

Assuming that condition (A) is satisfied, we write

$$S(z|q) = \begin{cases} w, & \text{if (B) holds for some } t \in \mathbb{N} \text{ and } w \in \{0,1\}^* \\ \perp, & \text{else} \end{cases} \tag{7.20}$$

$$T_S(z|q) = \begin{cases} t, & \text{if (B) holds for some } t \in \mathbb{N} \text{ and } w \in \{0,1\}^* \\ \perp, & \text{else} \end{cases} \tag{7.21}$$

When $S(z|q) = \perp$, we say that the machine does not halt on input (z,q).

The name "prefix-free Turing machine" is justified by the following property that the set of programs on which the machine halts is prefix-free.

Theorem 7.8: *For any machine $S \in \mathcal{S}$ and an oracle $q \in \{0,1\}^\#$, set*

$$\mathbf{A}_S(q) := \{z \in \{0,1\}^* : S(z|q) \neq \perp\} \tag{7.22}$$

is prefix-free.

Proof: A prefix-free Turing machine can move along the tape $(X_k)_{k \in \mathbb{N}}$ only in one direction. Hence, if it halts on a program z, then it cannot read the tape any further and halt on a program zu, where $|u| \geq 1$. $\quad\square$

The elements of sets $\mathbf{A}_S(q)$ will be called admissible or self-delimiting programs.

Subsequently, we will define the Kolmogorov complexity of a string as the length of the shortest program that computes the string.

Definition 7.12 *(Prefix-free Kolmogorov complexity)*: The prefix-free Kolmogorov complexity $K_S(w|q)$ of a string $w \in \{0,1\}^*$ given an oracle $q \in \{0,1\}^\#$ and a machine $S \in \mathcal{S}$ is defined as

$$\mathbf{H}_S(w|q) := \min_{z \in \{0,1\}^*} \{|z| : S(z|q) = w\} \tag{7.23}$$

We also use notation $\mathbf{H}_S(w) := \mathbf{H}_S(w|\lambda)$.

In parallel, we will introduce universal machines, which are machines that, given a special simulation program, can simulate any other machine.

Definition 7.13 *(Universal machine)*: A machine $S \in \mathcal{S}$ is called universal if for each machine $V \in \mathcal{S}$ there exists a string $u \in \{0,1\}^*$ such that

$$S(uz|q) = V(z|q) \tag{7.24}$$

for all $z \in \{0,1\}^*$ and $q \in \{0,1\}^\#$.

Theorem 7.9: *There exist universal machines.*

Proof: The formal proof is not complicated but tedious. Using a higher level language of description, the string u can be chosen as a machine-readable formal description of the machine $V = (Q', s', h', \delta')$. Given such a string u, machine S compiles a model of machine V and afterwards simulates the actions of machine V on input (z, q). $\square$

One of reasons for considering universal machines is that the Kolmogorov complexity for a universal machine is determined up to an additive constant. In the following, we will write $f(q) \overset{+}{<} g(q)$ and $f(q) \overset{+}{>} g(q)$ if there exists a constant $c \geq 0$ such that $f(q) \leq g(q) + c$ and $f(q) \geq g(q) + c$ holds respectively for all arguments q. We will write $f(q) \overset{+}{=} g(q)$ when we have both $f(q) \overset{+}{<} g(q)$ and $f(q) \overset{+}{>} g(q)$.

Theorem 7.10 *(Invariance theorem)*: *For any universal machine $S \in \mathcal{S}$ and any machine $V \in \mathcal{S}$, we have*

$$\mathbf{H}_S(w|q) \overset{+}{<} \mathbf{H}_V(w|q) \tag{7.25}$$

In particular, for any two universal machines $S, S' \in \mathcal{S}$, we have

$$\mathbf{H}_S(w|q) \overset{+}{=} \mathbf{H}_{S'}(w|q) \tag{7.26}$$

Proof: We have $S(uz|q) = V(z|q)$ for a certain string u. Hence, $\mathbf{H}_S(w|q) \leq \mathbf{H}_V(w|q) + |u|$. The second bound is a special case of the first bound. $\square$

Thus, for further considerations we will choose a certain universal machine as a reference to determine the prefix-free Kolmogorov complexity.

Definition 7.14 *(Prefix-free Kolmogorov complexity)*: Let $S \in \mathcal{S}$ be a certain reference universal machine. We will write

$$\mathbf{H}(w|q) := \mathbf{H}_S(w|q) \qquad (7.27)$$

We also use notation $\mathbf{H}(w) := \mathbf{H}(w|\lambda)$.

For the vast majority of strings, their Kolmogorov complexity is quite large since there are not so many short programs. A string w will be called (c, q)-incompressible if $\mathbf{H}(w|q) \geq |w| - c$. In view of the following result, at least a half of strings of length n is $(1, q)$-incompressible.

Theorem 7.11: *There exist at least $2^n - 2^{n-c} + 1$ distinct (c, q)-incompressible strings of length n.*

Proof: The number of (c, q)-incompressible strings of length n is greater than the difference between the number of all strings of length n and the number of distinct programs of length strictly smaller than $n - c$. There exist at most $2^{n-c} - 1$ distinct programs of length strictly smaller than $n - c$ and there exists 2^n distinct strings of length n. Hence follows the claim. $\square$

From the above result, we can see that compressibility is a property of only quite special strings. As we will see in Section 7.4, however, it is not possible to give a constructive example of a particular incompressible string.

Subsequently, analogously to the Kolmogorov complexity of strings conditioned on arbitrary sequences, we would like to discuss also the Kolmogorov complexity of some discrete objects conditioned on some nondiscrete objects. Examples of the discrete objects are natural numbers, strings over finite alphabets, and tuples thereof, such as rational numbers. Examples of nondiscrete objects are infinite sequences, real numbers, and tuples thereof, such as vectors, complex numbers, or real-valued functions of discrete objects. The distinction between the considered discrete objects and the nondiscrete objects is that the discrete objects are countably many and can be mapped one-to-one to finite sequences, whereas the considered nondiscrete objects are uncountably many but can be mapped one-to-one to infinite sequences.

Formally, let $\mathbb{A}$ be some set of discrete objects that can be mapped by a one-to-one function $\phi : \mathbb{A} \to \{0,1\}^*$ to strings and let $\mathbb{B}$ be some set of nondiscrete objects that can be mapped by a one-to-one function $\psi : \mathbb{B} \to \{0,1\}^{\#}$ to finite or infinite sequences. We have some freedom in choosing functions ϕ and ψ but they should be in some informal sense simple so that no pathologies arise in the following reasonings, concerning recursive and computable functions in particular. Subsequently, we will adopt these formal definitions of functions ϕ and ψ as well as sets $\mathbb{A}$ and $\mathbb{B}$.

Definition 7.15 *(Object encoding)*: Let $B(n)$ be the binary expansion of the natural number $n \in \mathbb{N}$. For a string $w \in \{0,1\}^*$, we also denote $\bar{w} := E(n)w$, where $n = |w| + 1$ and $E(n)$ is the Elias omega code word for number $n \in \mathbb{N}$. For $\chi :$ $\mathbb{A} \to \{0,1\}^*$, we also write $\bar{\chi}(a) := \bar{w} \iff \chi(a) = w$. We adopt these definitions:

(1) For strings $w \in \mathbb{A}_0 := \{0,1\}^*$, we use identity function $\phi_0(w) := w$.
(2) For natural numbers $n \in \mathbb{A}_1 := \mathbb{N}$, we use the binary expansion stripped of the redundant initial symbol, $\phi_1(n) := w \iff B(n) = 1w$.
(3) For rational numbers $q \in \mathbb{A}_2 := \mathbb{Q}$, we use

$$\phi_2(q) := \frac{s+1}{2}\bar{\phi}_1(n+1)\phi_1(m) \tag{7.28}$$

where $q = sn/m$, $s \in \{-1,1\}$, $n \in \mathbb{N} \cup \{0\}$, and $m \in \mathbb{N}$.
(4) For tuples $a = (a_1, \dots, a_n) \in \mathbb{A} := (\mathbb{A}_0 \cup \mathbb{A}_1 \cup \mathbb{A}_2)^*$, we use

$$\phi(a) := \bar{\phi}_1(n+1)\bar{\phi}_1(i_1)\bar{\phi}_{i_1}(a_1) \cdots \bar{\phi}_1(i_{n-1})\bar{\phi}_{i_{n-1}}(a_{n-1})\bar{\phi}_1(i_n)\phi_{i_n}(a_n) \tag{7.29}$$

where $a_k \in \mathbb{A}_{i_k}$.
(5) For discrete tuples treated as nondiscrete objects $a \in \mathbb{B}_0 := \mathbb{A}$, we use

$$\psi_0(a) := \bar{\phi}(a)00 \cdots \tag{7.30}$$

(6) For infinite sequences, $z \in \mathbb{B}_1 := \{0,1\}^{\mathbb{N}}$, we use identity function $\psi_1(z) := z$.
(7) For real numbers $r \in \mathbb{B}_2 := \mathbb{R}$, we use

$$\psi_2(r) := \frac{s+1}{2}\bar{\phi}_1(n+1)t_1 t_2 \cdots \tag{7.31}$$

where $r = sn + t$, $s \in \{-1,1\}$, $n \in \mathbb{N} \cup \{0\}$, $t = \sum_{k=1}^{\infty} 2^{-k}t_k$, and $t_i \in \{0,1\}$ are the least binary digits that satisfy this relationship.
(8) For finite nondiscrete tuples $b = (b_1, \dots, b_n) \in \mathbb{B} = (\mathbb{B}_0 \cup \mathbb{B}_1 \cup \mathbb{B}_2)^*$, we use

$$\psi(b) := 0\bar{\phi}_1(n+1)t_1 t_2 \cdots \tag{7.32}$$

where $b_k \in \mathbb{B}_{i_k}$, $\psi_{i_k}(b_k) = z_{i1}z_{i2} \cdots$ and $t_p = z_{ij}$ if $p = 2^i 3^j$ for some $i \in \{1, \dots, n\}$ and $j \in \mathbb{N}$, whereas $t_p = 0$ otherwise.

(9) For infinite nondiscrete tuples $b = (b_1, b_2, ...) \in \mathbb{B} = (\mathbb{B}_0 \cup \mathbb{B}_1 \cup \mathbb{B}_2)^\mathbb{N}$, we use

$$\psi(b) := 1t_1t_2 \cdots \tag{7.33}$$

where $b_k \in \mathbb{B}_{i_k}$, $\psi_{i_k}(b_k) = z_{i1}z_{i2}\cdots$ and $t_p = z_{ij}$ if $p = 2^i 3^j$ for some $i, j \in \mathbb{N}$, where as $t_p = 0$ otherwise.

We stress that under any reasonable convention the undefined value $\perp$ remains a special type of an object which is never encoded, i.e. $\perp \notin \mathbb{A}$ and $\perp \notin \mathbb{B}$.

Definition 7.16 *(Halting)*: For some fixed one-to-one mappings $\phi : \mathbb{A} \to \{0,1\}^*$ and $\psi : \mathbb{B} \to \{0,1\}^\#$ we will say that machine $S \in \mathcal{S}$ halts on a program $z \in \{0,1\}^*$ with an oracle $b \in \mathbb{B}$ and returns an object $a \in \mathbb{A}$ if $S(z|\psi(b)) = \phi(a)$. Overloading the notation, we will write this fact as

$$S(z|b) = a \tag{7.34}$$

When $S(z|\psi(b)) = \perp$, we will say that the machine does not halt on input (z, b) and we will write it as

$$S(z|b) = \perp \tag{7.35}$$

Analogously, we will adjust all previous definitions and notations in this section, admitting writing b instead of $\psi(b)$ and a instead of $\phi(a)$.

Now we can give some simple bounds for the prefix-free Kolmogorov complexity of discrete objects.

Theorem 7.12: *For any discrete object,*

$$\mathbf{H}(a|b) \overset{+}{<} \mathbf{H}_0(a) := |\bar{\phi}(a)| \tag{7.36}$$

Proof: Let $w = \phi(a)$ and $n = |w| + 1$. There is a machine $V \in \mathcal{S}$ which satisfies $V(E(n)w|b) = w$, where $E(n)$ is the Elias omega code word for number n, with length $|E(n)| = \log^* n$. The machine generates w as follows. First it reads the Elias omega code word for number n from the unidirectional tape. Then it copies string w from the unidirectional tape to the output tape. After copying the last symbol of w, the machine halts because it knows the length of w from reading the code word $E(n)$. Hence, we obtain the desired claim by $\mathbf{H}(a|b) \overset{+}{<} \mathbf{H}_V(a|b) = \mathbf{H}_0(a)$. $\square$

Theorem 7.13: *For discrete objects u and w we have inequalities:*

(1) $\mathbf{H}(w|u, b) \overset{+}{<} \mathbf{H}(w|b) \overset{+}{<} \mathbf{H}(u, w|b)$.

(2) $\mathbf{H}(u, w|b) \overset{+}{<} \mathbf{H}(u|b) + \mathbf{H}(w|u, b) \overset{+}{<} \mathbf{H}(u|b) + \mathbf{H}(w|b)$.

Proof:

(1) We have $\mathbf{H}(w|u, b) \overset{+}{<} \mathbf{H}(w|b)$ because a certain program that computes w given u and b has form: *Ignore u and execute the shortest program that computes w.*

Moreover, $\mathbf{H}(w|b) \overset{+}{<} \mathbf{H}(u, w|b)$ because a certain program that computes w has form: *Execute the shortest program that computes pair (u, w) and then compute w from (u, w).*

(2) We have $\mathbf{H}(u, w|b) \overset{+}{<} \mathbf{H}(u|b) + \mathbf{H}(w|u, b)$ because a certain program that computes pair (u, w) has form: *Execute the shortest program that computes u given b and the shortest program that computes w given u and b and from that compute pair (u, w).*

The second inequality follows from $\mathbf{H}(w|u, b) \overset{+}{<} \mathbf{H}(w|b)$. □

Now we define recursive discrete functions. A discrete function $f : \mathbb{B} \to \mathbb{A}$ is called total, whereas a discrete function $f : \mathbb{B} \to \mathbb{A} \cup \{\perp\}$ is called partial.

Definition 7.17 *(Recursive partial discrete function):* A partial discrete function $f : \mathbb{B} \to \mathbb{A} \cup \{\perp\}$ is called recursive if there exists a program $z \in \{0, 1\}^*$ for the reference universal Turing machine $S \in \mathcal{S}$ such that $S(z|b) = f(b)$ for all $b \in \mathbb{B}$.

Theorem 7.14 *(Argument raising):* *For a recursive partial discrete function $f :$ $\mathbb{A} \times \mathbb{B} \to \mathbb{A} \cup \{\perp\}$ there exist a string $u \in \{0, 1\}^*$ for the reference universal Turing machine $S \in \mathcal{S}$ such that $S(u\bar{\phi}(a)|b) = f(a, b)$ for all $a \in \mathbb{A}$ and $b \in \mathbb{B}$.*

Proof: There is a machine $V \in \mathcal{S}$ which reads the self-delimiting program $\bar{\phi}(a)$ from the unidirectional tape and then simulates machine S given program z and oracle $\psi(a, b)$, where z is some program from Definition 7.17. Thus, $V(\bar{\phi}(a)|b) = f(a, b)$. Since machine S is universal, there also exists a string $u \in \{0, 1\}^*$ such that $S(uw|b) = V(w|b)$ for all $w \in \{0, 1\}^*$ and $b \in \mathbb{B}$. Hence, $S(u\bar{\phi}(a)|b) = f(a, b)$ holds in particular. □

Recursive partial discrete functions can be also treated as discrete objects.

Definition 7.18 *(Object encoding):* For a recursive partial discrete function $f :$ $\mathbb{B} \to \mathbb{A} \cup \{\perp\}$, we use encoding $\phi_3(f) := z$, where z is some program from Definition 7.17. Subsequently, we extend the definitions of functions ϕ and ψ as well as sets $\mathbb{A}$ and $\mathbb{B}$ appropriately.

Having adopted this convention, we may demonstrate this bound.

Theorem 7.15: *For a recursive partial discrete function f and a discrete object w, we have*

$$H(f(w)|b) \overset{+}{<} H(w|b) + H(f|b) \tag{7.37}$$

Proof: The inequality follows from the fact that some program that computes $f(w)$ has form: *Execute the shortest program that computes object w, compile function f from its shortest description, and apply function f to object w.* □

There exists also nonrecursive functions. A celebrated example of a nonrecursive function is the solution of the halting problem.

Theorem 7.16 *(Halting problem):* *Total function*

$$h(z, b) := \begin{cases} 1, & \text{if } S(z|b) \neq \bot \\ 0, & \text{if } S(z|b) = \bot \end{cases} \tag{7.38}$$

is not recursive.

Proof: Suppose that function h is recursive. Then we may define the recursive partial function

$$g(z) := \begin{cases} 0, & \text{if } h(z, z) = 0 \\ \bot, & \text{else} \end{cases} \tag{7.39}$$

Since the partial function g is recursive, there exists a program e such that $S(e|b) = g(b)$ for all b. Let us inspect the value of $h(e, e)$. On the one hand, if $h(e, e) = 0$ then $g(e) = S(e, e) = \bot$ and so $h(e, e) \neq 0$. On the other hand, if $h(e, e) = 1$ then $g(e) = S(e|e) \neq \bot$ so $h(e, e) = 0$. Since we have obtained a contradiction then h cannot be recursive. □

When dealing with computability of real-valued functions, it is convenient to distinguish three important cases.

Definition 7.19 *(Computable real numbers and real functions):* We adopt these definitions:

(1) A real number $y \in \mathbb{R}$ is called lower semicomputable if there is a recursive function $A_1 : \mathbb{N} \to \mathbb{Q}$ which satisfies $A_1(k + 1) \geq A_1(k)$ and $\lim_{k \to \infty} A_1(k) = y$.

(2) A real number $y \in \mathbb{R}$ is called upper semicomputable if there is a recursive function $A_2 : \mathbb{N} \to \mathbb{Q}$ which satisfies $A_2(x, k + 1) \leq A_2(x, k)$ and $\lim_{k \to \infty} A_2(k) = y$.

(3) A real number $y \in \mathbb{R}$ is called computable if there is a recursive function $A_3 : \mathbb{N} \to \mathbb{Q}$ which satisfies $|y - A_3(k)| < 1/k$.

Analogously, we define:

(1) A real function $f : \mathbb{B} \to \mathbb{R}$ is called lower semicomputable if there is a recursive function $A_1 : \mathbb{B} \times \mathbb{N} \to \mathbb{Q}$ which satisfies $A_1(x, k+1) \geq A_1(x, k)$ and $\lim_{k \to \infty} A_1(x, k) = f(x)$ for all $x \in \mathbb{B}$.

(2) A real function $f : \mathbb{B} \to \mathbb{R}$ is called upper semicomputable if there is a recursive function $A_2 : \mathbb{B} \times \mathbb{N} \to \mathbb{Q}$ which satisfies $A_2(x, k+1) \leq A_2(x, k)$ and $\lim_{k \to \infty} A_2(x, k) = f(x)$ for all $x \in \mathbb{B}$.

(3) A real function $f : \mathbb{B} \to \mathbb{R}$ is called computable if there is a recursive function $A_3 : \mathbb{B} \times \mathbb{N} \to \mathbb{Q}$ which satisfies $|f(x) - A_3(x, k)| < 1/k$ for all $x \in \mathbb{B}$.

We have the following results.

Theorem 7.17: *A recursive function $f : \mathbb{B} \to \mathbb{Q}$ or a rational number $y \in \mathbb{Q}$ considered as a real function or a real number respectively are computable.*

Proof: We put $A_3(x, k) := f(x)$ or $A_3(k) := y$ respectively. □

Theorem 7.18: *A real function or a real number which is both upper and lower semicomputable is computable.*

Proof: We present a proof for the real function. The proof for the real number is analogous. Given the recursive functions $A_1, A_2 : \mathbb{B} \times \mathbb{N} \to \mathbb{Q}$ and a number $k \in \mathbb{N}$, there is a Turing machine which computes the approximations $A_1(x, n)$ and $A_2(x, n)$ for increasing n successively until $A_2(x, n) - A_1(x, n) < 1/k$ and then outputs $A_3(x, k) := \frac{1}{2}(A_2(x, n) + A_1(x, n))$. □

Finally, it can be shown that Kolmogorov complexity can be approximated from above but cannot be known exactly.

Theorem 7.19: *Function $\mathbb{A} \times \mathbb{B} \ni (a, b) \mapsto H(a|b) \in \mathbb{N}$ considered as a real function is upper semicomputable but it is not recursive.*

Proof:

(1) By Theorem 7.12, we have $\mathbf{H}(a|b) \leq \mathbf{H}_0(a) + c$ for some constant c. Let $S \in \mathcal{S}$ be the reference universal machine. We may define the time-bounded Kolmogorov complexity

$$\mathbf{H}^t(a|b) := \min_{z \in \{0,1\}^*} \{|z| : S(z|b) = a, z \leq \mathbf{H}_0(a) + c, \ T_S(z|b) \leq t\}$$

(7.40)

Function $\mathbf{H}^t(a|b)$ is recursive since there exists a Turing machine, which given b simulates the operation of machine S in parallel on all programs z shorter than $\mathbf{H}_0(a) + c$ and from this simulation it computes $\mathbf{H}^t(a|b)$. Moreover, we have $\mathbf{H}^{t+1}(a|b) \le \mathbf{H}^t(a|b)$ and $\lim_{t\to\infty}\mathbf{H}^t(a|b) = \mathbf{H}(a|b)$. Hence, function $(a, b) \mapsto \mathbf{H}(a|b)$ is upper semicomputable.

(2) Assume that function $(a, b) \mapsto \mathbf{H}(a|b)$ is recursive. Then we may define the total recursive function

$$g(l|b) := \min\{n \in \mathbb{N} : \mathbf{H}(n|b) \ge l\} \tag{7.41}$$

Since function g is recursive, there exists a string u such that $S(u\bar{\phi}(l)|b) = g(l|b)$ for all natural numbers l for the reference universal machine S. Hence, we obtain

$$l \le \mathbf{H}(n|b) \le |u| + \mathbf{H}_0(l) \tag{7.42}$$

whenever $n = g(l|b)$. Hence, we obtain a contradiction since inequality $l > |u| + \mathbf{H}_0(l)$ holds for sufficiently large l. Thus, function $(a, b) \mapsto \mathbf{H}(a|b)$ is not recursive. □

In the following three sections, we will prove more advanced results. First, we will exhibit some correspondences between Kolmogorov complexity and Shannon entropy. Second, we will shed some light onto limits of mathematics. Third, we will analyze the phenomenon of randomness from an algorithmic perspective.

7.3 Algorithmic Coding Theorems

The prefix-free Kolmogorov complexity resembles the pointwise Shannon entropy both in terms of its algebraic properties and its numerical value. The fundamental reason for these analogies is that the Shannon–Fano code for a given probability semidistribution is effectively computable if the semidistribution is effectively computable, even in the limit, or if the semidistribution is given by the oracle. This section will be devoted to a series of such results.

First of all, by an analogy to the chain rule for Shannon entropy

$$H(X, Y|Z) = H(X|Z) + H(Y|X, Z) \tag{7.43}$$

which follows for any random variables X, Y, and Z by Theorems 5.7 and 5.29, we may formulate this theorem, expressing a chain rule for Kolmogorov complexity.

Theorem 7.20 *(Chain rule):* *For discrete objects u and w, we have*

$$\mathbf{H}(u, w|b) \overset{+}{=} \mathbf{H}(u|b) + \mathbf{H}(w|u, \mathbf{H}(u|b), b) \tag{7.44}$$

We will use Theorem 7.20 to demonstrate theorems about facts and words in Section 8.4. In Theorem 7.20, independently due to Gács (1974) and Chaitin (1975a), it is easy to show that the left hand side is smaller than the right hand side. The proof of the converse inequality is harder but it rests on an interesting series of auxiliary theoretical concepts and results. In the following, we will demonstrate the entire reasoning. We will begin with the concept of recursively enumerable sets, or rather more generally, recursively enumerable set functions.

Definition 7.20 *(Recursively enumerable set function)*: A set function $\mathcal{E} : \mathbb{B} \to 2^{\mathbb{A}}$ is called recursively enumerable if there exists a total recursive function $f : \mathbb{N} \times \mathbb{B} \to \mathbb{A}$ such that $f(\mathbb{N}|b) = \mathcal{E}(b)$ for all $b \in \mathbb{B}$.

We will say informally that function f from the above definition enumerates set function $\mathcal{E}$. This phrasing motivates the following more precise construction.

Definition 7.21 *(Uniquely enumerating function)*: A partial recursive function $g : \mathbb{N} \times \mathbb{B} \to \mathbb{A} \cup \{\bot\}$ is called a uniquely enumerating function if for each $b \in \mathbb{B}$ there exists a set $\mathbb{X}_b^g = \{1, 2, \dots, N_b^g\}$ or $\mathbb{X}_b^g = \mathbb{N}$ such that:

(1) $g(i|b) \neq g(j|b)$ and $g(i|b) \neq \bot$ for $i \neq j$ and $i, j \in \mathbb{X}_b^g$.
(2) $g(i|b) = \bot$ for $i \notin \mathbb{X}_b^g$.

It is quite obvious that any recursively enumerable set function can be enumerated by a uniquely enumerating function.

Theorem 7.21: *If a set function $\mathcal{E} : \mathbb{B} \to 2^{\mathbb{A}}$ is recursively enumerable then there exists a uniquely enumerating function $g : \mathbb{N} \times \mathbb{B} \to \mathbb{A} \cup \{\bot\}$ such that $g(\mathbb{X}_b^g|b) = \mathcal{E}(b)$ for all $b \in \mathbb{B}$.*

Proof: Let $f : \mathbb{N} \times \mathbb{B} \to \mathbb{A}$ such that $f(\mathbb{N}|b) = \mathcal{E}(b)$ for all $b \in \mathbb{B}$ be a total recursive function. For a fixed $b \in \mathbb{B}$, let us define $n_1 = 1$ and let n_{i+1} be the smallest number such n that $f(n|b)$ is different from $f(n_1|b), \dots, f(n_i|b)$ if such a number exists or $n_{i+1} := \bot$ else. Then an example of a uniquely enumerating function is $g(i|b) = f(n_i|b)$ if $n_i \in \mathbb{N}$ and $g(i|b) = \bot$ if $n_i = \bot$. □

These simple observations have a surprising consequence. Namely, we can introduce a generalization of the inverse Kraft inequality (Theorem 7.5) for codes that can be both effectively encoded and decoded.

Theorem 7.22 *(Effective inverse Kraft inequality)*: *Suppose that a uniquely enumerating function $l : \mathbb{N} \times \mathbb{B} \to \mathbb{N} \cup \{\bot\}$ satisfies inequality*

$$\sum_{i \in \mathbb{X}_b^l} 2^{-l(i|b)} \leq 1 \tag{7.45}$$

for each $b \in \mathbb{B}$. Then there exist:

(1) *A partial recursive code* $C : \mathbb{N} \times \mathbb{B} \rightarrow \{0,1\}^* \cup \{\perp\}$ *such that the partial code* $C(\cdot|b)$ *is prefix-free for each* $b \in \mathbb{B}$ *and satisfies* $|C(i|b)| = l(i|b)$ *for all* $i \in \mathbb{X}_b^l$ *and* $C(i|b) = \perp$ *else.*

(2) *A prefix-free Turing machine* $V \in \mathcal{S}$ *such that*

$$V(z|b) = \begin{cases} i, & \text{if } z = C(i|b) \text{ for some } i \in \mathbb{X}_b^l \\ \perp, & \text{else} \end{cases} \tag{7.46}$$

Proof: The existence of the partial recursive code C follows from Theorem 7.5, since the construction in its proof is effective. Subsequently, the machine V may operate in the following way. Given a program z and an oracle b, it constructs code words $C(i|b)$ for increasing numbers i until it encounters a code word equal to program z and then it returns number i. An important technical detail of this algorithm is that program z can be read from the unidirectional tape $(X_k)_{k \in \mathbb{N}}$ incrementally while checking against code words $C(i|b)$. Namely, if $C(i|b) = c_1^n$ for some i then the machine V reads the tape $(X_k)_{k \in \mathbb{N}}$ until position k such that $X_1^{k-1} = c_1^{k-1}$ and $X_k \neq c_k$ or, if there is no such position, the machine V reads the tape $(X_k)_{k \in \mathbb{N}}$ until position n, detects the end of code word $C(i|b)$, returns number i, and halts. $\qquad \square$

As a corollary, we can bound Kolmogorov complexity in terms of apparently obscure recursively enumerable set functions that satisfy the Kraft inequality.

Theorem 7.23 (*Chaitin bound*): *Fix some* $b \in \mathbb{B}$. *Suppose that a set function* $\mathcal{E} :$ $\mathbb{B} \rightarrow 2^{\mathbb{A} \times \mathbb{N}}$ *is recursively enumerable and satisfies*

$$\sum_{(a,n) \in \mathcal{E}(b)} 2^{-n} \leq 1 \tag{7.47}$$

Then the prefix-free Kolmogorov complexity is bounded by the inequality

$$\mathbf{H}(a|b) \overset{+}{<} \min\{n \in \mathbb{N} : (a,n) \in \mathcal{E}(b)\} \tag{7.48}$$

Proof: Let g be a uniquely enumerating function for the set function $\mathcal{E}$. Subsequently, let $l(i|b) := n$ for $g(i|b) = (a,n)$ and $l(i|b) := \perp$ for $g(i|b) = \perp$. In view of Theorem 7.22, there exists a prefix-free Turing machine $V \in \mathcal{S}$ such that for all i where $l(i|b) \neq \perp$ there exists exactly one program z such that $V(z|b) = i$ and $|z| = l(i|b)$. Moreover, since function $g(i|b) = (a,n)$ is recursive, there exists a prefix-free Turing machine $U \in \mathcal{S}$ that outputs a instead of i, i.e. for i such that $g(i|b) = (a,n)$ there exists exactly one program z such that $U(z|b) = a$ and $|z| = n$. Hence, for the reference Turing machine we obtain the bound

$$\mathbf{H}(a|b) \overset{+}{<} \mathbf{H}_U(a|b) = \min\{n \in \mathbb{N} : (a,n) \in \mathcal{E}(b)\} \tag{7.49}$$
$$\square$$

Now we will introduce another ingredient for the proof of Theorem 7.20, which is the algorithmic probability. Strictly speaking, the algorithmic probability is a semidistribution over discrete objects, whose normalizing constant is called the halting probability.

Definition 7.22 (*Algorithmic and halting probabilities*): For a machine $S \in \mathcal{S}$, we define algorithmic probability as

$$\Pi_S(a|b) := \sum_{z \in \{0,1\}^*} 2^{-|z|} \mathbf{1}\{S(z|b) = a\} \tag{7.50}$$

and the halting probability as

$$\Omega_S(b) := \sum_{p \in \{0,1\}^*} 2^{-|z|} \mathbf{1}\{S(z|b) \neq \perp\} \tag{7.51}$$

If S is the reference machine, we will write

$$\Pi(a|b) := \Pi_S(a|b) \tag{7.52}$$

$$\Omega(a|b) := \Omega_S(b) \tag{7.53}$$

Theorem 7.24: *We have*

$$\sum_{a \in A} 2^{-H_S(a|b)} \leq \sum_{a \in A} \Pi_S(a|b) = \Omega_S(b) \in [0,1] \tag{7.54}$$

Proof: Set $A_S(b) := \{z \in \{0,1\}^* : S(z|b) \neq \perp\}$ is prefix-free by Theorem 7.8. Hence, the claim follows by the Kraft inequality (Theorem 7.4). □

It can be also shown that the algorithmic and the halting probabilities for a universal machine are not computable but are lower semicomputable. This result will be developed in the next section.

As for now, there are three important results for the algorithmic probabilities. The first one, called the coding theorem, applies the Chaitin bound (Theorem 7.23). The coding theorem asserts that the Kolmogorov complexity for the reference universal machine is bounded above by the pointwise entropy of the algorithmic probability for any prefix-free Turing machine.

Theorem 7.25 (*Coding theorem*): *For any machine $V \in \mathcal{S}$ we have*

$$H(a|b) \overset{+}{<} -\log \Pi_V(a|b) \tag{7.55}$$

Proof: Consider the set function

$$\mathcal{E}(b) := \{(a,n) : a \in \mathbb{A}, n \geq \lceil -\log \Pi_V(a|b)\rceil + 1\} \tag{7.56}$$

This set function is recursively enumerable since, simulating the execution on machine V of all programs of arbitrary lengths in parallel, we can enumerate all $(a, n) \in \mathcal{E}(b)$. This parallel simulation can be carried out for example in this way: In the ith step of the simulation, we iterate on $j \in \{1, 2, \dots, i - 1\}$ and we execute the jth step of computations for the $(j - i)$th program sorted in some effective fixed order. Moreover, for the set function $\mathcal{E}$, we obtain

$$\sum_{(a,n)\in\mathcal{E}(b)} 2^{-n} = \sum_{a\in\mathbb{A}} 2^{-\lceil -\log \Pi_V(a|b)\rceil} \leq \sum_{a\in\mathbb{A}} \Pi_V(a|b) \leq 1 \qquad (7.57)$$

Thus, in view of Theorem 7.23, we obtain

$$\mathbf{H}(a|b) \overset{+}{<} \min\{n \in \mathbb{N} : (a, n) \in \mathcal{E}(b)\} = \lceil -\log \Pi_V(a|b)\rceil + 1 \qquad (7.58)$$

$\square$

In the following, we will write $f(q) \overset{*}{<} g(q)$ and $f(q) \overset{*}{>} g(q)$ if there exists a constant $c \geq 1$ such that $f(q) \leq cg(q)$ and $f(q) \geq cg(q)$ holds respectively for all arguments q. We will write $f(q) \overset{*}{=} g(q)$ when we have both $f(q) \overset{*}{<} g(q)$ and $f(q) \overset{*}{>} g(q)$. It turns out that the algorithmic probability with respect to a reference universal machine roughly dominates algorithmic probabilities with respect to other machines.

Theorem 7.26 *(Coding theorem):* *For any machine $V \in \mathcal{S}$ we have*

$$\Pi(a|b) \overset{*}{=} 2^{-\mathbf{H}(a|b)} \overset{*}{>} \Pi_V(a|b) \qquad (7.59)$$

Proof: By the definition of algorithmic probabilities, we have $\Pi(a|b) \geq 2^{-\mathbf{H}(a|b)}$. Hence, the claim follows by Theorem 7.25. $\square$

The last important fact for the proof of Theorem 7.20 is that the algorithmic probability behaves almost like a complete probability distribution, namely, it is additive up to a multiplicative constant.

Theorem 7.27: *We have*

$$\Pi(u|b) \overset{*}{=} \sum_{w\in\mathbb{A}} \Pi(u, w|b) \qquad (7.60)$$

Proof: By $\mathcal{S}$ we will denote the reference machine. On the one hand, there exists a machine $V \in \mathcal{S}$ such that $V(p|b) = u$ if $S(p|b) = (u, w)$. Hence, by Theorem 7.26,

$$\Pi(u|b) \overset{*}{>} \Pi_V(u|b) \geq \sum_{w\in\mathbb{A}} \Pi(u, w|b) \qquad (7.61)$$

On the other hand, there exists a machine $V \in \mathcal{S}$ such that $V(p|b) = (u, u)$ if $S(p|b) = u$. Hence, by Theorem 7.26,

$$\sum_{w \in \mathbb{A}} \Pi(u, w|b) \geq \Pi(u, u|b) \stackrel{*}{>} \Pi_V(u, u|b) \geq \Pi(u|b) \tag{7.62}$$

$\square$

Now we may prove the chain rule for Kolmogorov complexity.

Proof of Theorem 7.20:   By S we will denote the reference machine. First we will prove the easier inequality

$$H(u, w|b) \stackrel{+}{<} H(u|b) + H(w|u, H(u|b), b) \tag{7.63}$$

Let z be the shortest program that satisfies $S(z|b) = u$ and let z' be the shortest program that satisfies $S(z'|u, H(u|b), b) = w$. Given z and b we may compute both u and $H(u|b)$ and whereas given z' and $(u, H(u|b), b)$ we may compute w. Thus, there exists a machine $V \in \mathcal{S}$ that satisfies $V(zz'|b) = (u, w)$. Hence, we obtain the claim.

Next, we will show the harder inequality

$$H(w|u, H(u|b), b) \stackrel{+}{<} H(u, w|b) - H(u|b) \tag{7.64}$$

By Theorems 7.26 and 7.27 we know that there exists a constant c such that for all u and w we have

$$2^{H(u|b)-c} \sum_{w \in \mathbb{A}} \Pi(u, w|b) \leq 1 \tag{7.65}$$

Consider now the set function

$$\mathcal{E}(u, m, b) := \{(w, n) : a \in \mathbb{A}, n \geq \lceil -\log \Pi(u, w|b) \rceil + 1 - m + c\} \tag{7.66}$$

This set function is recursively enumerable, by the argument similar as in the proof of Theorem 7.25. Moreover, by inequality (7.65), we obtain

$$\sum_{(w,n) \in \mathcal{E}(u, H(u|b), b)} 2^{-n} \leq 1 \tag{7.67}$$

Thus, in view of Theorems 7.23 and 7.26, we obtain

$$H(w|u, H(u|b), b) \stackrel{+}{<} \min\{n \in \mathbb{N} : (w, n) \in \mathcal{E}(u, H(u|b), b)\}$$
$$= \lceil -\log \Pi(u, w|b) \rceil + 1 - H(u|b) + c$$
$$\stackrel{+}{=} H(u, w|b) - H(u|b) \tag{7.68}$$

$\square$

Commenting upon Theorem 7.20, let us remark that the parallels between the chain rules (7.43) and (7.44) remain incomplete because in the algorithmic version there appears term $\mathbf{H}(w|u, \mathbf{H}(u|b), b)$ rather than $\mathbf{H}(w|u, b)$. Although $\mathbf{H}(w|u, \mathbf{H}(u|b), b)$ differs from $\mathbf{H}(w|u, b)$, in the next theorem we can see that expressions $\mathbf{H}(u, \mathbf{H}(u|b)|b)$ and $\mathbf{H}(u|b)$ are approximately equal.

Theorem 7.28: *We have*

$$\mathbf{H}(w, \mathbf{H}(w|b)|b) \stackrel{+}{=} \mathbf{H}(w|b) \tag{7.69}$$

Proof: From the shortest program that computes w given b, we may compute both w and $\mathbf{H}(w|b)$. Hence, $\mathbf{H}(w, \mathbf{H}(w|b)|b) \stackrel{+}{<} \mathbf{H}(w|b)$. On the hand, we have $\mathbf{H}(w, \mathbf{H}(w|b)|b) \stackrel{+}{>} \mathbf{H}(w|b)$ since given pair $(w, \mathbf{H}(w|b))$ we may compute w. □

We may also define an algorithmic analogue of mutual information.

Definition 7.23 *(Algorithmic mutual information)*: We define the algorithmic mutual information between discrete objects u and w given an object b as

$$\mathbf{I}(u; w|b) := \mathbf{H}(u|b) + \mathbf{H}(w|b) - \mathbf{H}(u, w|b) \tag{7.70}$$

Theorem 7.29: *We have*

$$\mathbf{I}(u; w|b) \stackrel{+}{=} \mathbf{I}(w; u|b) \stackrel{+}{>} 0 \tag{7.71}$$

Proof: We have $\mathbf{H}(u, w|b) \stackrel{+}{=} \mathbf{H}(w, u|b)$ since given pair (u, w) we may compute pair (w, u) and converse. Moreover, we have $\mathbf{H}(u, w|b) \stackrel{+}{<} \mathbf{H}(u|b) + \mathbf{H}(w|b)$ since given u and w, computed independently, we may compute the pair (u, w). □

Theorem 7.30: *We have*

$$\mathbf{I}(u; w|b) \stackrel{+}{=} \mathbf{H}(u|b) - \mathbf{H}(u|w, \mathbf{H}(w|b), b) \tag{7.72}$$

Proof: The claim follows by Theorem 7.20. □

Now we will derive some different links between Kolmogorov complexity and Shannon entropy. Let us recall that the coding theorem (Theorem 7.25) shows that the Kolmogorov complexity is bounded above by the pointwise entropy of the algorithmic probability. Using much simpler technique which applies the Shannon–Fano coding, we will show that the Kolmogorov complexity of a string given a semidistribution is bounded above by the pointwise Shannon entropy of the string for any semidistribution, see Gács et al. (2001).

Theorem 7.31: *Let p be a semidistribution on discrete objects, i.e. $p(w) \geq 0$ for $w \in \mathbb{A}$ and $\sum_{w \in \mathbb{A}} p(w) \leq 1$. We have*

$$\mathbf{H}(w|p) \overset{+}{<} - \log p(w) \tag{7.73}$$

Proof: For a semidistribution p, by Theorem 7.5 there exists a prefix-free Shannon–Fano code C_p such that $|C_p(w)| = \lceil - \log p(w) \rceil$. This code is effectively computable. Therefore, a certain machine $V \in \mathcal{S}$ may operate in the following way. Given a program z and the oracle p, it enumerates objects w and constructs code words $C_p(w)$ until it encounters a code word equal to program z and then it returns object w. As in the proof of Theorem 7.22, program z can be read from the unidirectional tape $(X_k)_{k \in \mathbb{N}}$ incrementally while checking against code words $C_p(w)$. In view of the above, we obtain inequalities

$$\mathbf{H}(w|p) \overset{+}{<} \mathbf{H}_V(w|p) = |C_p(w)| = \lceil - \log p(w) \rceil \tag{7.74}$$

$\square$

The above theorem combined with Theorem 7.24 and the general properties of semidistributions has the following corollary, which states that the Kolmogorov complexity of a string given a probability measure is close to the pointwise entropy of the string with a high probability.

Theorem 7.32: *Consider a stochastic process $(X_i)_{i \in \mathbb{N}}$ on a probability space $(\mathbb{X}^\mathbb{N}, \mathcal{X}^\mathbb{N}, P)$, where $X_j : \mathbb{X}^\mathbb{N} \ni (x_i)_{i \in \mathbb{N}} \mapsto x_j \in \mathbb{X}$ for some countable set $\mathbb{X}$. Let $P(x_1^n) := P(X_1^n = x_1^n)$ be the distribution of the process. We have a nonprobabilistic upper bound*

$$\mathbf{H}(x_1^n|P) \overset{+}{<} - \log P(x_1^n) + 2 \log n \tag{7.75}$$

for all x_1^n, and probabilistic lower bounds

$$\mathbf{E}[- \log P(X_1^n)] \leq \mathbf{E}\mathbf{H}(X_1^n|P) \tag{7.76}$$

$$P(\mathbf{H}(X_1^n|P) \leq - \log P(X_1^n) - m) \leq 2^{-m} \tag{7.77}$$

$$\lim_{n \to \infty} [\mathbf{H}(X_1^n|P) + \log P(X_1^n)] = \infty \text{ almost surely} \tag{7.78}$$

Proof: W observe that $2^{-\mathbf{H}(x_1^n|P)}$ and $\pi(n) P(x_1^n)$ where $\pi(n) := 6\pi^{-2}n^{-2}$ are semidistributions on strings of arbitrary length. Thus, the claims follow by Theorems 7.31 and 7.24 combined with Theorems 3.18–3.20. $\square$

Similarly, if we are interested in the unconditional Kolmogorov complexity, we note the following result for recursive measures.

Theorem 7.33: *Consider a stochastic process $(X_i)_{i \in \mathbb{N}}$ on a probability space $(\mathbb{X}^\mathbb{N}, \mathcal{X}^\mathbb{N}, P)$, where $X_j : \mathbb{X}^\mathbb{N} \ni (x_i)_{i \in \mathbb{N}} \mapsto x_j \in \mathbb{X}$ for some countable set $\mathbb{X}$. Let*

$P(x_1^n) := P(X_1^n = x_1^n)$ *be the distribution of the process. If P is a recursive function then we have a nonprobabilistic upper bound*

$$\mathbf{H}(x_1^n) \overset{+}{<} -\log P(x_1^n) + 2\log n + \mathbf{H}(P) \tag{7.79}$$

for all x_1^n. Moreover, for an arbitrary P we have probabilistic lower bounds

$$\mathbf{E}[-\log P(X_1^n)] \le \mathbf{E}\mathbf{H}(X_1^n) \tag{7.80}$$

$$P(\mathbf{H}(X_1^n) \le -\log P(X_1^n) - m) \le 2^{-m} \tag{7.81}$$

$$\lim_{n\to\infty} [\mathbf{H}(X_1^n) + \log P(X_1^n)] = \infty \text{ almost surely} \tag{7.82}$$

Proof: We observe that $2^{-\mathbf{H}(x_1^n)}$ and $\pi(n)P(x_1^n)$ where $\pi(n) := 6\pi^{-2}n^{-2}$ are semidistributions on strings of arbitrary length, whereas

$$\mathbf{H}(X_1^n) \overset{+}{<} \mathbf{H}(X_1^n|P) + \mathbf{H}(P) \tag{7.83}$$

since a recursive measure is a discrete object. Thus, the claims follow by Theorems 7.31 and 7.24 combined with Theorems 3.18–3.20. □

7.4 Limits of Mathematics

An important achievement of the twentieth century mathematics was the incompleteness theorem by Kurt Gödel. The Gödel incompleteness theorem sets some limits on the power of formal inference systems. A formal inference system is a finite collection of axioms and inference rules, which can be encoded as a tuple of strings – a discrete object. A formal inference system is called consistent if, within this system, i.e. using the axioms and inference rules of the system, it is not possible to prove both a statement and its negation. The system is called sound if only true propositions can be proved. Thus, a sound system is consistent, provided mathematics is consistent as a whole. According to the Gödel incompleteness theorem, for any sound formal inference system which is sufficiently expressible, there exist statements expressible in this system which are true but not provable within this system (Gödel, 1931).

An interesting version of the Gödel incompleteness theorem was proved by Gregory Chaitin. According to this version of the Gödel theorem, in any sound formal inference system it is not possible to prove that the Kolmogorov complexity of any particular string is substantially larger than the Kolmogorov complexity of this formal system (Chaitin, 1975b). This result reveals that unprovability of true propositions is a much more common phenomenon than suggested by the original proof of the incompleteness theorem by Gödel. Let us state our version of Chaitin's theorem formally.

Theorem 7.34 (*Chaitin incompleteness theorem*): *There exists a constant c such that for any sound formal inference system M propositions* $\mathbf{H}(w|M) \geq c$ *are unprovable in system M for any particular string w – if these statements are expressible in system M.*

Gregory Chaitin commented on the results in this vein as follows:

> I would like to able to say that if one has 10 pounds of axioms and a 20-pound theorem, then that theorem cannot be derived from those axioms. And I will argue that this approach to Gödel's theorem does suggest a change in the daily habits of mathematicians, and that Gödel's theorem cannot be shrugged away.
>
> (Chaitin, 1982)

In fact, as we have seen in Section 7.2, we have $\mathbf{H}(w|M) \overset{+}{>} \mathbf{H}(w) - \mathbf{H}(M)$. Thus, by Theorem 7.34 there exists a constant c' such that for any sound formal inference system M statements $\mathbf{H}(w) \geq c' + \mathbf{H}(M)$ are unprovable in M – if they are expressible in system M. Thus, there is something in the air about comparing the weights of formal systems and of provable propositions but we will not pursue this metaphor further.

Theorem 7.34 should be contrasted with Theorem 7.11, which states that there are $2^n - 2^{n-c} + 1$ distinct strings w of length n such that $\mathbf{H}(w) \geq |w| - c$. Although we can prove that there are so many incompressible strings, in view of Theorem 7.34, it is not possible to prove incompressibility of any explicitly given sufficiently long string. Thus, the only way to assert incompressibility of any explicitly given long string is to assume it as a new axiom in the inference system. To prove Theorem 7.34 we will apply a method similar to the proof of Theorem 7.19, i.e. uncomputability of Kolmogorov complexity.

Proof of Theorem 7.34: We observe that for a given formal inference system M we may effectively enumerate all proofs in the system and check their conclusions. In particular, there is a total recursive function π such that

$$\pi(n, l|M) := \begin{cases} w, & \text{if the } n\text{th proof in } M \text{ proves } \mathbf{H}(w|M) \geq l \text{ for a string } w \\ \lambda, & \text{else} \end{cases}$$

$$(7.84)$$

In consequence there exists a partial recursive function

$$f(l|M) := \begin{cases} \min\{n \in \mathbb{N} : \pi(n, l|M) \neq \lambda\}, & \text{if the set is not empty} \\ \bot, & \text{else} \end{cases}$$

$$(7.85)$$

Assume that the formal inference system is sound. Then function

$$g(l|M) := \begin{cases} \pi(f(l|M), l|M), & \text{if } f(l|M) \neq \bot \\ \bot, & \text{else} \end{cases} \tag{7.86}$$

is also a partial recursive function and returns some string $w = g(l|M)$ such that $\mathbf{H}(w|M) \geq l$, if there exists a proof of $\mathbf{H}(w|M) \geq l$. Since function g is recursive, there exists a string u such that $S(u\bar{\phi}(l)|M) = g(l|M)$ for all natural numbers l for the reference universal machine S. Hence, we obtain

$$l \leq \mathbf{H}(w|M) \leq |u| + \mathbf{H}_0(l) \tag{7.87}$$

whenever $g(l|M) = w$. Thus, propositions $\mathbf{H}(w|M) \geq c$ must be unprovable in system M where c is the smallest number such that $c > |u| + \mathbf{H}_0(c)$. □

Let us stress that the Chaitin incompleteness theorem (Theorem 7.34) holds only for strings w that we may write down explicitly. It is still possible to prove incompressibility of certain special strings that cannot be written down explicitly, such as the binary expansion of the halting probability, introduced in the previous section.

Definition 7.24 *(Halting probability)*: Let $S \in \mathcal{S}$ be the reference universal Turing machine. The unconditional halting probability is defined as

$$\Omega := \sum_{z \in \{0,1\}^*} 2^{-|z|}\mathbf{1}\{S(z|\lambda) \neq \bot\} \tag{7.88}$$

As shown in Theorem 7.24, we have $\Omega \in [0, 1]$. Subsequently, we will define the binary expansion of Ω.

Definition 7.25 *(Halting probability)*: We define the binary digits $\Omega_k \in \{0, 1\}$ of the halting probability Ω via equality

$$\Omega = \sum_{k=1}^{\infty} 2^{-k}\Omega_k \tag{7.89}$$

Digits Ω_k are given uniquely since the halting probability Ω is incompressible and thus uncomputable and irrational. Before we state this result, we will show a simpler fact, namely, that Ω can be approximated from below with an uncontrolled precision.

Theorem 7.35: *Number Ω is lower semicomputable.*

Proof: Let us simulate the execution on machine S of all programs of arbitrary lengths in parallel. This parallel simulation can be carried out for example in this way: In the ith step of the simulation, we iterate on $j \in \{1, 2, \dots, i-1\}$ and we execute the jth step of computations for the $(j-i)$th program sorted in some fixed effective order. In the beginning of the simulations, we set the approximation of number Ω as $\Omega' := 0$. When the simulated machine S halts for a certain program z, we improve the approximation by setting $\Omega' := \Omega' + 2^{-|z|}$. Approximation Ω' is an increasing function of the computation time. Hence, Ω is lower semicomputable. □

In the next step, we will introduce the concept of algorithmically incompressible sequences and real numbers.

Definition 7.26 (*Algorithmically incompressible sequence*): An infinite sequence $(r_k)_{k \in \mathbb{N}}$ or a real number $r = \sum_{k=1}^{\infty} 2^{-k} r_k$, where $r_k \in \{0, 1\}$, is called algorithmically incompressible if inequality $\mathbf{H}(r_1^n) \geq n - c$ holds for any n and some $c \geq 0$.

It is easy to see that an algorithmically incompressible sequence or, equivalently, an algorithmically incompressible real number are not computable.

Theorem 7.36: *If a real number is algorithmically incompressible then it is not computable.*

Proof: Suppose that a real number $r = \sum_{k=1}^{\infty} 2^{-k} r_k$ is computable. Then, by computability, there exists a recursive function $f(n)$ such that $|r - f(n)| < 2^{-n}$. Hence, there exists a recursive function $g(n)$ such that $g(n) = r_1^n$. Hence, by Theorem 7.15, we obtain

$$\mathbf{H}(r_1^n) = \mathbf{H}(g(n)) \overset{+}{<} \mathbf{H}(n) + \mathbf{H}(g) \overset{+}{<} \mathbf{H}_0(n) + \mathbf{H}(g) \tag{7.90}$$

which implies that r is not algorithmically incompressible. □

Subsequently, we will show that halting probability Ω is algorithmically incompressible despite it is lower semicomputable. As an important auxiliary result, we will demonstrate first that given Ω_1^n we may decide whether the reference universal machine halts on any program of length up to n.

Theorem 7.37 (*Halting problem*): *There exists a partial recursive function h such that*

$$h(z, \Omega_1^n) = \begin{cases} 1, & \text{if } S(z|\lambda) \neq \perp \\ 0, & \text{if } S(z|\lambda) = \perp \end{cases} \tag{7.91}$$

for any program $z \in \{0, 1\}^*$ such that $|z| \leq n$.

Proof: Let us simulate the execution of machine S in parallel on all programs of length up to n. In the beginning of the simulations, we set the approximation of number Ω as $\Omega' := 0$. When the simulated machine S halts for a certain program z, we improve the approximation by setting $\Omega' := \Omega' + 2^{-|z|}$. At any instant, we have $\Omega' \leq \Omega$. But at a certain instant, the approximation Ω' becomes greater than $\sum_{k=1}^n 2^{-k} \Omega_k$. Since then we have $\sum_{k=1}^n 2^{-k} \Omega_k \leq \Omega' \leq \Omega < \sum_{k=1}^n 2^{-k} \Omega_k + 2^{-n}$, then it becomes evident that machine S will not halt on any other program z of length $|z| \leq n$. Hence, we obtain a finite list of programs of lengths up to n with an information which of them halts and which does not. From this we may compute function $h(z, \Omega_1^n)$. $\square$

Some simple corollary of the above result is that given Ω_1^n we may also decide which mathematical statements of length up to $n - c$ are true or false. This property is metaphorically called philosophers' stone.

Theorem 7.38 (Philosophers' stone): *There exist a constant c and a partial recursive function g such that*

$$g(\varphi, \Omega_1^n) = \begin{cases} 1, & \text{if } \varphi(n) \text{ holds for all } n \in \mathbb{N} \\ 0, & \text{else} \end{cases} \tag{7.92}$$

for any recursive predicate of natural numbers $\varphi : \mathbb{N} \to \{true, false\}$ such that $\mathbf{H}_0(\varphi) \leq n - c$.

Proof: There is a recursive partial function f which given the program for computing predicate φ searches for the smallest n such that $\varphi(n)$ does not hold. Namely, we have

$$f(\varphi) := \begin{cases} \min\{n \in \mathbb{N} : \varphi(n) \text{ is false}\}, & \text{if the set is not empty} \\ \bot, & \text{else} \end{cases} \tag{7.93}$$

For the reference machine S, we have $S(u\bar{\phi}(\varphi)|\lambda) = f(\varphi)$ for a certain string u. Hence, by Theorem 7.37, there exists a partial recursive function g that satisfies (7.92) if $\mathbf{H}_0(\varphi) \leq n - |u|$. $\square$

In the sequel, we will show that the halting probability is algorithmically incompressible – and thus not computable and irrational.

Theorem 7.39: *Number Ω is algorithmically incompressible.*

Proof: By Theorem 7.37, we can enumerate, given string Ω_1^n, all programs of length up to n for which the reference universal machine S halts. For any string w which is not computed by these programs, we have $\mathbf{H}(w) > n$. Thus, we can construct a recursive function g which returns the first such string w given an argument Ω_1^n. Hence, by Theorem 7.15, we obtain

$$n < H(g(\Omega_1^n)) \overset{+}{<} H(\Omega_1^n) + H(g) \tag{7.94}$$

which is tantamount to Ω being algorithmically incompressible. $\qquad\square$

In the next section, we will see that halting probability Ω is a special case of a much more common object, namely, algorithmically random sequences.

7.5 Algorithmic Randomness

Considering preformal conceptions about a random infinite sequence (with respect to a given probability measure), it is intuitive to require that the set of random sequences have probability one and consist only of those sequences that cannot be differentiated from any set of sequences typical of that measure by any effective statistical means. In particular, sequences such as 000000 ..., 111111 ..., or 010101 ... should not be considered random with respect to the Lebesgue measure, i.e. the uniform measure on infinite sequences, since they can be generated by simple deterministic algorithms. Beforehand, it is not obvious that the concept of a random infinite sequence can be successfully formalized and that this can be done for an arbitrary probability measure.

Historically, the formal concept of an algorithmically random sequence was developed by Per Martin-Löf by means of algorithmic statistical tests (Martin-Löf, 1966). It was proved later by Claus-Peter Schnorr (in 1974, unpublished) that this approach is equivalent to a definition which applies Kolmogorov complexity like in Theorem 7.32. Since we want to demonstrate the Schnorr theorem – in a version for general probability measures, we will discuss the algorithmic tests first.

To state necessary preliminaries, let us return to the setting of Theorem 7.32. Consider a stochastic process $(X_i)_{i\in\mathbb{N}}$ on a probability space $(\mathbb{X}^{\mathbb{N}}, \mathcal{X}^{\mathbb{N}}, P)$, where $X_j : \mathbb{X}^{\mathbb{N}} \ni (x_i)_{i\in\mathbb{N}} \mapsto x_j \in \mathbb{X}$ for some countable set $\mathbb{X}$. To make the notation shorter, in the following, we will write infinite sequences as $x_1^\infty = (x_i)_{i\in\mathbb{N}}$, where $x_i \in \mathbb{X}$. For strings $w \in \mathbb{X}^*$ and subsets $L \subseteq \mathbb{X}^*$, called formal languages, will also write generalized cylinder sets as

$$[L] := \{x_1^\infty \in \mathbb{X}^{\mathbb{N}} : x_1^n \in L \text{ for some } n \geq 0\} \tag{7.95}$$
$$[w] := \{x_1^\infty \in \mathbb{X}^{\mathbb{N}} : x_1^n = w \text{ for some } n \geq 0\} \tag{7.96}$$

and abbreviate the notation

$$P(L) := P(X_1^\infty \in [L]) \tag{7.97}$$
$$P(w) := P(X_1^\infty \in [w]) \tag{7.98}$$

so that we obtain the familiar identity $P(x_1^n) = P(X_1^n = x_1^n)$ in particular. Equating probability measures with their values on the cylinder sets (which determine them uniquely), we also note that, according to our earlier conventions, the set of nondiscrete objects $\mathbb{B}$ contains all probability measures P of stochastic processes $(X_i)_{i \in \mathbb{N}}$, where $X_i : \Omega \to \mathbb{X}$ for a countable $\mathbb{X}$.

Having made these remarks, we may introduce Martin-Löf and Solovay tests.

Definition 7.27 (*Martin-Löf and Solovay tests*): Let $U : \mathbb{B} \to 2^{\mathbb{N} \times \mathbb{X}^*}$ be a recursively enumerable set function. We write

$$U_m(b) := \{w : (m, w) \in U(b)\} \tag{7.99}$$

The set function U is called a Martin-Löf test for a probability measure P if for all $m \in \mathbb{N}$ sets $U_m(P)$ are prefix-free and

$$P(U_m(P)) \leq 2^{-m} \tag{7.100}$$

In contrast, the set function U is called a Solovay test for a probability measure P if for all $m \in \mathbb{N}$ sets $U_m(P)$ are prefix-free and

$$\sum_{m=1}^{\infty} P(U_m(P)) < \infty \tag{7.101}$$

Borrowing names from mathematical statistics, formal languages $U_m(P)$ and their cylinder sets $[U_m(P)]$ will be called critical regions, whereas numbers 2^{-m} will be called significance levels. Thus, we may say that a Martin-Löf test U for measure P is an arbitrary sequence of significance levels 2^{-m} and critical regions $U_m(P)$ that are jointly recursively enumerable. For a Solovay test, we relax the requirement that significance levels be of form 2^{-m}.

The above definition requires a certain illustration. Consider the following example. Let P_θ be the probability measure of the Bernoulli(θ) process, i.e. $P_\theta(x_1^n) = \prod_{i=1}^n \theta^{x_i}(1 - \theta)^{1-x_i}$ where $x_i \in \{0, 1\}$ and $\theta \in [0, 1]$. By the Hoeffding inequality (3.99), we have

$$P_\theta\left(\left\{x_1^\infty : \left|\frac{1}{m}\sum_{i=1}^m x_i - \theta\right| \geq t\right\}\right) \leq 2\exp(-2mt^2) \tag{7.102}$$

Putting $t = m^{-1/2}\log m$, we obtain $P_\theta(U_m(P_\theta)) \leq 2m^{-2}$ for

$$U_m(P_\theta) := \left\{x_1^m : \left|\frac{1}{m}\sum_{i=1}^m x_i - \theta\right| \geq \sqrt{\frac{\log m}{m}}\right\} \tag{7.103}$$

Hence,

$$U(P_\theta) := \{(m, w) : w \in U_m(P_\theta)\} \tag{7.104}$$

is a Solovay test for measure P_θ. Strictly speaking, to prove that U is a Solovay test, we need to additionally show that parameter θ can be computed with a sufficient precision given measure P_θ in the particular form in which we represent it on the tape of the Turing machine. We leave this problem as an exercise.

Let us observe that, for a given measure P, critical regions $[U_m(P)]$ are sets of approximately nonrandom sequences. In particular, if condition (7.100) is satisfied then the set of sequences x_1^∞ that are not contained in the intersection $\bigcap_{m=1}^\infty [U_m(P)]$ of all critical regions has full measure P, whereas if condition (7.101) is satisfied then the set of sequences x_1^∞ that are contained in only finitely many critical regions $[U_m(P)]$ has also full measure P – by the Borel–Cantelli lemma (Theorem 3.12). This motivates the following definition.

Definition 7.28 *(Passing Martin-Löf and Solovay tests):* We will say that:

(1) Sequence x_1^∞ passes a Martin-Löf test U for measure P if sequence x_1^∞ is not contained in the intersection $\bigcap_{m=1}^\infty [U_m(P)]$ of all critical regions;
(2) Sequence x_1^∞ passes a Solovay test U for measure P if sequence x_1^∞ is contained in only finitely many critical regions $[U_m(P)]$.

For example, for the Solovay test (7.103) and the Bernoulli measure P_θ, a sequence x_1^∞ passes the test if

$$\left| \frac{1}{m} \sum_{i=1}^m x_i - \theta \right| < \sqrt{\frac{\log m}{m}} \tag{7.105}$$

for all but finitely many m. Since the set of such sequences x_1^∞ has full measure P_θ, we may suppose that a proper subset of these sequences can be naturally called algorithmically random sequences with respect to P_θ. Intuitively these algorithmically random sequences should pass any imaginable Solovay or Martin-Löf test. Since there are only countably many Solovay or Martin-Löf tests, the set of sequences that pass all the tests still has full measure P_θ.

In fact, passing any Martin-Löf or Solovay test corresponds to a very natural property. The following Schnorr theorem states that a sequence x_1^∞ passes any imaginable Solovay or Martin-Löf test for measure P if and only if the conditional Kolmogorov complexity of prefixes x_1^n is essentially larger than $-\log P(x_1^n)$. The Schnorr theorem is usually presented for recursive probability measures. Here we present a version for general measures, also uncomputable ones.

Theorem 7.40 *(Schnorr theorem):* *Let P be a probability measure. The following conditions are equivalent:*

1. *Sequence x_1^∞ satisfies $\lim_{n\to\infty}[\mathbf{H}(x_1^n|P) + \log P(x_1^n)] = \infty$.*

2. *Sequence x_1^∞ passes any Solovay test V for probability measure P.*
3. *Sequence x_1^∞ passes any Martin-Löf test U for probability measure P.*
4. *Sequence x_1^∞ satisfies $\inf_{n\in\mathbb{N}}[\mathbf{H}(x_1^n|P) + \log P(x_1^n)] > -\infty$.*

Proof: We will demonstrate implications (1) $\implies$ (4) $\implies$ (3) $\implies$ (2) $\implies$ (1), which is sufficient to show the equivalence.

(1) $\implies$ (4): Obviously, statistic $\mathbf{H}(x_1^n|P) + \log P(x_1^n)$ is bounded below if it tends to infinity. (In that case we have $P(x_1^n) \neq 0$ for all n.)

(4) $\implies$ (3): Suppose by contradiction that $x_1^\infty \in \bigcap_{m=1}^\infty [U_m(P)]$ for a certain Martin-Löf test U. We will show that x_1^∞ satisfies $\inf_{n\in\mathbb{N}}[\mathbf{H}(x_1^n|P) + \log P(x_1^n)] = -\infty$. We obtain

$$\sum_{m=2}^\infty \sum_{w\in U_{m^2}(P)} 2^{m-\lceil-\log P(w)\rceil} \le \sum_{m=2}^\infty \sum_{w\in U_{m^2}(P)} 2^{m+\log P(w)}$$

$$= \sum_{m=2}^\infty 2^m P(U_{m^2}(P)) \le \sum_{m=2}^\infty 2^{m-m^2} \le 1 \quad (7.106)$$

Since set function U is recursively enumerable, in view of Theorem 7.23, we obtain

$$\mathbf{H}(w|P) \overset{+}{<} \lceil-\log P(w)\rceil - m \quad (7.107)$$

for any $w \in U_{m^2}(P)$ and $m \ge 2$. Since $x_1^\infty \in \bigcap_{m=1}^\infty [U_m(P)]$, there exists a sequence of prefixes $x_1^{n(m)}$ such that $x_1^{n(m)} \in U_{m^2}(P)$. This yields

$$\mathbf{H}(x_1^{n(m)}|P) + \log P(x_1^{n(m)}) \overset{+}{<} -m \quad (7.108)$$

for an arbitrary m.

(3) $\implies$ (2): Suppose by contradiction that x_1^∞ is contained in infinitely many critical regions $[V_m(P)]$ for a certain Solovay test V. We will show that $x_1^\infty \in \bigcap_{m=1}^\infty [U_m(P)]$ for a certain Martin-Löf test U. For $\sum_{m=1}^\infty P(V_m(P)) \le C$, the test U is constructed via

$$[U_m(b)] := \{y_1^\infty : y_1^\infty \text{ belongs to at least in } 2^m C \text{ of } [V_i(b)]\} \quad (7.109)$$

Then x_1^∞ belongs to all sets $[U_m(P)]$. To show that U of this form is a Martin-Löf test, we notice the following. We can choose some prefix-free sets $U_m(b)$ generating the sets $[U_m(b)]$ in such a way that set function U is recursively enumerable. Moreover, we notice that

$$C \ge \sum_{i=1}^\infty P(V_i(P)) \ge \sum_{i=1}^\infty P(X_1^\infty \in [V_i(P)] \cap [U_m(P)])$$

$$= \sum_{i=1}^\infty \sum_{w\in U_m(P)} P(X_1^\infty \in [V_i(P)] \cap [w])$$

$$\geq 2^m C \sum_{w \in U_m(P)} P(w) = 2^m C P(U_m(P)) \tag{7.110}$$

Hence, $P(U_m(P)) \leq 2^{-m}$ and U is a Martin-Löf test.

(2) $\Rightarrow$ (1): Suppose by contradiction that x_1^∞ satisfies

$$\liminf_{n \to \infty} [\mathbf{H}(x_1^n|P) + \log P(x_1^n)] < \infty \tag{7.111}$$

We will show that x_1^∞ belongs to infinitely many critical regions $[V_m(P)]$ for a certain Solovay test V. We observe that if $\liminf_{n \to \infty} [\mathbf{H}(x_1^n|P) + \log P(x_1^n)] < \infty$ then sequence x_1^∞ belongs to infinitely many sets $[V_m(P)]$ if we put

$$V_m(P) := \{y_1^m : \mathbf{H}(y_1^m|P) + \log P(y_1^m) \leq N\} \tag{7.112}$$

where N is a certain finite number. For this choice of $V_m(P)$, set function V is recursively enumerable and sets $V_m(P)$ are prefix-free. Thus, it remains to prove the second condition for the Solovay test. Denote

$$W(y_1^m|P) = \frac{2^{-\mathbf{H}(y_1^m|P)}}{P(y_1^m)} \tag{7.113}$$

Then by the Markov inequality (Theorem 3.11) and by the Kraft inequality for Kolmogorov complexity (Theorem 7.24), we obtain

$$\sum_{m=1}^\infty P(V_m(P)) = \sum_{m=1}^\infty P(W(X_1^m|P) \geq 2^{-N})$$

$$\leq 2^N \sum_{m=1}^\infty \mathbf{E} W(X_1^m|P)$$

$$\leq 2^N \sum_{m=1}^\infty \sum_{y_1^m} 2^{-\mathbf{H}(y_1^m|P)} \leq 2^N < \infty \tag{7.114}$$

$\square$

Thus, algorithmically random sequences in the Martin-Löf sense, also called 1-random sequences, can be defined in the following way.

Definition 7.29 *(Algorithmically random sequences):* We say that a sequence x_1^∞ is 1-random or algorithmically random in the Martin-Löf sense with respect to a probability measure P when it satisfies conditions enumerated in Theorem 7.40. The set of such sequences is denoted as $\mathcal{R}_P$.

There are also stronger concepts of algorithmic randomness such as n-randomness (Downey and Hirschfeldt, 2010).

It follows easily that the set of Martin-Löf random sequences has the full measure.

Theorem 7.41: *We have $P(\mathcal{R}_P) = 1$.*

Proof: Follows by Theorem 7.32. $\square$

In view of the Schnorr theorem, statistic $\mathbf{H}(x_1^n|P) + \log P(x_1^n)$ either tends to $+\infty$ or gets arbitrarily close to $-\infty$. In particular, we obtain that all algorithmically incompressible sequences are 1-random with respect to the Lebesgue measure over binary sequences, i.e. $P(x_1^n) = 2^{-n}$ where $x_i \in \{0, 1\}$. Thus, the Chaitin halting probability Ω is 1-random in the with respect to the Lebesgue measure by Theorem 7.39 and the Schnorr theorem. As a corollary of this fact, the relative frequencies of 0's and 1's in the binary expansion of Ω are equal to $1/2$ by the relationship (7.105).

Algorithmic randomness is a subject of intense research in recent years. Some fundamental problem is strengthening various theorems of form "$\varphi(x_1^\infty)$ for P-almost all x_1^∞" to claims of form "$\varphi(x_1^\infty)$ for all $x_1^\infty \in \mathcal{R}_P$," where $\mathcal{R}_P$ is the set of 1-random sequences for a relevant measure P. Interesting examples include the Birkhoff ergodic theorem (Bienvenu et al., 2012; Franklin et al., 2012), ergodic decomposition (Hoyrup, 2013), the problem of superefficient estimators (Vovk, 2009), and the Lambalgen theorem about decomposition of the set of jointly random sequences (van Lambalgen, 1987; Vovk and V'yugin, 1993, 1994; V'yugin, 2007; Takahashi, 2008, 2011), which is of a substantial importance for foundations of Bayesian statistics.

Problems

7.1 Let $C : \mathbb{N} \to \{0, 1\}^*$ be a uniquely decodable code for natural numbers. Show that $|C(n)| \geq \log n + \log\log n$ for infinitely many n.

7.2 Show that there exists a uniquely decodable code $C : \{0, 1\}^n \to \{0, 1\}^*$ such that

$$|C(x_1^n)| \leq 2 + \lfloor \log(n + 1)\rfloor + \left\lfloor \log\left(\sum_{i=1}^{n} x_i\right)\right\rfloor \tag{7.115}$$

7.3 Consider the entropy $\mathcal{H}(p)$ of a two-point distribution defined in (5.111). Let $(X_i)_{i \in \mathbb{N}}$ be a Bernoulli process. Show that

$$0 \leq H(X_i) - \mathbf{E}\mathcal{H}\left(\frac{\sum_{i=1}^{n} X_i}{n}\right) \leq \frac{c_1 \log n + c_2}{n} \tag{7.116}$$

for certain constants c_1 and c_2. What are the optimal c_1 and c_2?

7.4 In this task we will consider a certain restricted minimal grammar-based code. The general idea of a minimal grammar-based code is to encode a given string as the minimal context-free grammar that generates the string as its only production. This can be done when imposing various constraints on the shape of admissible context-free grammars. Here we will assume that the admissible grammars are flat, i.e. nonterminal symbols can appear only in the definition of the start symbol. The respective minimal flat grammars correspond to parsing the given string into a sequence of some optimal words. In the task, you will be requested to show that the minimal flat grammar-based code is universal and satisfies an analogue of inequality (6.86) for the prediction by partial matching (PPM) probability.

Consider the set strings $\mathbb{X}^*$ over a finite alphabet $\mathbb{X} = \{x_1, x_2, \dots, x_D\}$. Denote the set of finite sequences of strings as $\mathbb{X}^{**}$, i.e. for each $V \in \mathbb{X}^{**}$ there exist a number $n \in \mathbb{N} \cup \{0\}$ and strings $w_i \in \mathbb{X}^*$ such that $V = (w_1, w_2, \dots, w_n)$. Let $E : \mathbb{N} \to \{0, 1\}^*$ and $B : \mathbb{X} \cup \{\$, \#\} \to \{0, 1\}^*$ be prefix-free codes such that $B(x_i) := E(i)$, $B(\$) := E(D + 1)$, and $B(\#) := E(D + 2)$. Define codes

$$B^* : \mathbb{X}^* \ni w = (y_1, y_2, \dots, y_{|w|}) \mapsto B(y_1)B(y_2) \cdots B(y_{|w|}) \quad (7.117)$$

$$B'' : \mathbb{X}^{**} \ni V = (w_1, w_2, \dots, w_{|V|})$$
$$\mapsto B^*(w_1)B(\$)B^*(w_2)B(\$) \cdots B(\$)B^*(w_{|V|}) \quad (7.118)$$

Subsequently, for $V = (w_1, w_2, \dots, w_{|V|}) \in (\mathbb{X} \cup \{\$, \#\})^{**}$ and $P = (n_1, n_2, \dots, n_k) \in \{1, 2, \dots, |V|\}^*$ we define mappings

$$\phi(P|V) := w_{n_1} w_{n_2}, \dots, w_{n_k} \quad (7.119)$$

$$E^*(P) := E(n_1)E(n_2) \cdots E(n_k) \quad (7.120)$$

Now for $w \in \mathbb{X}^*$ we define code

$$C(w) := B''(V)B(\#)E^*(P)B(\#) \quad (7.121)$$

where $V = (w_1, w_2, \dots, w_{|V|}) \in \mathbb{X}^{**}$ and $P \in \mathbb{N}^*$ are chosen in such a way that the length of code word $C(w)$ is minimal given constraint

$$w = \phi(P|(x_1, x_2, \dots, x_D, \$, \#, w_1, w_2, \dots, w_{|V|})) \quad (7.122)$$

For these optimal V and P we will write $V(w) := V$ and $P(w) := P$.

(a) Show that $C : \mathbb{X}^* \ni w \mapsto C(w) \in \{0, 1\}^*$ is a prefix-free code.
(b) Can we choose code $E : \mathbb{N} \to \{0, 1\}^*$ such that $Q(w) := 2^{-|C(w)|}$ is a universal semidistribution?
(c) Let $L(u) := \max\{|s| : u = u_1 s u_2 = u_1' s u_2' \text{ for } u_1 \neq u_1'\}$ be the maximal repetition of string u. Show that

$$|C(u)| + |C(v)| - |C(uv)| \leq [\max_{i \leq D+2} |E(i)|][|V(uv)| + 1][L(uv) + 1]$$

$$(7.123)$$

Remark: Inequality (7.123) is an analogue of inequality (6.86) for the PPM probability and it was demonstrated by Dębowski (2011) in the context of proving a theorem about facts and words for minimal grammar-based codes (see Theorems 8.22 and 8.23 later). Both inequalities (7.123) and (6.86) relate the mutual information pertaining to a universal code and the cardinality of the code's vocabulary. In contrast to formula (6.86), formula (7.123) contains maximal repetition $L(uv)$ rather than term $\log|uv|$. Moreover, the minimal code C from formula (7.123) probably cannot be computed in a polynomial time, see Charikar et al. (2005). This second factor makes the theorem about facts and words for code C, as discussed by Dębowski (2011), difficult to apply. Potentially, however, this theorem can be interesting for a linguistic interpretation since the words of code C may resemble orthographic words more than the PPM words, see de Marcken (1996). As an open problem, one could consider proving inequality (7.123) for some versions of code C applying local minimization, see Kieffer and Yang (2000), Charikar et al. (2005), and Ochoa and Navarro (2019).

7.5 Show that all computable functions $\mathbb{R} \to \mathbb{R}$ are continuous.

7.6 Show that for natural numbers n and m we have
$$|\mathbf{H}(m+n) - \mathbf{H}(m)| \le 2\log n + c \qquad (7.124)$$

7.7 Let $n = |w|$ for a string $w \in \{0,1\}^*$. Show that:
(a) $\mathbf{H}(w) \overset{+}{<} n + 2\log n$;
(b) $\mathbf{H}(w|n) \overset{+}{<} n$.

7.8 For a natural number n define $\mathbf{H}^+(n) = \max\{\mathbf{H}(m) : m \le n\}$. Show that $\mathbf{H}^+(n) \overset{+}{=} \log n + \mathbf{H}(\lceil \log n \rceil)$.

7.9 Consider strings $w, u \in \{0,1\}^*$. Assume that $\mathbf{H}(u) \ge |u| - c$ for a fixed number c. Show that $\mathbf{H}(u, w) \overset{+}{>} |u| + \mathbf{H}(w|u)$.

7.10 Having pair $(u, \mathbf{H}(u))$ we may effectively enumerate all programs of the smallest length that compute a discrete object u. Let u^* be the first such program in some fixed enumeration. Prove that:
a) $\mathbf{H}(w|u, \mathbf{H}(u)) \overset{+}{=} \mathbf{H}(w|u^*)$;
b) $\mathbf{H}(w^*) \overset{+}{=} |w^*| = \mathbf{H}(w)$.

7.11 Show that $\mathbf{H}(\mathbf{H}(\mathbf{H}(u))|u, \mathbf{H}(u)) \overset{+}{=} 0$.

7.12 We have $\sum_{u \in \mathbb{A}} 2^{-H(u|w)} \leq 1$ by the Kraft inequality. Show that

$$\sum_{w \in \mathbb{A}} 2^{-H(u|w)} = \infty \tag{7.125}$$

7.13 Consider the Lebesgue measure over binary sequences, i.e. $P(x_1^n) = 2^{-n}$ where $x_i \in \{0, 1\}$. Show that sequence y_1^∞ is algorithmically random with respect to P if and only if y_m^∞ is algorithmically random for any $m > 1$.

7.14 Let a probability measure P satisfy $P(A) = 1$, where A is a countable set of infinite sequences. Show that $\mathcal{R}_P = A$.

7.15 Show that for a stationary process $(X_i)_{i \in \mathbb{Z}}$ over a finite alphabet and positive growing sequences $(k_n)_{n \in \mathbb{N}}$ and $(l_n)_{n \in \mathbb{N}}$ of natural numbers such that $\lim_{n \to \infty}(k_n + l_n) = \infty$, we have

$$\lim_{n \to \infty} \frac{\mathbf{H}(X_{-k_n}^{l_n} | \alpha)}{k_n + l_n + 1} = h_F \text{ almost surely} \tag{7.126}$$

where h_F is the Shannon entropy rate of the random ergodic measure of the process and object α is fixed. For $k_n = -1, l_n = n$, and $\alpha = \lambda$, this result was shown by Brudno (1982).

7.16 Many theorems in probability can be stated as "$\varphi(\omega)$ holds for μ-almost all ω," where φ is a random proposition and μ is a probability measure. As we have discussed in Section 7.5, there are related stronger statements "$\varphi(\omega)$ for all $\omega \in \mathcal{R}_\mu$," where $\mathcal{R}_\mu$ is the set of Martin-Löf random points for measure μ. Of course, we have $\mu(\mathcal{R}_\mu^c) = 0$, which implies $\varphi(\omega)$ for μ-almost all ω if $\varphi(\omega)$ holds for all $\omega \in \mathcal{R}_\mu$. Does $\varphi(\omega)$ for μ-almost all ω imply $\varphi(\omega)$ for all $\omega \in \mathcal{R}_\mu$ automatically if we assume some sort of generalized computability of random proposition φ and measure μ? How general notions of computability can we consider?

8

Power Laws for Information

This chapter opens the exposition of more original research, done mostly by the author of this book. In the chapters to follow, we would like to discuss stochastic processes that satisfy certain power laws common to natural language. As we have identified in Chapter 1, there are several such power laws possibly exhibited by texts in natural language: power-law growth of block mutual information (1.132), power-law logarithmic growth of maximal repetition (1.133), and power-law decay of mutual information between individual characters (1.134). Plausibly, there may be more such laws. It is important to understand to what extent these laws are mathematical consequences of one another. For possible applications in engineering and artificial intelligence, it is also vital to understand general constructions of stochastic processes that satisfy these power laws as well as to comprehend reasons why these laws are satisfied.

We will begin, in this chapter, with a discussion of power laws that are connected to the growth of mutual information between two adjacent blocks of a random text. In the context of natural language modeling, the power-law growth of block mutual information is called the relaxed Hilberg hypothesis, as it was discussed in Section 1.9. A large body of our research concerned relating this hypothesis to other plausible statistical patterns of natural language. Some natural candidates have been the famous Zipf law concerning the distribution of words in texts and its corollary, Herdan's law, stating a power-law growth of the vocabulary size with the text length.

The author of this book was convinced already in 2000 that such a link should be true. However, it took him a long time as a nonprofessional mathematician to absorb enough formal knowledge to express his intuitions as sound theorems – the theorems about facts and words, which were half-formally introduced in Section 1.12. The heuristic idea of such results was stated by Dębowski (2006), whereas fully formal proofs were presented by Dębowski (2011, 2018a). Although the general wording of the theorems about facts and words is not complicated, namely, there should be more distinct words in a text than independent facts, it was a long

Information Theory Meets Power Laws: Stochastic Processes and Language Models, First Edition. Łukasz Dębowski.
© 2021 John Wiley & Sons, Inc. Published 2021 by John Wiley & Sons, Inc.

journey to identify and fill all gaps in the proofs. We have realized that, in order to present the whole reasoning, its motivations, and conceptual preliminaries, we need to write this book.

With the motivations described in Chapter 1 and the formal preliminaries described in Chapters 2–7, we can now proceed to the core results of this book, stated as follows. In Section 8.1, we will investigate a convenient concept of the Hilberg exponent of a real number sequence, which is the exponent of the tightest power-law upper bound of the sequence. Hilberg exponents were mentioned in the half-formal Section 1.9. We will formulate a few general statements concerning this concept. In particular, we will bound the Hilberg exponent for an increasing sequence of random variables with the Hilberg exponents for sequences of their medians and expectations.

In Section 8.2, we will apply the above results to pointwise and expected mutual information between increasing blocks drawn from a stationary process. We will see that the Hilberg exponent for pointwise mutual information is almost surely constant for ergodic processes. Moreover, this exponent is less than the Hilberg exponent for expected mutual information and is equal to it if the Hilberg exponent for the variance of mutual information is sufficiently small. This statement resembles somehow the Shannon–McMillan–Breiman (SMB) theorem discussed in Chapter 6, and for this reason, it will be called the second order SMB theorem.

Section 8.3 is a conceptual preparation for Section 8.4. We will seek to understand the anatomy of strongly nonergodic processes, half-formally mentioned in Section 1.11 and formally defined in Section 5.6, and a related concept of perigraphic processes, also half-formally analyzed in Section 1.11. It turns out that these notions can be related to the phenomenon of keywords and the topical variation of texts in natural language. We can ask what happens when the description of the topic of an infinite text is infinitely complex, necessarily describable by a sequence of algorithmically or probabilistically random bits. Let us call these bits algorithmic or probabilistic facts, respectively. It turns out that if there are infinitely many independent probabilistic facts for a text generated by a stationary probability measure, then the respective process is strongly nonergodic. We exhibit a simple example of such processes, called Santa Fe processes. These processes satisfy a stronger property, namely, the number of facts that are described in a finite text grows like a power of the text length. Motivated by this example, we define perigraphic processes as having this general property.

In Section 8.4, we formulate and demonstrate the theorems about facts and words half-formally mentioned in Section 1.12. These propositions link the Hilberg exponents for three sorts of quantities: the number of independent facts that are described in a finite text drawn from a stationary process, the mutual information between two adjacent text blocks of the same length, and the number of distinct words detected in the text by the prediction by partial matching (PPM) algorithm introducedin Section 6.3. It turns out that those Hilberg exponents are dominated in order that we have named them, i.e. a random text must contain roughly more distinct words than independent facts. There are two versions of this proposition, one for algorithmic facts and another for probabilistic facts. Seen from a linguistic perspective, these theorems link the relaxed Hilberg hypothesis, introducedin Section 1.9 with some sort of Herdan's laws for words and facts, introduced in Section 1.3.

8.1 Hilberg Exponents

In this section, we will study some general properties of power-law growth of sequences of random variables induced by stationary stochastic processes. Suppose that we have a real function of strings $S : \mathbb{X}^* \to \mathbb{R}$ and a certain stationary stochastic process $(X_i)_{i \in \mathbb{Z}}$, where $X_i : \Omega \to \mathbb{X}$. Suppose also that function $S(X_1^n)$ grows asymptotically linearly. To be a bit more specific, let us assume that there exist random limits

$$\lim_{n \to \infty} \frac{S(X_{-n+1}^0)}{n} = \lim_{n \to \infty} \frac{S(X_1^n)}{n}$$

$$= \lim_{n \to \infty} \frac{S(X_{-n+1}^n)}{2n} = s((X_i)_{i \in \mathbb{Z}}) \text{ almost surely} \tag{8.1}$$

A function $S : \mathbb{X}^* \to \mathbb{R}$ that satisfies (8.1) will be called an extensive function. By a generalization of the SMB theorem noticed in Problem 6.3, some example of an extensive function is

$$S(x_1^n) = -\log P(X_1^n = x_1^n) \tag{8.2}$$

or, by Problem 7.15, we can also put

$$S(x_1^n) = \mathbf{H}(x_1^n | \alpha) \tag{8.3}$$

where $\mathbf{H}(w | \alpha)$ is the prefix Kolmogorov complexity of string w given an object α. In both cases,

$$s((X_i)_{i \in \mathbb{Z}}) = h_F \tag{8.4}$$

is the Shannon entropy rate of the random ergodic measure.

Knowing of the asymptotic relationship (8.1), we may be subsequently interested in characterizing the asymptotic behavior of function

$$J(x_{-n+1}^0; x_1^n) := S(x_{-n+1}^0) + S(x_1^n) - S(x_{-n+1}^n) \tag{8.5}$$

which will be called the excess function. In case of (8.2) and (8.3), the excess function equals the respective pointwise mutual information. Of course, equalities (8.1) imply

$$\lim_{n\to\infty} \frac{J(X^0_{-n+1}; X^n_1)}{n} = 0 \text{ almost surely} \tag{8.6}$$

Hence, the excess function $J(X^0_{-n+1}; X^n_1)$ is a sublinear function of the block length n. In particular, the excess function may exhibit an asymptotic power-law growth, the phenomenon of our present interest. However, we should be aware that the excess function $J(X^0_{-n+1}; X^n_1)$ may grow by leaps and bounds so we need an appropriately general theory.

While characterizing the behavior of the excess function $J(X^0_{-n+1}; X^n_1)$, it is insightful to look first at the average case. Assume that there exist expectations

$$\mathfrak{S}(n) := \mathbf{E}\, S(X^n_1) \tag{8.7}$$

$$\mathfrak{s} := \mathbf{E}\, s((X_i)_{i\in\mathbb{Z}}) \tag{8.8}$$

$$\mathfrak{J}(n) := \mathbf{E}\, J(X^0_{-n+1}; X^n_1) \tag{8.9}$$

This happens both for (8.2) and (8.3). For example, for (8.2), we obtain the Shannon entropy $\mathfrak{S}(n) = H(X^n_1)$ and the mutual information $\mathfrak{J}(n) = I(X^0_{-n+1}; X^n_1)$.

By stationarity of $(X_i)_{i\in\mathbb{Z}}$, the expected excess function satisfies

$$\mathfrak{J}(n) = 2\mathfrak{S}(n) - \mathfrak{S}(2n) \tag{8.10}$$

whereas both for (8.2) and (8.3) by the SMB theorem (Theorem 6.9) we obtain the asymptotic expected rates

$$\lim_{n\to\infty} \frac{\mathfrak{S}(n)}{n} = \mathfrak{s} \tag{8.11}$$

$$\lim_{n\to\infty} \frac{\mathfrak{J}(n)}{n} = 0 \tag{8.12}$$

Let us assume that the expected extensive function $\mathfrak{S} : \mathbb{N} \to \mathbb{R}$ satisfies

$$\mathfrak{S}(n) \geq n\mathfrak{s} \tag{8.13}$$

This happens both for (8.2) and (8.3). As shown in the following proposition, in this case, it turns out that the expected excess function $\mathfrak{J} : \mathbb{N} \to \mathbb{R}$ must be positive for infinitely many n.

Theorem 8.1 (*Excess bound*): *For a function* $\mathfrak{S} : \mathbb{N} \to \mathbb{R}$, *define* $\mathfrak{J}(n) := 2\mathfrak{S}(n) - \mathfrak{S}(2n)$. *If* $\lim_{n\to\infty} \mathfrak{S}(n)/n = \mathfrak{s}$ *for a* $\mathfrak{s} \in \mathbb{R}$ *and* $\mathfrak{S}(n) \geq n\mathfrak{s}$ *for all but finitely many* n *then*

$$\mathfrak{J}(n) \geq 0 \tag{8.14}$$

for infinitely many n.

Theorem 8.1 was originally demonstrated by Dębowski (2011). Let us note that for (8.2) we have $\mathfrak{J}(n) \geq 0$ for all n. In contrast, for (8.3) we may have $\mathfrak{J}(n) < 0$, unless the extensive function $S : \mathbb{X}^* \to \mathbb{R}$ is subadditive, i.e. unless the excess function is positive,

$$J(x^0_{-n+1}; x^n_1) \geq 0 \tag{8.15}$$

Then, condition (8.14) is, of course, satisfied for all n.

Proof of Theorem 8.1: We have the identity

$$\sum_{k=0}^{m-1} \frac{\mathfrak{J}(2^k n)}{2^{k+1}} = \mathfrak{S}(n) - n \times \frac{\mathfrak{S}(2^m n)}{2^m n} \tag{8.16}$$

For m tending to infinity, the above implies

$$\sum_{k=0}^{\infty} \frac{\mathfrak{J}(2^k n)}{2^{k+1}} = \mathfrak{S}(n) - n\mathfrak{s} \tag{8.17}$$

Putting $n = 2^p$, we obtain

$$\sum_{k=p}^{\infty} \frac{\mathfrak{J}(2^k)}{2^{k+1}} \geq 0 \tag{8.18}$$

for all but finitely many p. If $\mathfrak{J}(n) \geq 0$ held only for finitely many n, then the sum in (8.18) would be strictly negative for all sufficiently large p. Hence, $\mathfrak{J}(n) \geq 0$ must hold for infinitely many n. □

Suppose now that the expected extensive function satisfies

$$\mathfrak{S}(n) \geq n^\beta + n\mathfrak{s} \tag{8.19}$$

for a certain $\beta \in (0, 1)$. Then, applying Theorem 8.1 to function $\mathfrak{S}'(n) = \mathfrak{S}(n) - n^\beta$ we obtain inequality

$$\mathfrak{J}(n) \geq (2 - 2^\beta)n^\beta \tag{8.20}$$

for infinitely many n. Thus, we have obtained an asymptotic power-law bound for the expected excess function $\mathfrak{J} : \mathbb{N} \to \mathbb{R}$. In general, the asymptotic power-law growth may be captured by the following concept of the Hilberg exponent.

Definition 8.1 *(Hilberg exponent):* The Hilberg exponent of a function $\mathfrak{J} : \mathbb{N} \to \mathbb{R}$ is defined as

$$\text{hilb } \mathfrak{J}(n) := \limsup_{n \to \infty} \frac{\log^+ \mathfrak{J}(n)}{\log n} \tag{8.21}$$

where $\log^+ x := \log(x + 1)$ for $x \geq 0$ and $\log^+ x := 0$ for $x < 0$.

Definition 8.1 generalizes a definition from Dębowski (2015a, 2018a), originally introduced only for the pointwise and expected mutual information. The name "Hilberg exponent" was motivated by Hilberg's hypothesis about the power-law growth of block mutual information (1.132). In particular, if $An^\beta \leq \mathfrak{J}(n) \leq Bn^\beta$ for all n and some $A, B > 0$, then $\mathrm{hilb}_{n\to\infty} \mathfrak{J}(n) = \beta$. Property (8.12) implies also

$$\mathrm{hilb}_{n\to\infty} \mathfrak{J}(n) \in [0, 1] \tag{8.22}$$

Let us compare the Hilberg exponents for functions $\mathfrak{S}(n) - n\mathfrak{s}$ and $\mathfrak{J}(n)$. They turn out to be equal if the expected excess function $\mathfrak{J} : \mathbb{N} \to \mathbb{R}$ is bounded below by a constant.

Theorem 8.2 *(Excess bound): For a function* $\mathfrak{S} : \mathbb{N} \to \mathbb{R}$, *define* $\mathfrak{J}(n) := 2\mathfrak{S}(n) - \mathfrak{S}(2n)$. *If* $\lim_{n\to\infty} \mathfrak{S}(n)/n = \mathfrak{s}$ *for a* $\mathfrak{s} \in \mathbb{R}$ *then*

$$\mathrm{hilb}_{n\to\infty} (\mathfrak{S}(n) - n\mathfrak{s}) \leq \mathrm{hilb}_{n\to\infty} \mathfrak{J}(n) \tag{8.23}$$

with an equality if $\mathfrak{J}(n) \geq -C$ *for all but finitely many n and some* $C > 0$.

Proof: Write $\delta = \mathrm{hilb}_{n\to\infty} \mathfrak{J}(n)$. Since the left hand of (8.23) side is less than 1, it is sufficient to prove this inequality for $\delta < 1$. Observe then that $\mathfrak{J}(n) \leq n^{\delta+\epsilon}$ for all but finitely many n for any $\epsilon > 0$. Then for $\epsilon < 1 - \delta$, by the telescope sum (8.17), we obtain for sufficiently large n that

$$\mathfrak{S}(n) - n\mathfrak{s} \leq \sum_{k=0}^{\infty} \frac{n^{\delta+\epsilon} 2^{k(\delta+\epsilon)}}{2^{k+1}} \leq n^{\delta+\epsilon} \sum_{k=0}^{\infty} 2^{(\delta+\epsilon-1)k-1} = \frac{n^{\delta+\epsilon}}{2(1 - 2^{\delta+\epsilon-1})} \tag{8.24}$$

Since ϵ can be taken arbitrarily small, we obtain (8.23).

Now assume that $\mathfrak{J}(n) \geq -C$ for all but finitely many n. By the telescope sum (8.17), we have $[\mathfrak{J}(n) - C]/2 \leq \mathfrak{S}(n) - n\mathfrak{s}$ for sufficiently large n. Hence,

$$\delta \leq \mathrm{hilb}_{n\to\infty} (\mathfrak{S}(n) - n\mathfrak{s}) \tag{8.25}$$

Thus, we obtain the equality in (8.23). □

Subsequently, we will investigate the Hilberg exponents for some sequences of real random variables. We will present four propositions which restate some results from Dębowski (2015a, 2018a) with certain modifications. They relate the Hilberg exponent for an approximately growing sequence of bounded below random variables to the Hilberg exponent for their expectations. In the following, for a probability space $(\Omega, \mathcal{J}, P)$, we call a measurable transformation $T : \Omega \to \Omega$ stationary or ergodic if the probability measure P is stationary or ergodic with respect to T, respectively.

Theorem 8.3: *Consider a sequence of real random variables $(J(n))_{n\in\mathbb{N}}$. If T is a stationary ergodic transformation of the probability space and $J(n + 1) \geq J(n) \circ T$ for all $n \in \mathbb{N}$, then random variable $\operatorname{hilb}_{n\to\infty} J(n)$ is almost surely constant.*

Proof: Denote random variable $Y := \operatorname{hilb}_{n\to\infty} J(n)$. By the hypothesis $J(n + 1) \geq J(n) \circ T$, we obtain

$$Y = \operatorname*{hilb}_{n\to\infty} J(n + 1) \geq \operatorname*{hilb}_{n\to\infty} J(n) \circ T = Y \circ T \text{ almost surely} \qquad (8.26)$$

But T is stationary, so random variables Y and $Y \circ T$ have the same distribution. Combining these two observations yields

$$P(Y \geq q) = P(Y \circ T \geq q) = P(Y \geq q, Y \circ T \geq q) \qquad (8.27)$$

Hence, $P(A_q) = 1$ for $A_q = ((Y - q)(Y \circ T - q) \geq 0)$. Hence, $P(Y = Y \circ T) = \bigcap_{q\in\mathbb{Q}} A_q = 1$. But T is ergodic, so the latter means that Y is almost surely constant. $\qquad \qquad \square$

To state some further results, we need to introduce the essential infimum and the essential supremum of a real random variable.

Definition 8.2 *(Essential infimum and supremum):* Let X be a real random variable. We define the essential infimum and the essential supremum as

$$\operatorname{ess\,inf} X := \sup\{r \in \mathbb{R} : P(X \leq r) = 0\} \qquad (8.28)$$

$$\operatorname{ess\,sup} X := \inf\{r \in \mathbb{R} : P(X \geq r) = 0\} \qquad (8.29)$$

Theorem 8.4: *Consider a sequence of real random variables $(J(n))_{n\in\mathbb{N}}$. If inequalities*

$$J(m) \geq J(n) - C \log n \qquad (8.30)$$

$$J(n) \geq -C \qquad (8.31)$$

are satisfied for all $m \geq n \geq 1$ and some $C > 0$ then

$$\operatorname{ess\,sup} \operatorname*{hilb}_{n\to\infty} J(n) \leq \operatorname*{hilb}_{n\to\infty} \mathbf{E}\, J(n) \qquad (8.32)$$

Proof: For $2^k \leq n < 2^{k+1}$, we have

$$J(2^k) - Ck \leq J(n) \leq J(2^{k+1}) + C(k + 1) \qquad (8.33)$$

From this sandwich bound, we obtain that the Hilberg exponent for $J(n)$ can be almost surely defined on a sparse subsequence,

$$\text{hilb } J(n) = \limsup_{k \to \infty} \frac{\log^+ J(2^k)}{k} \quad \text{almost surely} \qquad (8.34)$$

(where $\limsup$ is indexed by $n \to \infty$ below hilb)

As we will show now, condition (8.34) is sufficient to obtain inequality (8.32). Denote the Hilberg exponent for $\mathbf{E} J(n)$ as

$$\delta := \limsup_{n \to \infty} \frac{\log^+ \mathbf{E} J(n)}{\log n} \qquad (8.35)$$

Now take an $\epsilon > 0$. Since $J(n) \geq -C$ hence from the Markov inequality (Theorem 3.11), we have

$$\sum_{k=1}^{\infty} P\left(\frac{J(2^k) + C}{2^{k(\delta+\epsilon)}} \geq 1 \right) \leq \sum_{k=1}^{\infty} \frac{\mathbf{E} J(2^k) + C}{2^{k(\delta+\epsilon)}}$$

$$\leq A + \sum_{k=1}^{\infty} \frac{2^{k(\delta+\epsilon/2)}}{2^{k(\delta+\epsilon)}} < \infty \qquad (8.36)$$

where $A < \infty$. Hence, by the Borel–Cantelli lemma (Theorem 3.12),

$$\text{hilb } J(n) = \limsup_{k \to \infty} \frac{\log^+ J(2^k)}{k} \leq \delta + \epsilon \quad \text{almost surely} \qquad (8.37)$$

Since we can take ϵ arbitrarily small then we obtain the claim. $\qquad \square$

We recall that $\mathbf{M}X := \sup\left\{ r : P(Y < r) \leq \frac{1}{2} \right\} \geq \inf\left\{ r : P(Y > r) \leq \frac{1}{2} \right\}$ stands for the median of a real random variable X. Some generalization of this concept are quantiles

$$\mathbf{Q}_p X := \sup\{ r : P(X < r) \leq p \} \geq \inf\{ r : P(X > r) \leq 1 - p \} \qquad (8.38)$$

for $p \in (0, 1)$. We have $\mathbf{M}X = \mathbf{Q}_{1/2}X$, $\text{ess sup } X = \sup_{0 < p < 1} \mathbf{Q}_p X$, and $\text{ess inf } X = \inf_{0 < p < 1} \mathbf{Q}_p X$, see Problem 8.1.

Theorem 8.5: *Consider a sequence of real random variables* $(J(n))_{n \in \mathbb{N}}$ *such that* $J(n) \geq -C$ *for some* $C > 0$. *For any* $0 < p < 1$, *we have*

$$\text{hilb } \mathbf{Q}_p J(n) \leq \text{ess sup hilb } J(n) \qquad (8.39)$$

Proof: Let $(X_n)_{n \in \mathbb{N}}$ be a sequence of real random variables such that

$$\limsup_{n \to \infty} X_n \leq a \quad \text{almost surely} \qquad (8.40)$$

By the Riesz theorem (Theorem 3.27) for random variables $(X_n - a)\mathbf{1}\{X_n > a\}$, condition (8.40) implies that for any $\epsilon > 0$ we have

$$\lim_{n\to\infty} P(X_n > a + \epsilon) = 0 \qquad (8.41)$$

Now denote $\gamma := \text{ess sup hilb}_{n\to\infty} J(n)$. Then,

$$\text{hilb } J(n) = \limsup_{n\to\infty} \frac{\log^+(J(n) + C)}{\log n} \leq \gamma \text{ almost surely} \qquad (8.42)$$

Consequently, for any $\epsilon > 0$, we obtain

$$\lim_{n\to\infty} P\left(\frac{\log^+(J(n) + C)}{\log n} \geq \gamma + \epsilon\right) = 0 \qquad (8.43)$$

Hence, for $0 < p < 1$ we have

$$\forall_{\epsilon>0} \; \exists_{N>0} \; \forall_{n>N} \; \mathbf{Q}_p\left[\frac{\log^+(J(n) + C)}{\log n}\right] \leq \gamma + \epsilon \qquad (8.44)$$

Since function $\log^+ x$ is strictly growing for $x \geq 0$ and $J(n) + C \geq 0$, we have

$$\mathbf{Q}_p\left[\frac{\log^+(J(n) + C)}{\log n}\right] = \left[\frac{\log^+(\mathbf{Q}_p J(n) + C)}{\log n}\right] \qquad (8.45)$$

Resuming these two observations, we obtain

$$\text{hilb } \mathbf{Q}_p J(n) = \limsup_{n\to\infty} \frac{\log^+(\mathbf{Q}_p J(n) + C)}{\log n} \leq \gamma + \epsilon \qquad (8.46)$$

for any $\epsilon > 0$. Thus, we have derived the claim. $\qquad\square$

For a real random variable X, we denote the variance $\text{Var } X := \mathbf{E}(X - \mathbf{E}X)^2$.

Theorem 8.6: *Consider a sequence of real random variables $(J(n))_{n\in\mathbb{N}}$. We have*

$$\text{hilb}_{n\to\infty}\sqrt{\text{Var } J(n)} < \text{hilb}_{n\to\infty} \mathbf{E} J(n) \implies \text{hilb}_{n\to\infty} \mathbf{M}J(n) \geq \text{hilb}_{n\to\infty} \mathbf{E} J(n) \qquad (8.47)$$

Proof: Applying the median–mean inequality (3.45), we obtain inequalities

$$\text{hilb}_{n\to\infty} \mathbf{E} J(n) \leq \text{hilb}_{n\to\infty} [\mathbf{M}J(n) + \sqrt{\text{Var}J(n)}] \leq \max\left\{\text{hilb}_{n\to\infty}\mathbf{M}J(n), \text{hilb}_{n\to\infty}\sqrt{\text{Var}J(n)}\right\} \qquad (8.48)$$

which imply the claim. $\qquad\square$

In the following sections, we will see some applications of the above results. Theorems 8.4 and 8.5 can be further strengthened, see Problem 8.3.

8.2   Second Order SMB Theorem

In this section, we will derive a sublinear analogue of the SMB theorem (Theorem 6.1) for the block mutual information. The material in this section comes from Dębowski (2015a) with some modifications. The preliminaries are as follows. For an extensive function $S_Q : \mathbb{X}^* \to \mathbb{R}$ and a stationary stochastic process $(X_i)_{i\in\mathbb{Z}}$, where $X_i : \Omega \to \mathbb{X}$, let us consider the pointwise excess function

$$J_Q(x^0_{-n+1};x^n_1) := S_Q(x^0_{-n+1}) + S_Q(x^n_1) - S_Q(x^n_{-n+1}) \tag{8.49}$$

and the random variables

$$J_Q(n) := J_Q(X^0_{-n+1};X^n_1) \tag{8.50}$$

Some particular extensive functions which we will consider in this section are the prefix-free Kolmogorov complexity conditioned on an object α,

$$S_{|\alpha}(x^n_1) := \mathbf{H}(x^n_1|\alpha) \tag{8.51}$$

and the pointwise entropy of the distribution of the process,

$$S_P(x^n_1) := -\log P(X^n_1 = x^n_1) \tag{8.52}$$

For these functions, let us denote the Hilberg exponents

$$\delta_Q := \operatorname*{hilb}_{n\to\infty} \mathbf{E}\, J_Q(n) \tag{8.53}$$

$$\gamma_Q := \operatorname*{hilb}_{n\to\infty} J_Q(n) \tag{8.54}$$

$$\epsilon_Q := \operatorname*{hilb}_{n\to\infty} \sqrt{\operatorname{Var} J_Q(n)} \tag{8.55}$$

Exponent γ_Q is a real random variable, whereas exponents δ_Q and ϵ_Q are real numbers.

First, we will prove the theorem which links the expected and the random Hilberg exponents for Kolmogorov complexity. Although the Kolmogorov complexity itself is uncomputable, the Hilberg exponents for Kolmogorov complexity can be evaluated in some cases and enjoy a few nice properties. These properties stem from the fact that function $J_{|\alpha}(n)$ up to an additive constant equals the algorithmic mutual information

$$J_{|\alpha}(n) \overset{+}{=} \mathbf{I}(X^0_{-n+1};X^n_1|\alpha)$$
$$= \mathbf{H}(X^0_{-n+1}|\alpha) + \mathbf{H}(X^n_1|\alpha) - \mathbf{H}(X^0_{-n+1},X^n_1|\alpha) \tag{8.56}$$

which is a positive and growing function in some approximation.

Theorem 8.7: *Consider a discrete-valued stationary process* $(X_i)_{i \in \mathbb{Z}}$ *and the extensive function (8.51). We have*

$$\operatorname{ess\,sup} \gamma_{|\alpha} \leq \delta_{|\alpha} \qquad (8.57)$$

with an equality if $\epsilon_{|\alpha} < \delta_{|\alpha}$*. Moreover, exponent* $\gamma_{|\alpha}$ *is almost surely constant for an ergodic process over a finite alphabet.*

Proof:
1. The first claim, i.e. inequality (8.57) follows by Theorems 8.4–8.6 since, as we will prove, we have

$$J_{|\alpha}(m) \geq J_{|\alpha}(n) - C \log n \qquad (8.58)$$

$$J_{|\alpha}(n) \geq -C \qquad (8.59)$$

for $m \geq n \geq 1$. First, for strings u and v, we denote the algorithmic mutual information

$$\mathbf{I}(u; v|\alpha) := \mathbf{H}(u|\alpha) + \mathbf{H}(v|\alpha) - \mathbf{H}(u, v|\alpha) \qquad (8.60)$$

By Theorem 7.15, for $|u| = |v|$, we obtain

$$\mathbf{H}(uv|\alpha) \overset{+}{=} \mathbf{H}(u, v|\alpha) \qquad (8.61)$$

whereas for $|u| \neq |v|$, we have

$$\mathbf{H}(uv|\alpha) \overset{+}{<} \mathbf{H}(u, v|\alpha) \overset{+}{<} \mathbf{H}(uv|\alpha) + \mathbf{H}(|v|) \qquad (8.62)$$

Put $a = X_{-m+1}^{-n}$, $b = X_{-n+1}^{0}$, $c = X_1^n$, and $d = X_{n+1}^m$. Since $\mathbf{H}(n) \overset{+}{<} 2 \log n$, hence

$$J_{|\alpha}(m) + 4 \log n \overset{+}{>} \mathbf{I}(a, b; c, d|\alpha) \qquad (8.63)$$

whereas by Theorem 7.29,

$$J_{|\alpha}(n) \overset{+}{=} \mathbf{I}(b; c|\alpha) \overset{+}{>} 0 \qquad (8.64)$$

It suffices now to relate $\mathbf{I}(a, b; c, d|\alpha)$ and $\mathbf{I}(b; c|\alpha)$. In fact, using identity $\mathbf{H}(u, v|\alpha) \overset{+}{=} \mathbf{H}(u|\alpha) + \mathbf{H}(v|u, \mathbf{H}(u|\alpha), \alpha)$ from Theorem 7.20, we can show

$$\begin{aligned} \mathbf{I}(a, b; c, d|\alpha) &\overset{+}{=} \mathbf{I}(a, b; c|\alpha) + \mathbf{I}(a, b; d|c, \mathbf{H}(c|\alpha), \alpha) \\ &\overset{+}{>} \mathbf{I}(a, b; c|\alpha) \\ &\overset{+}{=} \mathbf{I}(b; c|\alpha) + \mathbf{I}(a; c|b, \mathbf{H}(b|\alpha), \alpha) \\ &\overset{+}{>} \mathbf{I}(b; c|\alpha) \qquad (8.65) \end{aligned}$$

which completes the proof.

2. Now, we will prove that $\gamma_{|\alpha}$ is almost surely constant for an ergodic process over a finite alphabet. For $t > 0$, from the shortest program that computes x_1^n, we can construct a program that computes x_{t+1}^{t+n}, whose length exceeds the length of the program for x_1^n no more than $\mathbf{H}(x_{n+1}^{n+t}|\alpha) + C$, where $C > 0$. Analogously, from the shortest program that computes x_{t+1}^{t+n}, we can construct a program that computes x_1^n, whose length exceeds the length of the program for x_{t+1}^{t+n} no more than $\mathbf{H}(x_1^t|\alpha) + C$. This yields

$$|\mathbf{H}(x_1^n|\alpha) - \mathbf{H}(x_{t+1}^{t+n}|\alpha)| \leq \max\{\mathbf{H}(x_1^t|\alpha), \mathbf{H}(x_{n+1}^{n+t}|\alpha)\} + C \tag{8.66}$$

Thus, for the shift transformation (4.1), we obtain

$$|J_{|\alpha}(n) - J_{|\alpha}(n) \circ T|$$
$$\leq 3\max\{\mathbf{H}(X_{-n+1}|\alpha), \mathbf{H}(X_1|\alpha), \mathbf{H}(X_{n+1}|\alpha)\} + 3C \tag{8.67}$$

Hence, for a finite alphabet, random variable $\gamma_{|\alpha}$ is invariant, i.e. $\gamma_{|\alpha} = \gamma_{|\alpha} \circ T$ almost surely. Since the process is ergodic, it means that $\gamma_{|\alpha}$ must be almost surely constant. $\qquad\square$

In the following, we will show that the Hilberg exponents for the Kolmogorov complexity and the pointwise entropy are almost surely equal if we condition the Kolmogorov complexity on the probability measure of the process. The result follows by Theorem 7.32.

Theorem 8.8: *Consider a discrete-valued stationary process $(X_i)_{i \in \mathbb{Z}}$ and the extensive function (8.51) with $\alpha = P$, where P is the probability measure of the process. We have*

1. $\delta_{|P} = \delta_P$,
2. $\gamma_{|P} = \gamma_P$ *almost surely,*
3. $\epsilon_{|P} = \epsilon_P$.

Proof: We note that $S_{|P}(x_1^n) = \mathbf{H}(x_1^n|P)$ and, in the following, we apply some partial results from Theorem 7.32.

1. By (7.75) and (7.76), we obtain

$$|\mathbf{E}J_{|P}(n) - \mathbf{E}J_P(n)| \leq 4\log n + 2C \tag{8.68}$$

Thus, $\delta_{|P} = \delta_P$.
2. By (7.75) and (7.78), we obtain

$$|J_{|P}(n) - J_P(n)| \leq 4\log n + 2C \tag{8.69}$$

for all but finitely many n almost surely. Thus, $\gamma_{|P} = \gamma_P$ holds almost surely.

3. By (7.75) and (7.77), we obtain

$$P(|J_{|P}(n) - J_P(n)| \geq 4\log n + 2C + m) \leq 2^{-m} \tag{8.70}$$

Subsequently, this yields

$$\mathbf{E}\,(J_{|P}(n) - J_P(n))^2 \leq (4\log n + 2C)^2 + \sum_{m=0}^{\infty} (4\log n + 2C + m + 1)^2 2^{-m}$$

$$\leq A(\log n)^2 + B \tag{8.71}$$

for some constants $A, B > 0$. Hence, equality $\epsilon_{|P} = \epsilon_P$ follows by the relation

$$|\sqrt{\mathrm{Var}X} - \sqrt{\mathrm{Var}Y}| \leq \sqrt{\mathrm{Var}(X - Y)} \leq \sqrt{\mathbf{E}\,(X - Y)^2} \tag{8.72}$$

□

Theorems 8.7 and 8.8 can be combined into a second-order analogue of the SMB theorem (Theorem 6.1). The original idea of the SMB theorem was to relate the asymptotic growth of the pointwise and the expected entropies for an ergodic process P. In contrast, relating the random Hilberg exponent γ_P and the expected Hilberg exponent δ_P means relating the rate of growth of the pointwise and expected mutual informations, which are differences of the respective entropies. This is a somewhat subtler effect than the SMB theorem, hence our "second-order" terminology.

Theorem 8.9 *(Second order SMB theorem): Consider a discrete-valued stationary process* $(X_i)_{i \in \mathbb{Z}}$ *and the extensive function (8.52). We have*

$$\mathrm{ess\,sup}\,\gamma_P \leq \delta_P \tag{8.73}$$

with an equality if $\epsilon_P < \delta_P$. *Moreover, exponent* γ_P *is almost surely constant for an ergodic process over a finite alphabet.*

Proof: The claim follows by Theorems 8.7 and 8.8. □

Hilberg exponents δ_P and γ_P can be effectively computed for some simple processes such as Bernoulli processes or Santa Fe processes, to be discussed in the next section. The reader is referred to Dębowski (2015a), see also Problem 8.4.

8.3 Probabilistic and Algorithmic Facts

In Definitions 4.3, 4.5, and 4.6, we have subsequently defined ergodic processes as such that have a trivial invariant σ-field, whereas in Definition 5.16, we have defined strongly nonergodic processes as those whose invariant σ-field contains a nonatomic sub-σ-field. In this section, we will present a different characterization of ergodic and strongly nonergodic processes which is based on a reflection upon natural language, sketched in the half-formal Section 1.11. However, we will present the ideas in the opposite direction than in Section 1.11, i.e. we will discuss the probabilistic approach first and seek for its algorithmic counterpart afterward.

As a prerequisite, we recall that according to the Birkhoff ergodic theorem (Theorem 4.6), a process is ergodic if and only if the relative frequency of any finite subsequence in a random sample converges to a certain constant in the long run. Consider now the following thought experiment. Suppose that we select a random book from a library. Yaglom and Yaglom (1983) observed that there is hardly any book that contains a large number of occurrences of both word *lemma* and word *love*. Simply put, there are some keywords that are specific to particular topics of texts and the relative frequencies of these keywords are higher for books concerning some topics, whereas they are lower for books concerning other topics. In other words, the relative frequencies of keywords are random variables with values depending on the particular topic of the randomly selected book. Since keywords are just some particular strings, we may conclude that the stochastic process that models arbitrary books in natural language should be nonergodic.

The above thought experiment provides another perspective onto nonergodic processes. Counting the relative frequencies of keywords, such as *lemma* or *love*, we can effectively recognize the topic of a random text. In general, we can imagine quite arbitrary methods of topic identification. According to the following Theorems 8.10 and 8.11, a process is nonergodic exactly when we can distinguish in the limit at least two random topics in it. Functions $f : \mathbb{X}^* \to \{0, 1, 2\}$ and $g : \mathbb{N} \times \mathbb{X}^* \to \{0, 1, 2\}$ appearing in these statements serve that goal. Precisely, they assume values 0 or 1 when we can identify some partial description of the topic, whereas they take value 2 when we are not certain which partial topic description should be assigned to the text.

In the following, let $\mathcal{J}$ be the invariant σ-field of the process $(X_i)_{i \in \mathbb{Z}}$ for the shift operation (4.1) and let $\sigma_P(\mathcal{A})$ be the completion of a σ-field $\mathcal{A}$.

Theorem 8.10: *Assume that $(X_i)_{i \in \mathbb{Z}}$ is a stationary process. A random bit Z is $\sigma_P(\mathcal{J})$-measurable if and only if it satisfies condition*

$$\lim_{n \to \infty} P(f(X_{j+1}^{j+n}) = Z) = 1 \tag{8.74}$$

for any $j \in \mathbb{Z}$ and some function $f : \mathbb{X}^* \to \{0, 1, 2\}$.

Proof: In the following, T denotes the shift operation (4.1), whereas $\mathcal{G}_j^k$ is the σ-field generated by random variables $(X_i)_{j \leq i \leq k}$.

Assume first that Z is $\sigma_P(\mathcal{J})$-measurable. Hence, by Theorem 5.34, variable Z is also $\sigma_P(\mathcal{G}_1^\infty)$-measurable. Hence, by Theorem 5.37, $\lim_{n \to \infty} P(f(X_1^k) = Z) = 1$ for some function f. Finally, stationarity of $(X_i)_{i \in \mathbb{Z}}$ and $\sigma_P(\mathcal{J})$-measurability of Z imply that the probabilities $P(f(X_{j+1}^{j+n}) = Z)$ do not depend on j. So random variable Z satisfies condition (8.74).

As for the converse, assume that Z satisfies (8.74). By Theorem 5.37, variable Z is $\sigma_P(\mathcal{G}_1^\infty)$-measurable. Consider random variables $Z_n = f(X_1^n)$. By condition (8.74), $\lim_{n \to \infty} P(Z_n \circ T^j = Z) = 1$, for all $j \in \mathbb{Z}$, and hence $\lim_{n \to \infty} P(Z_n = Z \circ T^{-j}) = 1$ by stationarity of $(X_i)_{i \in \mathbb{Z}}$. The latter implies $Z = Z \circ T^{-j}$ almost surely and thus Z is $\sigma_P(\mathcal{J})$-measurable. □

Now let us introduce this definition.

Definition 8.3 *(Probabilistic fact)*: A random bit Z satisfying condition (8.74) will be called a probabilistic fact.

A probabilistic fact tells which of two topics the infinite text generated by the stationary process is about. It is a kind of a random switch which is preset before we start scanning the infinite text, compare this statement with a similar wording by Gray and Davisson (1974b). Consequently, a process is nonergodic when there exists at least one probabilistic fact which is not constant.

Theorem 8.11: *A stationary process $(X_i)_{i \in \mathbb{Z}}$ is nonergodic if and only if there exists a random bit Z which satisfies condition (8.74) and has distribution*

$$P(Z = 1) = 1 - P(Z = 0) \in (0, 1) \tag{8.75}$$

Proof: By Definition 4.6, process $(X_i)_{i \in \mathbb{Z}}$ is ergodic if and only if there does not exists a $\sigma_P(\mathcal{J})$-measurable binary random variable Z with distribution (8.75). Hence, the claim follows by Theorem 8.10. □

As for books in natural language, we may have an intuition that the pool of available book topics is extremely large and contains many more topics than just two. For this reason, we may need not a single probabilistic fact Z but rather a sequence of probabilistic facts $Z_1, Z_2, \dots$ to specify the topic of a random book completely. Formally, a stationary processes requiring an infinite sequence of independent uniformly distributed probabilistic facts to describe its topic turns out to be strongly nonergodic.

Theorem 8.12: *A stationary process $(X_i)_{i \in \mathbb{Z}}$ is strongly nonergodic if and only if condition*

$$\lim_{n \to \infty} P(g(k, X_{j+1}^{j+n}) = Z_k) = 1 \tag{8.76}$$

holds for all $j \in \mathbb{Z}$ and $k \in \mathbb{N}$, some function $g : \mathbb{N} \times \mathbb{X}^* \to \{0, 1, 2\}$, and some fair-coin process $(Z_k)_{k \in \mathbb{N}}$.

Proof: By Definition 5.16, process $(X_i)_{i \in \mathbb{Z}}$ is strongly nonergodic if and only if $\sigma_P(\mathcal{J})$ contains a nonatomic sub-σ-field. By Theorem 3.5 this holds if and only if there exists a $\sigma_P(\mathcal{J})$-measurable fair-coin process $(Z_k)_{k \in \mathbb{N}}$. Hence, the claim follows by Theorem 8.10. □

Theorem 8.12 was established by Dębowski (2009).

As we have stated above, for a strongly nonergodic process, there is an infinite number of independent probabilistic facts $(Z_k)_{k \in \mathbb{N}}$ with a uniform distribution on the alphabet $\{0, 1\}$. Formally, these probabilistic facts can be assembled into a single real random variable $T = \sum_{k=1}^{\infty} 2^{-k} Z_k$, which is uniformly distributed on the unit interval $[0, 1]$, i.e. according to the Lebesgue measure. The value of random variable T identifies the topic of a random infinite text. Thus, for a strongly nonergodic process, we have a continuum of available topics which can be incrementally identified from any sufficiently long text.

As shown in Theorem 5.42, the mixture Bernoulli process (5.114) is some example of a strongly nonergodic process. In this case, the random parameter Y defined in Eq. (5.116) plays the role of the random variable $T = \sum_{k=1}^{\infty} 2^{-k} Z_k$. As another example, we will construct now some strongly nonergodic processes for which condition (8.76) can be verified directly and easily. The example comes under an umbrella of Santa Fe processes. The first mention of Santa Fe processes appears in articles (Dębowski, 2009, 2011).

Definition 8.4 *(Mixture Zipfian Santa Fe process)*: Let $(K_i)_{i \in \mathbb{Z}}$ be an independent identically distributed (IID) process assuming values in natural numbers with distribution

$$P(K_i = k) = \frac{k^{-\alpha}}{\zeta(\alpha)}, \quad \alpha > 1 \tag{8.77}$$

where $\zeta(\alpha) := \sum_{k=1}^{\infty} k^{-\alpha}$ is the Riemann zeta function, and let $(Z_k)_{k \in \mathbb{N}}$ be a fair-coin process independent from process $(K_i)_{i \in \mathbb{Z}}$. Then, the mixture Zipfian Santa Fe process $(X_i)_{i \in \mathbb{Z}}$ with an exponent α is defined as

$$X_i := (K_i, Z_{K_i}) \tag{8.78}$$

Theorem 8.13: *Mixture Zipfian Santa Fe processes are strongly nonergodic.*

Proof: Mixture Zipfian Santa Fe process are strongly nonergodic since condition (8.76) holds for example for

$$g(k, x_1^n) = \begin{cases} 0, & \text{if } \exists_{1 \le i \le n} x_i = (k, 0) \text{ and } \neg \exists_{1 \le i \le n} x_i = (k, 1) \\ 1, & \text{if } \exists_{1 \le i \le n} x_i = (k, 1) \text{ and } \neg \exists_{1 \le i \le n} x_i = (k, 0) \\ 2, & \text{else} \end{cases} \tag{8.79}$$

Simply speaking, function $g(k, \cdot)$ returns 0 or 1 when an unambiguous value of the second constituent can be read off from pairs $x_i = (k, \cdot)$ and returns 2 when there is some ambiguity. Condition (8.76) is satisfied since

$$P(g(k, X_{j+1}^{j+n}) = Z_k) = P(K_i = k \text{ for some } j + 1 \le i \le j + n)$$

$$= 1 - (1 - P(K_i = k))^n \xrightarrow[n \to \infty]{} 1 \tag{8.80}$$

$\square$

Some property that distinguishes the mixture Zipfian Santa Fe processes from the mixture Bernoulli process (5.114) is a power-law growth of the number of inferrable probabilistic facts, defined as follows.

Definition 8.5 *(Set of inferrable probabilistic facts):* For a strongly nonergodic stationary process $(X_i)_{i \in \mathbb{Z}}$, let $(Z_k)_{k \in \mathbb{N}}$ be a fair-coin process that satisfies condition (8.76). The set of probabilistic facts inferrable from block x_1^n is defined as

$$U(X_1^n) := \{l \in \mathbb{N} : g(k, X_1^n) = Z_k \text{ for all } k \le l\} \tag{8.81}$$

In other words, we have $U(X_1^n) = \{1, 2, \dots, l\}$ where l is the largest number such that $g(k, X_1^n) = Z_k$ for all $k \le l$.

Theorem 8.14: *For the mixture Zipfian Santa Fe process with an exponent α and function (8.79),*

$$\operatorname*{hilb}_{n \to \infty} \#U(X_1^n) = 1/\alpha \text{ almost surely} \tag{8.82}$$

$$\operatorname*{hilb}_{n \to \infty} \mathbf{E} \#U(X_1^n) = 1/\alpha \tag{8.83}$$

Proof: First, we obtain

$$P(\#U(X_1^n) \le m_n) \le \sum_{k=1}^{m_n} P(g(k, X_1^n) \ne Z_k) = \sum_{k=1}^{m_n} [1 - P(K_i = k)]^n$$

$$\le m_n \left[1 - \frac{m_n^{-\alpha}}{\zeta(\alpha)} \right]^n \le m_n \exp(-nm_n^{-\alpha}/\zeta(\alpha)) \tag{8.84}$$

Put now $m_n = n^{1/\alpha - \epsilon}$ for an $\epsilon > 0$. It is easy to observe that

$$\sum_{n=1}^{\infty} P(\#U(X_1^n) \le m_n) < \infty \tag{8.85}$$

Hence, by the Borel–Cantelli lemma (Theorem 3.12), we obtain $\#U(X_1^n) > m_n$ for all but finitely many n almost surely.

Second, we obtain

$$P(\#U(X_1^n) \ge M_n) \le \frac{n!}{(n - M_n)!} \prod_{k=1}^{M_n} P(K_i = k)$$

$$\le \left(\frac{1}{M_n!}\right)^{\alpha} \left(\frac{n}{\zeta(\alpha)}\right)^{M_n} \le \left(\frac{e^{\alpha} n}{M_n^{\alpha} \zeta(\alpha)}\right)^{M_n} \tag{8.86}$$

where we use inequality $k! \ge (k/e)^k$. Put now $M_n = e(2n/\zeta(\alpha))^{1/\alpha}$. We obtain

$$P(\#U(X_1^n) \ge M_n) \le \left(\frac{1}{2}\right)^{M_n} \tag{8.87}$$

so we obtain

$$\sum_{n=1}^{\infty} P(\#U(X_1^n) \ge M_n) < \infty \tag{8.88}$$

Hence, the Borel–Cantelli lemma (Theorem 3.12) yields $\#U(X_1^n) < M_n$ for all but finitely many n almost surely. Combining this result with the previous result yields equality (8.82).

To obtain equality (8.83), we notice that $\#U(X_1^n)$ is a positive growing function of n and we invoke Theorem 8.4 for the lower bound, whereas for the upper bound we observe that

$$\mathbf{E} \, \#U(X_1^n) \le M_n + n P(\#U(X_1^n) \ge M_n) \tag{8.89}$$

where the last term decays according to the stretched exponential bound (8.87) for $M_n = e(2n/\zeta(\alpha))^{1/\alpha}$. □

As it was motivated in Section 1.11, we may ask whether the idea of the power-law growth of the number of inferrable facts can be translated somehow to the case of ergodic processes. Some straightforward method to apply is to replace the fair-coin process $(Z_k)_{k\in\mathbb{N}}$ with a certain algorithmically incompressible sequence of particular binary digits $(z_k)_{k\in\mathbb{N}}$. Subsequently, let us try to generalize condition (8.76) as

$$\lim_{n \to \infty} P(g(k, X_{j+1}^{j+n}) = z_k) = 1 \tag{8.90}$$

In fact, condition (8.90) is trivially satisfied for any stationary process for a certain recursive function $g : \mathbb{N} \times \mathbb{X}^* \to \{0, 1, 2\}$ and an algorithmically incompressible sequence $(z_k)_{k\in\mathbb{N}}$. Let $(\mathbf{\Omega}_k)_{k\in\mathbb{N}}$ be the binary expansion of the halting probability $\mathbf{\Omega} = \sum_{k=1}^{\infty} 2^{-k}\mathbf{\Omega}_k$. By Theorem 7.35, the halting probability $\mathbf{\Omega}$ is lower semicomputable so there exists a recursive function $\omega : \mathbb{N}\times\mathbb{N} \to \{0,1\}$ such that $\lim_{n\to\infty}\omega(k,n) = \mathbf{\Omega}_k$. Putting $(z_k)_{k\in\mathbb{N}} = (\mathbf{\Omega}_k)_{k\in\mathbb{N}}$ and $g(k,X_{j+1}^{j+n}) = \omega(k,n)$, we recover condition (8.90).

In spite of this negative result, let us define the set of algorithmic facts inferrable from an arbitrary stationary process.

Definition 8.6 *(Set of inferrable algorithmic facts)*: For a strongly nonergodic stationary process $(X_i)_{i\in\mathbb{Z}}$, let $(z_k)_{k\in\mathbb{N}}$ be an algorithmically incompressible sequence that satisfies condition (8.90) for a certain recursive function $g : \mathbb{N}\times\mathbb{X}^* \to \{0,1,2\}$. The set of algorithmic facts inferrable from block x_1^n is defined as

$$\mathbf{U}(X_1^n) := \{l \in \mathbb{N} : g(k,X_1^n) = z_k \text{ for all } k \leq l\} \tag{8.91}$$

Subsequently, we will call a process perigraphic if the expected number of algorithmic facts which can be inferred from a finite text sampled from the process grows asymptotically like a power of the text length.

Definition 8.7 *(Perigraphic process)*: A stationary discrete process $(X_i)_{i\in\mathbb{Z}}$ is called perigraphic if

$$\operatorname*{hilb}_{n\to\infty} \mathbf{E} \,\#\mathbf{U}(X_1^n) > 0 \tag{8.92}$$

for some recursive function $g : \mathbb{N}\times\mathbb{X}^* \to \{0,1,2\}$ and an algorithmically incompressible sequence $(z_k)_{k\in\mathbb{N}}$.

The concept of a perigraphic process was introduced by Dębowski (2018a).

Let us also adjust the definition of the Santa Fe process.

Definition 8.8 *(Individual Zipfian Santa Fe process)*: Let $(K_i)_{i\in\mathbb{Z}}$ be an IID process assuming values in natural numbers with the power-law distribution (8.77), and let $(z_k)_{k\in\mathbb{N}}$ be an arbitrary algorithmically incompressible sequence of binary digits. Then, the individual Zipfian Santa Fe process $(X_i)_{i\in\mathbb{Z}}$ with an exponent α is defined as

$$X_i = (K_i, z_{K_i}) \tag{8.93}$$

Theorem 8.15: *Individual Zipfian Santa Fe processes are ergodic.*

Proof: Individual Zipfian Santa Fe process are IID processes, which are mixing and ergodic by Theorem 4.14. □

Theorem 8.16: *For the individual Zipfian Santa Fe process with an exponent α and function (8.79),*

$$\operatorname*{hilb}_{n\to\infty} \#\mathbf{U}(X_1^n) = 1/\alpha \text{ almost surely} \tag{8.94}$$

$$\operatorname*{hilb}_{n\to\infty} \mathbf{E} \,\#\mathbf{U}(X_1^n) = 1/\alpha \tag{8.95}$$

Proof: We obtain the proof by plugging $(z_k)_{k\in\mathbb{N}}$ instead of $(Z_k)_{k\in\mathbb{N}}$ in the proof of Theorem 8.14. □

According to the above proposition, individual Zipfian Santa Fe processes are perigraphic. But as we will show in the next section, not every process is perigraphic since all perigraphic processes have nonrecursive distributions.

8.4 Theorems About Facts and Words

Consider the Shannon mutual information $I(X_1^n; X_{n+1}^{2n})$ between two blocks of an infinite random text and let us consider its limit, the excess entropy

$$E = \lim_{n\to\infty} I(X_1^n; X_{n+1}^{2n}) = I(X_{-\infty}^0; X_0^\infty) \tag{8.96}$$

where the second equality was established in Theorem 5.33. In Theorem 5.40, we have shown that $E = \infty$ for any strongly nonergodic process. Intuitively, we may expect this result since an infinite sequence of independent random bits $(Z_k)_{k\in\mathbb{N}}$ can be learned given either the infinite past or the infinite future of $(X_i)_{i\in\mathbb{Z}}$.

Subsequently, Theorem 5.40 can be strengthened in form of an inequality between the mutual information and the expected number of facts that can be inferred from a finite random text. Moreover, using Theorem 6.16 for the PPM probabilities, we can demonstrate a reverse inequality between the mutual information and the expected number of PPM words detected in the random text. Chaining these two results together, we obtain statements which we call theorems about facts and words, which were mentioned half-formally in the introductory Section 1.12. These statements say that the number of inferrable independent facts must be roughly smaller than the number of detected PPM words. The theorems hold both in a probabilistic and an algorithmic setting.

The preliminaries are as follows. Let us restrict ourselves to processes over a finite alphabet. We will compare the Shannon entropy $H(X_1^n)$, the Kolmogorov complexity $\mathbf{H}(X_1^n)$, and the PPM probability of a random block x_1^n. We recall notation $\langle \mathrm{PPM}(X_1^n)\rangle := -\log \mathrm{PPM}(X_1^n)$. Since the PPM probability is a recursive distribution then by (7.79), we have

$$\mathbf{H}(X_1^n) \overset{+}{<} \langle \mathrm{PPM}(X_1^n)\rangle + 2\log n + \mathbf{H}(\mathrm{PPM}) \tag{8.97}$$

whereas by (7.80), we have

$$H(X_1^n) \le \mathbf{E}\,\mathbf{H}(X_1^n) \tag{8.98}$$

Hence, by Theorem 5.12 (existence of the entropy rate) and Theorem 6.7 (universality of the PPM probability), we obtain equality of rates

$$h = \lim_{n\to\infty} \frac{H(X_1^n)}{n} = \lim_{n\to\infty} \frac{\mathbf{E}\,\mathbf{H}(X_1^n)}{n} = \lim_{n\to\infty} \frac{\mathbf{E}\,\langle \mathrm{PPM}(X_1^n)\rangle}{n} \tag{8.99}$$

Moreover, for a stationary process, the mutual information satisfies

$$I(X_1^n; X_{n+1}^{2n}) = 2H(X_1^n) - H(X_1^{2n}) \ge 0 \tag{8.100}$$

$$\mathbf{E}\,\mathbf{I}(X_1^n; X_{n+1}^{2n}) = 2\,\mathbf{E}\,\mathbf{H}(X_1^n) - \mathbf{E}\,\mathbf{H}(X_1^{2n}) \overset{+}{>} 0 \tag{8.101}$$

Hence, by Theorem 8.2, we obtain a network of inequalities

$$\underset{n\to\infty}{\mathrm{hilb}}\,[H(X_1^n) - hn] = \underset{n\to\infty}{\mathrm{hilb}}\,I(X_1^n; X_{n+1}^{2n})$$

$$\le$$

$$\underset{n\to\infty}{\mathrm{hilb}}\,[\mathbf{E}\,\mathbf{H}(X_1^n) - hn] = \underset{n\to\infty}{\mathrm{hilb}}\,\mathbf{E}\,\mathbf{I}(X_1^n; X_{n+1}^{2n})$$

$$\le$$

$$\underset{n\to\infty}{\mathrm{hilb}}\,[\mathbf{E}\,\langle \mathrm{PPM}(X_1^n)\rangle - hn] \le \underset{n\to\infty}{\mathrm{hilb}}\,\mathbf{E}\,[2\langle \mathrm{PPM}(X_1^n)\rangle - \langle \mathrm{PPM}(X_1^{2n})\rangle] \tag{8.102}$$

We also note this auxiliary statement.

Theorem 8.17: *For a random variable S with values in natural numbers,*

$$H(S) \le \log \frac{\pi^2}{6} + 2\log \mathbf{E}\,S \tag{8.103}$$

$$\mathbf{H}(S) \overset{+}{<} 2\log S \tag{8.104}$$

Proof: Let $\pi(n) := 6\pi^{-2}n^{-2}$. By the Gibbs inequality (Theorem 3.18) and the Jensen inequality (Theorem 3.14), we have

$$0 \le \sum_{s=1}^{\infty} P(S = s)\log \frac{P(S = s)}{\pi(s)}$$

$$\le \log \frac{\pi^2}{6} + 2\,\mathbf{E}\,\log S - H(S) \le \log \frac{\pi^2}{6} + 2\log \mathbf{E}\,S - H(S) \tag{8.105}$$

On the other hand, by Theorem 7.31, since function $\pi(n)$ is computable, we obtain

$$\mathbf{H}(S) \stackrel{+}{=} \mathbf{H}(S|\pi) \stackrel{+}{<} -\log \pi(S) \stackrel{+}{<} 2\log S \tag{8.106}$$

□

Having these results, we may formulate the main statements of this section. The first one is the theorem about probabilistic facts and mutual information.

Theorem 8.18 (*Probabilistic facts and mutual information*): Let $(X_i)_{i \in \mathbb{Z}}$ be a *stationary strongly nonergodic process, where $X_i : \Omega \to \mathbb{X}$ are random variables over a finite alphabet $\mathbb{X}$. We have inequality*

$$\operatorname*{hilb}_{n \to \infty} \mathbf{E} \, \#U(X_1^n) \le \operatorname*{hilb}_{n \to \infty} I(X_1^n; X_{n+1}^{2n}) \tag{8.107}$$

Proof: Let us write $S_n := \#U(X_1^n)$. Observe that

$$H(Z_1^{S_n}|S_n) = -\sum_{s,w} P(S_n = s, Z_1^s = w) \log P(Z_1^s = w|S_n = s)$$

$$\ge -\sum_{s,w} P(S_n = s, Z_1^s = w) \log \frac{P(Z_1^s = w)}{P(S_n = s)}$$

$$= -\sum_{s,w} P(S_n = s, Z_1^s = w) \log \frac{2^{-s}}{P(S_n = s)}$$

$$= \mathbf{E}\, S_n - H(S_n) \ge \mathbf{E}\, S_n - 2\log(\mathbf{E}\, S_n + 1) - \log \frac{\pi^2}{6} \tag{8.108}$$

where the last row of inequalities follows by Theorem 8.17. Hence, by the inequality

$$H(X|Y) \le H(X|f(Y)) \tag{8.109}$$

for a measurable function f and by the chain rule for Shannon entropy (7.43), we obtain

$$H(X_1^n) - H(X_1^n|Z_1^\infty) \ge H(X_1^n|S_n) - H(X_1^n|Z_1^\infty, S_n) - H(S_n)$$

$$\ge H(X_1^n|S_n) - H(X_1^n|Z_1^{S_n}, S_n) - H(S_n)$$

$$= I(X_1^n; Z_1^{S_n}|S_n) - H(S_n)$$

$$\ge H(Z_1^{S_n}|S_n) - H(Z_1^{S_n}|X_1^n, S_n) - H(S_n)$$

$$= H(Z_1^{S_n}|S_n) - H(S_n)$$

$$\ge \mathbf{E}\, S_n - 4\log(\mathbf{E}\, S_n + 1) - 2\log \frac{\pi^2}{6} \tag{8.110}$$

Now we observe that

$$H(X_1^n|Z_1^\infty) \ge H(X_1^n|X_{n+1}^\infty) = hn \tag{8.111}$$

by Theorem 5.34. Hence, we have

$$H(X_1^n) - H(X_1^n|Z_1^\infty) \le H(X_1^n) - hn \tag{8.112}$$

Chaining inequalities (8.110) and (8.112) and network (8.102), we obtain (8.107). □

There is also an algorithmic version of the above theorem, which can be applied to perigraphic processes in particular. Once we replace probabilistic facts with algorithmic facts, we have to replace the Shannon mutual information with the algorithmic mutual information.

Theorem 8.19 (*Algorithmic facts and mutual information*): *Let* $(X_i)_{i\in\mathbb{Z}}$ *be a stationary process, where* $X_i : \Omega \to \mathbb{X}$ *are random variables over a finite alphabet* $\mathbb{X}$. *We have inequality*

$$\operatorname{hilb}_{n\to\infty} \mathbf{E}\,\#\mathbf{U}(X_1^n) \le \operatorname{hilb}_{n\to\infty} \mathbf{E}\,\mathbf{I}(X_1^n; X_{n+1}^{2n}) \tag{8.113}$$

Proof: Let us write $S_n := \#\mathbf{U}(X_1^n)$. Observe that there exists a constant $c > 0$ depending on z_1^∞ such that

$$\mathbf{H}\left(z_1^{S_n}|S_n\right) \overset{+}{>} \mathbf{H}\left(z_1^{S_n}\right) - \mathbf{H}(S_n)$$
$$\overset{+}{=} S_n - c - \mathbf{H}(S_n) \tag{8.114}$$

$$\mathbf{H}\left(\mathbf{H}\left(z_1^{S_n}\right)\right) \overset{+}{<} 2\log\left(\mathbf{H}\left(z_1^{S_n}\right) + 1\right)$$
$$\overset{+}{<} 2\log(S_n + 1) \tag{8.115}$$

where the first row of inequalities follows by the algorithmic incompressibility of z_1^∞, whereas the second row of inequalities follows by Theorem 8.17. Moreover, for any a recursive function f, we have

$$\mathbf{H}(x|y) \overset{+}{<} \mathbf{H}(x|f(y)) + \mathbf{H}(f) \tag{8.116}$$

Hence, applying the chain rule for Kolmogorov complexity (7.44), we obtain

$$\mathbf{H}(X_1^n) - \mathbf{H}(X_1^n|z_1^\infty) \overset{+}{>} \mathbf{H}(X_1^n|S_n) - \mathbf{H}(X_1^n|z_1^\infty, S_n) - \mathbf{H}(S_n)$$
$$\overset{+}{>} \mathbf{H}(X_1^n|S_n) - \mathbf{H}(X_1^n|z_1^{S_n}, S_n) - \mathbf{H}(S_n)$$
$$\overset{+}{>} \mathbf{I}(X_1^n; z_1^{S_n}|S_n) - \mathbf{H}(\mathbf{H}(z_1^{S_n})) - \mathbf{H}(S_n)$$
$$\overset{+}{>} \mathbf{H}(z_1^{S_n}|S_n) - \mathbf{H}(z_1^{S_n}|X_1^n, \mathbf{H}(X_1^n|S_n), S_n)$$

$$- \mathbf{H}(\mathbf{H}(z_1^{S_n})) - \mathbf{H}(S_n)$$

$$\overset{+}{>} \mathbf{H}(z_1^{S_n}|S_n) - \mathbf{H}(g) - \mathbf{H}(\mathbf{H}(z_1^{S_n})) - \mathbf{H}(S_n)$$

$$\overset{+}{>} S_n - c - \mathbf{H}(g) - \mathbf{H}(\mathbf{H}(z_1^{S_n})) - 2\mathbf{H}(S_n)$$

$$\overset{+}{>} S_n - 6\log(S_n + 1) - c - \mathbf{H}(g) \tag{8.117}$$

where g is the recursive function from Definition 8.6. Hence, by the Jensen inequality (Theorem 3.14),

$$\mathbf{E}\,\mathbf{H}(X_1^n) - \mathbf{E}\,\mathbf{H}(X_1^n|z_1^\infty) \overset{+}{>} \mathbf{E}\,S_n - 6\log(\mathbf{E}\,S_n + 1) - c - \mathbf{H}(g) \tag{8.118}$$

Now we observe that

$$\mathbf{E}\,\mathbf{H}(X_1^n|z_1^\infty) \geq H(X_1^n) \geq hn \tag{8.119}$$

by Theorems 3.18 and 7.24. Hence, we derive

$$\mathbf{E}\,\mathbf{H}(X_1^n) - \mathbf{E}\,\mathbf{H}(X_1^n|z_1^\infty) \leq \mathbf{E}\,\mathbf{H}(X_1^n) - hn \tag{8.120}$$

Chaining inequalities (8.118) and (8.120) and network (8.102), we obtain (8.113). □

Applying inequalities (8.118) and (8.119) from the proof of the above theorem, we can also show that perigraphic processes have nonrecursive distributions in general. This result was mentioned in the previous section.

Theorem 8.20: *Any perigraphic process has a nonrecursive distribution.*

Proof: Assume that a perigraphic process $(X_i)_{i\in\mathbb{Z}}$ has a recursive distribution $P(x_1^n) = P(X_1^n = x_1^n)$. By inequalities (8.118) and (8.119), we have

$$\operatorname*{hilb}_{n\to\infty} \mathbf{E}\,\#\mathbf{U}(X_1^n) \leq \operatorname*{hilb}_{n\to\infty} [\mathbf{E}\,\mathbf{H}(X_1^n) - H(X_1^n)] \tag{8.121}$$

Since for a recursive distribution P we have inequality (7.79) then

$$\operatorname*{hilb}_{n\to\infty} \mathbf{E}\,\#\mathbf{U}(X_1^n) = 0 \tag{8.122}$$

Since we have obtained a contradiction with the assumption that the process is perigraphic, measure P cannot be recursive. □

Now we come back to universal semidistributions and the PPM probabilities. Using Theorem 6.16, we can demonstrate an opposite inequality between the Shannon or algorithmic mutual information and the expected number of distinct PPM words that can be detected in a random text.

Theorem 8.21 *(Mutual information and PPM words):* *Let $(X_i)_{i\in\mathbb{Z}}$ be a stationary process, where $X_i : \Omega \to \mathbb{X}$ are random variables over a finite alphabet $\mathbb{X}$. Let*

$G(X_1^n)$ *be the PPM order of block* x_1^n *and let* $V_G(X_1^n)$ *be the PPM vocabulary of block* x_1^n. *There holds inequality*

$$\text{hilb}_{n\to\infty} I(X_1^n; X_{n+1}^{2n}) \leq \text{hilb}_{n\to\infty} \mathbf{E}\, \mathbf{I}(X_1^n; X_{n+1}^{2n})$$

$$\leq \text{hilb}_{n\to\infty} \mathbf{E}\, [G(X_1^n) + \#V_G(X_1^n)] \qquad (8.123)$$

Proof: By Theorem 6.17 and stationarity, we obtain

$$\text{hilb}_{n\to\infty} \mathbf{E}\, [2\langle \text{PPM}(X_1^n)\rangle - \langle \text{PPM}(X_1^{2n})\rangle] \leq \text{hilb}_{n\to\infty} \mathbf{E}\, [G(X_1^n) + \#V_G(X_1^n)] \quad (8.124)$$

Hence, inequality (8.123) follows by network (8.102). □

Combining Theorems 8.18 and 8.21, we obtain a neat statement which links the expected number of independent facts that can be inferred from a random text with the expected number of distinct PPM words that can be detected in a random text. Namely, the number of inferred facts is roughly smaller than the number of the detected PPM words, if the PPM order does not grow too fast.

Theorem 8.22 *(Probabilistic facts and PPM words):* *Consider a strongly noner-godic stationary process* $(X_i)_{i\in\mathbb{Z}}$, *where* $X_i : \Omega \to \mathbb{X}$ *and the alphabet* $\mathbb{X}$ *is finite. We have inequalities*

$$\text{hilb}_{n\to\infty} \mathbf{E}\, \#\, U(X_1^n) \leq \text{hilb}_{n\to\infty} I(X_1^n; X_{n+1}^{2n})$$

$$\leq \text{hilb}_{n\to\infty} \mathbf{E}\, [G(X_1^n) + \#V_G(X_1^n)] \qquad (8.125)$$

Proof: The claim follows by conjunction of Theorems 8.18 and 8.21. □

The above theorem has also an algorithmic version, for ergodic and perigraphic processes in particular.

Theorem 8.23 *(Algorithmic facts and PPM words):* *Consider a stationary pro-cess* $(X_i)_{i\in\mathbb{Z}}$, *where* $X_i : \Omega \to \mathbb{X}$ *and the alphabet* $\mathbb{X}$ *is finite. We have inequalities*

$$\text{hilb}_{n\to\infty} \mathbf{E}\, \#\mathbf{U}(X_1^n) \leq \text{hilb}_{n\to\infty} \mathbf{E}\, \mathbf{I}(X_1^n; X_{n+1}^{2n})$$

$$\leq \text{hilb}_{n\to\infty} \mathbf{E}\, [G(X_1^n) + \#V_G(X_1^n)] \qquad (8.126)$$

Proof: The claim follows by conjunction of Theorems 8.19 and 8.21. □

The theorems about facts and words in the above versions were proved by Dębowski (2018a). A weaker analogue of Theorems 8.18, and 8.22 for minimal grammar-based codes rather than the PPM probability, with inequality lim inf $\leq$ lim sup rather than lim sup $\leq$ lim sup, was stated by Dębowski (2011), see Problem 7.4. Dębowski was not aware at that time of Theorem 8.2. Using Theorem 8.2, his result for the minimal grammar-based codes can be strengthened as well. We will not discuss this issue, however, since the respective minimal grammar-based codes probably cannot be computed in a reasonable time, see Charikar et al. (2005).

Let us recall the empirical observations from Section 1.12. In Figure 1.6, we have presented the growth of the PPM order $G(X_1^n)$ and the number of PPM words $\#V_G(x_1^n)$ for Shakespeare's First Folio/35 plays and a random permutation of the text characters. For the plays of Shakespeare, we seem to have a stepwise power law growth of $\#V_G(x_1^n)$ with the Hilberg exponent close to 0.8, whereas for the random permutation of characters we seem to obtain the Hilberg exponent equal to 0. Noticing the difference between these observations, we might suppose that the number of inferrable probabilistic or algorithmic facts for texts in natural language also obeys a power-law growth. Formally, however, we cannot make such a deduction since inequalities (8.125) and (8.126) can be strict.

The strictness of inequalities (8.125) and (8.126) probably can be related to the following observations, see Problem 5.6. First, the Gács–Körner common information between random variables X and Y is

$$C_{\mathrm{GK}}(X;Y) := \sup_{W\,:\,W=\,f(X)=g(Y)} H(W) \tag{8.127}$$

Gács and Körner (1973) showed that common information $C_{\mathrm{GK}}(X;Y)$ satisfies inequality $C_{\mathrm{GK}}(X;Y) \leq I(X;Y)$, which is usually strict. In our application, expectation $\mathbf{E}\,\#U(X_1^n)$ may be close to the Gács–Körner common information between block x_1^n and the persistent topic $(Z_k)_{k\in\mathbb{N}}$. Hence, $\mathbf{E}\,\#U(X_1^n)$ may be much smaller than mutual information $I(X_1^n;X_{n+1}^\infty)$, which is close to mutual information $I(X_1^n;X_{n+1}^{2n})$. Second, a similar idea might be possibly applied to the to the upper bound mutual information $I(X_1^n;X_{n+1}^{2n})$. Namely, the modified Wyner common information between random variables X and Y is

$$C_{\mathrm{MW}}(X;Y) := \inf_{W\,:\,X\perp\!\!\!\perp Y\,|\,W} H(W) \tag{8.128}$$

Wyner (1975) proved that common information $C_{\mathrm{MW}}(X;Y)$ satisfies inequality $C_{\mathrm{MW}}(X;Y) \geq I(X;Y)$, which is usually strict. In our application, expectation $\mathbf{E}\,\#V_G(X_1^{2n})$ may be close to the modified Wyner common information between blocks x_1^n and X_{n+1}^{2n}, which is larger than mutual information $I(X_1^n;X_{n+1}^{2n})$.

Let us also note that the theorems about facts and words have been stated only for processes over a finite alphabet, whereas Santa Fe processes, our basic examples of strongly nonergodic and perigraphic processes, are processes over a countably infinite alphabet. Thus, in Chapter 11, we will construct some strongly nonergodic and perigraphic processes over a finite alphabet via an operation called quasiperiodic coding, to be discussed in Section 10.2. This operation will be applied to Santa Fe processes in particular, so these processes are indeed convenient guiding examples in our theory.

Problems

8.1 Show that

$$\text{ess sup } X = \sup_{0<p<1} \mathbf{Q}_p X \tag{8.129}$$

$$\text{ess inf } X = \inf_{0<p<1} \mathbf{Q}_p X \tag{8.130}$$

8.2 Show that for $f(n), g(n) \geq 0$ we have

$$\operatorname*{hilb}_{n\to\infty} [f(n) + g(n)] \leq \max\left\{\operatorname*{hilb}_{n\to\infty} f(n), \operatorname*{hilb}_{n\to\infty} g(n)\right\} \tag{8.131}$$

$$\operatorname*{hilb}_{n\to\infty} [f(n)g(n)] \leq \operatorname*{hilb}_{n\to\infty} f(n) + \operatorname*{hilb}_{n\to\infty} g(n) \tag{8.132}$$

8.3 Show that the claims of Theorems 8.4 and 8.5 hold also for $J(n) \geq -A(n)$ and $J(m) - J(n) \geq -A(n)$, $m \geq n$, where $\operatorname{hilb}_{n\to\infty} A(n) = 0$.

8.4 Show that:
(a) $\delta_P = \gamma_P = 0$ almost surely for the Bernoulli(θ) process.
(b) $\delta_P = \gamma_P = 1/\alpha$ almost surely for the mixture Zipfian Santa Fe process with an exponent α.
Having solved the task, see Dębowski (2015a) for the solution.

8.5 Let $A(\gamma), B(\gamma) \geq 0$ be positive real functions of an arbitrary argument γ. Suppose that for $n \in \mathbb{N}$ we may define

$$C(n) := \min_{\gamma}[A(\gamma) + nB(\gamma)] \tag{8.133}$$

$$\Gamma(n) := \arg\min_{\gamma}[A(\gamma) + nB(\gamma)] \tag{8.134}$$

Show that:
(a) $\lim_{n\to\infty} C(n)/n = \lim_{n\to\infty} B(\Gamma(n))/n = b := \inf_{\gamma} B(\gamma)$;
(b) $\lim_{n\to\infty} A(\Gamma(n))/n = 0$;

(c) $\text{hilb}_{n\to\infty}[C(n) - bn] = \text{hilb}_{n\to\infty}A(\Gamma(n))$;

(d) $0 \le A(\Gamma(n)) \le 2C(n) - C(2n) \le A(\Gamma(2n))$.

What if we replace term $nB(\gamma)$ in (8.133) and (8.134) with a subadditive or a superadditive function $B(n|\gamma)$ of the argument n? Can these observations be somehow related to the theorems about facts and words and Problem 8.4?

8.6 Consider an arbitrary infinite sequence $(x_i)_{i\in\mathbb{N}}$ over a countable alphabet. Show that

$$\text{hilb}_{n\to\infty} \min \{\#\mathbf{U}(x_1^n), \#\mathbf{U}(x_{n+1}^{2n})\} \le \text{hilb}_{n\to\infty} \mathbf{I}(x_1^n; x_{n+1}^{2n}) \qquad (8.135)$$

8.7 Define the Kolmogorov order of a string

$$\mathbf{M}(x_1^n) := \min\{k : (n-k)h_k(x_1^n) \le \mathbf{H}(x_1^n)\} \qquad (8.136)$$

where $h_k(x_1^n)$ is the empirical entropy. Argue that $\mathbf{M}(x_1^n) \le L(x_1^n) + 1$, where $L(x_1^n)$ is the maximal repetition. Denote the respective Kolmogorov vocabulary as $V_{\mathbf{M}}(x_1^n) := V_{\mathbf{M}(x_1^n)}(x_1^n)$. Consider an arbitrary infinite sequence $(x_i)_{i\in\mathbb{N}}$ over a finite alphabet. Show that

$$\text{hilb}_{n\to\infty} \mathbf{I}(x_1^n; x_{n+1}^{2n}) \le \text{hilb}_{n\to\infty} [\mathbf{M}(x_1^n) + \#V_{\mathbf{M}}(x_1^n)] \qquad (8.137)$$

8.8 Consider a stationary ergodic Mth order Markov process $(X_i)_{i\in\mathbb{Z}}$. Show that

$$(n - M)h_M(X_1^n) \le -\log P(X_1^n) \qquad (8.138)$$

holds for the empirical entropy $h_k(x_1^n)$ of a string x_1^n.

8.9 Let a sequence $(x_i)_{i\in\mathbb{N}}$ be Martin-Löf random with respect to a stationary ergodic Mth order Markov measure. Prove inequalities

$$\limsup_{n\to\infty} \mathbf{M}(x_1^n) \le M \le \liminf_{n\to\infty} G(x_1^n) \qquad (8.139)$$

for the Kolmogorov order $\mathbf{M}(x_1^n)$, the Markov order M, and the PPM order $G(X_1^n)$.

Remark: Can these inequalities be strengthened as equalities or generalized to an arbitrary stationary ergodic measure? Denote the consistent estimator of the Markov order by Morvai and Weiss (2005) as $M(x_1^n)$ and define the respective vocabulary $V_M(x_1^n) := V_{M(x_1^n)}(x_1^n)$. If we could prove that

$$\liminf_{n\to\infty} [M(x_1^n) - \mathbf{M}(x_1^n)] \ge 0 \qquad (8.140)$$

for every sequence $(x_i)_{i\in\mathbb{N}}$ which is Martin-Löf random with respect to a stationary ergodic measure over a finite alphabet then we would obtain a pointwise version of the theorem about mutual information and words, namely,

$$\operatorname*{hilb}_{n\to\infty} \mathbf{I}(x_1^n; x_{n+1}^{2n}) \le \operatorname*{hilb}_{n\to\infty} [M(x_1^n) + \#V_M(x_1^n)] \tag{8.141}$$

8.10 Consider a stationary ergodic process $(X_i)_{i\in\mathbb{Z}}$. Define

$$\beta_+ := \operatorname*{hilb}_{n\to\infty} [\mathbf{E}\,\mathbf{H}(X_1^n) - hn] \tag{8.142}$$

$$\beta_- := \operatorname*{hilb}_{n\to\infty} [H(X_1^n) - hn] \tag{8.143}$$

Construct some examples of stationary ergodic processes such that:

(a) $\beta_- = \beta_+ = 0$;
(b) $\beta_- = \beta_+ = 1$;
(c) $0 < \beta_- = \beta_+ < 1$;
(d) $\beta_- = 0$ and $\beta_+ = 1$;
(e) $\beta_- = 0$ and $0 < \beta_+ < 1$;
(f) $0 < \beta_- < 1$ and $\beta_+ = 1$.

Compare your constructions with examples of processes presented in Chapter 11.

9

Power Laws for Repetitions

In the previous chapter, we have analyzed power laws connected to the growth of block mutual information (1.132). We have seen that these laws arise naturally if the persistent topic of a random text is infinitely complex. Equivalently, we may say that in this case, there exists an infinitely complex world that is repeatedly described by the random text. In contrast, in this chapter, we would like to study the power-law logarithmic growth of maximal repetition (1.133), which also seems to arise in texts created by humans, as introduced in Section 1.13.

The power-law logarithmic growth of maximal repetition was discovered by Dębowski (2012b, 2015b) accidentally. The initial motivation for this research was confirming a hypothesis of Dębowski (2011) that the maximal repetition for natural language grows logarithmically, which was connected to the theorem about facts and words originally stated for minimal grammar-based codes, see Problem 7.4. A converse statement turned out to be likely true for natural language. Consequently, we have asked the question what kind of stochastic processes can exhibit a power-law logarithmic growth of maximal repetition and in what ways they can resemble natural language. This idea inspired further papers (Dębowski, 2017, 2018b), where we constructed random hierarchical association (RHA) processes and demonstrated some general bounds for the maximal repetition in terms of conditional Rényi entropy rates.

In fact, the maximal repetition had been earlier researched by Erdös and Rényi (1970), Shields (1992), and Szpankowski (1993b). Whereas for finite-state processes, the maximal repetition grows logarithmically, Shields (1992) has shown that in general, the maximal repetition can grow arbitrarily fast for stationary processes, and this growth may hold regardless of the value of the Shannon entropy rate. Complementing this result, Szpankowski (1993b) showed that the logarithmic rate of growth of maximal repetition is equal to the so called collision entropy rate for some subclass of processes with a finite excess entropy. This result is less general than the partial result of Dębowski (2018b). Thus, we can

Information Theory Meets Power Laws: Stochastic Processes and Language Models, First Edition.
Łukasz Dębowski.
© 2021 John Wiley & Sons, Inc. Published 2021 by John Wiley & Sons, Inc.

see that we need both more general and stronger results so as to understand what happens in natural language.

As it was sketched in Section 1.13, our intuition is that the infinitely complex world repeatedly described in texts in natural language may emerge in the limit as a result of human tendency to imitate previously encountered chunks of text. More generally, we suppose that the process being perigraphic and the power-law growth of block mutual information can be a mathematical consequence of the power-law logarithmic growth of maximal repetition or rather of a related phenomenon. We have not succeeded in proving this linking statement yet, and, in this chapter, we will present a bundle of preliminary propositions that might help in future research. Some of these results have not been discussed in the literature so far, whereas some have been beautified.

Thus, Section 9.1 is devoted to Rényi–Arimoto entropies, a generalization of the conditional Shannon entropy. The Rényi–Arimoto entropies do not satisfy the chain rule and the polymatroid identities like the Shannon entropy but they enjoy a number of other nice relationships. They are parameterized with an extended real number parameter, and they are decreasing with respect to this parameter. They are also monotonic with respect to their other arguments.

In Section 9.2, we consider generalized entropy rates defined as the upper limits of the rates of Rényi–Arimoto block entropies for a stationary process. If the process satisfies a certain condition stronger than finite excess entropy then the respective generalized entropy rates exist as the proper limits and do not change their value if we impose conditioning on the infinite future. It is an important open question whether this result can be generalized to processes with an infinite excess entropy such as, hypothetically, the natural language.

Recurrence times are the topic of Section 9.3. The recurrence time is the first position in the stationary process on which the value of the initial block of the specified length reoccurs. Recurrence times are a classical topic in ergodic theory and information theory. We recall some received results such as the Kac lemma and the Kontoyiannis bounds for recurrence times in terms of the block probability. Subsequently, using these propositions, we prove the Wyner–Ziv/ Ornstein–Weiss (WZ/OW) theorem, which shows that the rate of recurrence times equals the Shannon entropy rate of a stationary ergodic process almost surely and in expectation.

Section 9.4 treats the concept of subword complexity. Subword complexity is a function that counts how many distinct substrings of a given length are contained in a given string. Subword complexity can be linked with recurrence times. In particular for a stationary process, the expected subword complexity equals the expectation of the recurrence time trimmed to the length of the considered string. Given this result, we can express Jensen gaps for the recurrence time, such as the variance, as series of the expected subword complexity. Using other simple techniques, we can also upper bound the expected subword complexity by the block entropy of the process.

In Section 9.5, we introduce two maximal lengths: the longest match length and the maximal repetition. The longest match length is smaller than the maximal repetition and grows essentially like the logarithm of the window length divided by the Shannon entropy rate, by the WZ/OW theorem proved in Section 9.3. In contrast, it was discussed in the literature that the maximal repetition can grow faster. For processes with a finite excess entropy, satisfying the condition mentioned in Section 9.2, it was shown that the maximal repetition grows essentially like the logarithm of the string length divided by the collision entropy rate. We show two ideas of generalizing this result to arbitrary stationary processes which fail.

To be more positive, in Section 9.6, we discuss a few bounds for maximal repetition that work for arbitrary stationary processes. We reformulate the problem of bounding the maximal repetition using entropy powers and Hilberg exponents. We show that the power-law logarithmic growth of maximal repetition corresponds to a power-law growth of an entropy function if the Hilberg exponent of the respective entropy power of the maximal repetition is approximately constant and bounded away from zero and infinity. We show that this quantity is indeed constant for Rényi–Arimoto entropies, and we present a few positive bounds which apply Kontoyiannis inequalities for recurrence times. These bounds do not fully constrain the growth of maximal repetition, however, and should be strengthened in the future.

9.1 Rényi–Arimoto Entropies

The general goal of this chapter is to furnish some theory for growth rates of maximal repetition. We will approach this topic gradually, moving through various related concepts, whose relationship to our main subject will become clear in the end. To begin, in this section, we will define and investigate properties of generalized entropies called Rényi entropies and Rényi–Arimoto conditional entropies. These quantities generalize the concepts of Shannon entropy and Shannon conditional entropy, respectively. We will link them to maximal repetition ultimately in Section 9.6.

The unconditional Rényi entropies are defined as follows, see Rényi (1961).

Definition 9.1 *(Rényi entropies)*: For a parameter $\gamma \in (0,1) \cup (1,\infty)$, the Rényi entropy of a simple random variable $X : \Omega \to \mathbb{X}$ is defined as

$$H_\gamma(X) := -\frac{1}{\gamma - 1} \log \left[\sum_{x \in \mathbb{X}} p(x)^\gamma \right] = -\frac{\gamma}{\gamma - 1} \log \left[\sum_{x \in \mathbb{X}} p(x)^\gamma \right]^{1/\gamma} \tag{9.1}$$

where $p(x) := P(X = x)$, whereas for $\gamma \in \{0, 1, \infty\}$, we define the Rényi entropyas

$$H_\gamma(X) := \lim_{\delta \to \gamma} H_\delta(X) \tag{9.2}$$

As a definition of the generalized entropy, Rényi proposed the first formula in (9.1), which is trivially equal to the second, more decorated formula. The second formula, however, motivates a straightforward generalization of the Rényi entropy to a conditional quantity called the Rényi–Arimoto entropy.

It took some time before this reasonable definition of a generalized conditional entropy was proposed. The nontrivial achievement was to insert the expectation operator in such a place that would guarantee the most interesting algebraic properties of the resulted object. Consequently, the conditional entropy will be generalized as follows, see Fehr and Berens (2014).

Definition 9.2 (*Rényi–Arimoto conditional entropies*): For a parameter $\gamma \in (0, 1) \cup (1, \infty)$, the Rényi–Arimoto conditional entropy of a simple random variable $X : \Omega \to \mathbb{X}$ given an arbitrary random variable Y is defined as

$$H_\gamma(X|Y) := -\frac{\gamma}{\gamma - 1} \log \mathbf{E} \left[\sum_{x \in \mathbb{X}} p(x|Y)^\gamma \right]^{1/\gamma} \tag{9.3}$$

where $p(x|Y) := P(X = x|Y)$, whereas for $\gamma \in \{0, 1, \infty\}$, we define the Rényi–Arimoto conditional entropy as

$$H_\gamma(X|Y) := \lim_{\delta \to \gamma} H_\delta(X|Y) \tag{9.4}$$

It can be easily seen that $H_\gamma(X), H_\gamma(X|Y) \geq 0$ and $H_\gamma(X) = H_\gamma(X|r)$ for a constant r. Some special cases of Rényi and Rényi–Arimoto entropies have their names, mentioned in the following proposition. Two of them, the Shannon and the Hartley entropy, have been already encountered.

Theorem 9.1: *Some special cases of Rényi and Rényi–Arimoto entropies are:*

1. *Hartley entropy:*

$$H_0(X) = \log \# \{x \in \mathbb{X} : p(x) > 0\} \tag{9.5}$$

$$H_0(X|Y) = \log \text{ess sup} \# \{x \in \mathbb{X} : p(x|Y) > 0\} \tag{9.6}$$

2. *Shannon entropy:*

$$H_1(X) = H(X) := - \sum_{x : p(x) > 0} p(x) \log p(x) \tag{9.7}$$

$$H_1(X|Y) = H(X|Y) := \mathbf{E} \left[- \sum_{x : p(x|Y) > 0} p(x|Y) \log p(x|Y) \right] \tag{9.8}$$

3. *Collision entropy:*

$$H_2(X) = -\log \sum_{x \in \mathbb{X}} p(x)^2 \tag{9.9}$$

$$H_2(X|Y) = -2 \log \mathbf{E} \left[\sum_{x \in \mathbb{X}} p(x|Y)^2 \right]^{1/2} \tag{9.10}$$

4. *Min-entropy:*

$$H_\infty(X) = -\log \max_{x \in \mathbb{X}} p(x) \tag{9.11}$$

$$H_\infty(X|Y) = -\log \mathbf{E} \max_{x \in \mathbb{X}} p(x|Y) \tag{9.12}$$

Proof: It is straightforward that the unconditional expressions are special cases of the conditional expressions so it is sufficient to derive the latter ones.

1. We have $\lim_{\gamma \to 0} p(x|Y)^\gamma = \mathbf{1}\{p(x|Y) > 0\}$. Let $Z := \#\{x \in \mathbb{X} : p(x|Y) > 0\}$. Then

$$H_0(X|Y) = \lim_{\gamma \to 0} \log \left[\mathbf{E} \, Z^{1/\gamma} \right]^\gamma = \log \operatorname{ess\,sup} Z \tag{9.13}$$

2. By the L'Hôpital rule, we obtain

$$H_1(X|Y) = \lim_{\gamma \to 1} \frac{d \left(\log \mathbf{E} \left(\sum_{x \in \mathbb{X}} p(x|Y)^\gamma \right)^{1/\gamma} \right)}{d\gamma} : \frac{d(1 - \gamma)}{d\gamma} \tag{9.14}$$

The first derivative can be developed as

$$\frac{d}{d\gamma} \ln \mathbf{E} \left(\sum_{x \in \mathbb{X}} p(x|Y)^\gamma \right)^{1/\gamma} = \frac{d}{d\gamma} \ln \mathbf{E} \, \exp \left(\frac{1}{\gamma} \ln \sum_{x \in \mathbb{X}} p(x|Y)^\gamma \right)$$

$$= \frac{\mathbf{E} \left(\sum_{x \in \mathbb{X}} p(x|Y)^\gamma \right)^{1/\gamma} \left(-\frac{1}{\gamma^2} \ln \sum_{x \in \mathbb{X}} p(x|Y)^\gamma + \frac{1}{\gamma} \frac{\sum_{x \in \mathbb{X}} p(x|Y)^\gamma \ln p(x|Y)}{\sum_{x \in \mathbb{X}} p(x|Y)^\gamma} \right)}{\mathbf{E} \left(\sum_{x \in \mathbb{X}} p(x|Y)^\gamma \right)^{1/\gamma}}$$

$$\xrightarrow[\gamma \to 1]{} \mathbf{E} \left(\sum_{x \in \mathbb{X}} p(x|Y) \ln p(x|Y) \right) \tag{9.15}$$

Consequently, we obtain the claim.

3. These equalities follow by a direct application of formulas (9.1) and (9.3).
4. We obtain

$$H_\infty(X|Y) = -\log \mathbf{E} \lim_{\gamma \to \infty} \left(\sum_{x \in \mathbb{X}} p(x|Y)^\gamma \right)^{1/\gamma} = -\log \mathbf{E} \max_{x \in \mathbb{X}} p(x|Y) \tag{9.16}$$

$\square$

It is a general fact that entropy $H_\gamma(X|Y)$ decreases if we increase parameter γ. This result requires a certain preparation and care. Let $q : \mathbb{X} \to \mathbb{R}$ be a probability distribution. Let us denote the γ-norm of this distribution as

$$||q||_\gamma := \left[\sum_{x\in\mathbb{X}} q(x)^\gamma\right]^{1/\gamma} \tag{9.17}$$

For $\gamma \in (0,1) \cup (1,\infty)$, we have

$$\frac{d}{d\gamma} \ln ||q||_\gamma = -\frac{1}{\gamma^2} \ln\left[\sum_{x\in\mathbb{X}} q(x)^\gamma\right] + \frac{1}{\gamma} \frac{\sum_{x\in\mathbb{X}} q(x)^\gamma \ln q(x)}{\sum_{x\in\mathbb{X}} q(x)^\gamma}$$

$$= \frac{1}{\gamma^2} \sum_{x\in\mathbb{X}} \frac{q(x)^\gamma}{\sum_{y\in\mathbb{X}} q(y)^\gamma} \ln \frac{q(x)^\gamma}{\sum_{y\in\mathbb{X}} q(y)^\gamma} \leq 0 \tag{9.18}$$

since the Shannon entropy is positive. Hence, $||q||_\gamma$ is a decreasing function of γ. This relationship is called the inequality on means. Somewhat surprisingly, unconditional Rényi entropy $H_\gamma(q) := -\frac{\gamma}{\gamma-1} \log ||q||_\gamma$ is also a decreasing function of γ, in spite of the negative sign, since

$$\frac{d}{d\gamma}\left(-\frac{\gamma}{\gamma-1} \ln ||q||_\gamma\right) = \frac{1}{(\gamma-1)^2} \ln\left[\sum_{x\in\mathbb{X}} q(x)^\gamma\right] - \frac{1}{(\gamma-1)} \frac{\sum_{x\in\mathbb{X}} q(x)^\gamma \ln q(x)}{\sum_{x\in\mathbb{X}} q(x)^\gamma}$$

$$= -\frac{1}{(\gamma-1)^2} \sum_{x\in\mathbb{X}} \frac{q(x)^\gamma}{\sum_{y\in\mathbb{X}} q(y)^\gamma} \ln \frac{q(x)^{\gamma-1}}{\sum_{y\in\mathbb{X}} q(y)^\gamma} \leq 0 \tag{9.19}$$

where the Kullback–Leibler (KL) divergence is also positive.

Now we can state the theorem by Fehr and Berens (2014) about monotonicity of Rényi–Arimoto entropies.

Theorem 9.2: *We have $H_\gamma(X|Y) \geq H_\delta(X|Y)$ for $\gamma \leq \delta$.*

Proof: It is sufficient to prove our claim for $0 < \gamma \leq \delta < 1$ and $1 < \gamma \leq \delta < \infty$ and conditional entropies. In both cases, we can denote $\epsilon = \frac{\gamma-1}{\gamma} \frac{\delta}{\delta-1}$ and write

$$2^{-H_\gamma(X|Y)} = \left(\mathbf{E}\, 2^{-\frac{\gamma-1}{\gamma} H_\gamma(p(\cdot|Y))}\right)^{\frac{\gamma}{\gamma-1}} \leq \left(\mathbf{E}\, 2^{-\frac{\gamma-1}{\gamma} H_\delta(p(\cdot|Y))}\right)^{\frac{\gamma}{\gamma-1}}$$

$$= \left(\mathbf{E}\, 2^{-\frac{\delta-1}{\delta} H_\delta(p(\cdot|Y))\epsilon}\right)^{\frac{\gamma}{\gamma-1}} \leq \left(\mathbf{E}\, 2^{-\frac{\delta-1}{\delta} H_\delta(p(\cdot|Y))}\right)^{\frac{\gamma}{\gamma-1}\epsilon}$$

$$= \left(\mathbf{E}\, 2^{-\frac{\delta-1}{\delta} H_\delta(p(\cdot|Y))}\right)^{\frac{\delta}{\delta-1}} = 2^{-H_\delta(X|Y)} \tag{9.20}$$

where the first inequality follows by (9.19) and the second inequality follows by the Jensen inequality (Theorem 3.14). Hence, we obtain the claim. $\qquad \square$

Yet we have a simple upper bound for some entropies in terms of the min-entropy, which is the smallest of all generalized entropies.

Theorem 9.3: *We have $H_\gamma(X|Y) \leq \frac{\gamma}{\gamma-1} H_\infty(X|Y)$ for $\gamma > 1$.*

Proof: For $\gamma > 1$, we obtain

$$H_\gamma(X|Y) = -\frac{\gamma}{\gamma-1} \log \mathbf{E} \left[\sum_{x \in \mathbb{X}} p(x|Y)^\gamma \right]^{1/\gamma}$$

$$\leq -\frac{\gamma}{\gamma-1} \log \mathbf{E} \max_{x \in \mathbb{X}} p(x|Y) = \frac{\gamma}{\gamma-1} H_\infty(X|Y) \qquad (9.21)$$

$\square$

As shown by Fehr and Berens (2014), the Rényi–Arimoto conditional entropy does not satisfy the exact chain rule like the Shannon conditional entropy, but it satisfies the following weaker inequalities:

$$H_\gamma(X|Z, Y) \leq H_\gamma(X|Z) \leq H_\gamma(X, Y|Z) \leq H_0(X|Z) + H_\gamma(Y|X, Z) \qquad (9.22)$$

We will demonstrate now the first two inequalities.

Theorem 9.4: *We have $H_\gamma(X|Z, Y) \leq H_\gamma(X|Z)$.*

Proof: We have $\mathbf{E}\left[p(x|Z, Y)|Z\right] = p(x|Z)$. Hence, for $\gamma \in (0, 1) \cup (1, \infty)$, we obtain

$$H_\gamma(X|Z, Y) = -\frac{\gamma}{\gamma-1} \log \mathbf{E}\, \mathbf{E}\left[\left[\sum_{x \in \mathbb{X}} p(x|Z, Y)^\gamma \right]^{1/\gamma} \Bigg| Z \right]$$

$$\leq -\frac{\gamma}{\gamma-1} \log \mathbf{E} \left[\sum_{x \in \mathbb{X}} \mathbf{E}\left[p(x|Z, Y)|Z\right]^\gamma \right]^{1/\gamma} = H_\gamma(X|Z) \qquad (9.23)$$

where the inequality follows by the repeated use of the Jensen inequality (Theorem 3.14) for function $f(x) = (x^\gamma + (1-x)^\gamma)^{1/\gamma}$. The other cases of γ can be obtained by continuity.

$\square$

Theorem 9.5: *We have $H_\gamma(X|Z) \leq H_\gamma(X, Y|Z)$.*

Proof: For $\gamma \in (0, 1) \cup (1, \infty)$, we obtain

$$H_\gamma(X|Z) = -\frac{\gamma}{\gamma-1} \log \mathbf{E} \left[\sum_{x \in \mathbb{X}} p(x|Z)^\gamma \right]^{1/\gamma}$$

$$\leq -\frac{\gamma}{\gamma-1} \log \mathbf{E} \left[\sum_{x \in \mathbb{X}} p(x|Z)^\gamma \sum_{y \in \mathbb{Y}} p(y|x, Z)^\gamma \right]^{1/\gamma}$$

$$= -\frac{\gamma}{\gamma-1} \log \mathbf{E} \left[\sum_{(x,y) \in \mathbb{X} \times \mathbb{Y}} p(x, y|Z)^\gamma \right]^{1/\gamma} = H_\gamma(X, Y|Z) \qquad (9.24)$$

The other cases follow by continuity.

$\square$

9.2 Generalized Entropy Rates

In this section, we will investigate simple properties of Rényi block entropies and Rényi entropy rates, tracing some ideas from Section 5.2 such as the Fekete lemma. In contrast to the Shannon entropy rates, the existence of Rényi entropy rates as proper limits is guaranteed only for a subclass of stationary processes. As it will become clear, this class of processes has a finite excess entropy, so it may be too small as for the needs of statistical language modeling.

Let us consider a stationary process $(X_i)_{i \in \mathbb{Z}}$ over a finite alphabet $\mathbb{X}$. By analogy to finite sequences $x_j^k := (x_j, x_{j+1}, \dots, x_k)$ of symbols $x_i \in \mathbb{X}$, we will also write infinite sequences $x_j^\infty = (x_j, x_{j+1}, \dots)$. A similar convention is applied to random variables. Recall the Shannon block entropy $H(n) := H(X_1^n)$ and the Shannon entropy rate $h := H(X_1 | X_2^\infty)$. Formally, we may also define the conditional Shannon block entropy $H^{\mathrm{cond}}(n) := H(X_1^n | X_{n+1}^\infty) = hn$. By analogy we may consider the following Rényi block entropies.

Definition 9.3 (*Block Rényi entropy*): For a stationary process $(X_i)_{i \in \mathbb{Z}}$ over a finite alphabet and $\gamma \in [0, \infty]$, let us define the block Rényi entropies

$$H_\gamma(n) := H_\gamma(X_1^n) \qquad\qquad (9.25)$$
$$H_\gamma^{\mathrm{cond}}(n) := H_\gamma(X_1^n | X_{n+1}^\infty) \qquad\qquad (9.26)$$

By the results of the previous section, $H_\gamma(n)$ and $H_\gamma^{\mathrm{cond}}(n)$ are increasing functions of n and decreasing functions of γ. Moreover, we have $H_\gamma(n) \geq H_\gamma^{\mathrm{cond}}(n)$, $H_\gamma(n) \leq \frac{\gamma}{\gamma-1} H_\infty(n)$, and $H_\gamma^{\mathrm{cond}}(n) \leq \frac{\gamma}{\gamma-1} H_\infty^{\mathrm{cond}}(n)$.

Recall that for a stationary process, the Shannon entropy is subadditive, $H(n + m) \leq H(n) + H(m)$, and hence by the Fekete lemma (Theorem 5.11), for the process over a finite alphabet, the Shannon entropy rate h can be equivalently defined as the limit

$$h = \lim_{n \to \infty} \frac{H(n)}{n} = \lim_{n \to \infty} \frac{H^{\mathrm{cond}}(n)}{n} \qquad\qquad (9.27)$$

By analogy, we may consider the following rates of Rényi block entropies:

$$h_\gamma := \limsup_{n \to \infty} \frac{H_\gamma(n)}{n} \qquad\qquad (9.28)$$

$$h_\gamma^{\mathrm{cond}} := \limsup_{n \to \infty} \frac{H_\gamma^{\mathrm{cond}}(n)}{n} \qquad\qquad (9.29)$$

In the second turn, we may ask whether the respective upper limits exist as proper limits. It turns out that the proper limit exists always for the largest of all Rényi entropies.

Theorem 9.6: *For a stationary process* $(X_i)_{i \in \mathbb{Z}}$ *over a finite alphabet, we have*

$$h_0 = \lim_{n \to \infty} \frac{H_0(n)}{n} = \inf_{n \in \mathbb{N}} \frac{H_0(n)}{n} \tag{9.30}$$

Proof: Observe that $\{x_1^{n+m} : P(X_1^{n+m} = x_1^{n+m}) > 0\} \subseteq \{x_1^n : P(X_1^n = x_1^n) > 0\} \times \{x_1^m : P(X_1^m = x_1^m) > 0\}$. Hence $H_0(n + m) \le H_0(n) + H_0(m)$ and thus (9.30) holds by the Fekete lemma (Theorem 5.11). □

Limit h_0 will be called the Hartley entropy rate.

The behavior of other block Rényi entropies seems to be less regular, unless we make some pretty strong assumptions. Consider for example conditions

$$K := \sup_{n,m \ge 1} \operatorname{ess\,sup} \frac{P(X_1^{n+m})}{P(X_1^n)P(X_{n+1}^{n+m})} < \infty \tag{9.31}$$

$$L := \inf_{n,m \ge 1} \operatorname{ess\,inf} \frac{P(X_1^{n+m})}{P(X_1^n)P(X_{n+1}^{n+m})} > 0 \tag{9.32}$$

Clearly, if condition (9.31) is satisfied then process $(X_i)_{i \in \mathbb{Z}}$ is finitary, i.e. its excess entropy is finite. Thus, condition (9.31) is probably too strong as for probabilistic modeling of natural language. But as we will see in Section 11.1, condition (9.31) is satisfied by hidden Markov processes with a finite number of hidden states. Similar statement holds also for condition (9.32). Therefore, it makes a certain sense to contemplate conditions (9.31) and (9.32) for a while.

In fact, it can be shown that condition (9.31) is sufficient for the existence of proper Rényi entropy rates. The following result is due to Pittel (1985).

Theorem 9.7: *For a stationary process* $(X_i)_{i \in \mathbb{Z}}$ *over a finite alphabet which satisfies condition (9.31) and* $\gamma \in (0, \infty]$, *we have*

$$h_\gamma = \lim_{n \to \infty} \frac{H_\gamma(n)}{n} \tag{9.33}$$

Proof: Write $p(x_1^n) := P(X_1^n = x_1^n)$. For $\gamma \in (0,1) \cup (1, \infty)$, observe that condition (9.31) implies

$$\left(\sum_{x_1^{n+m}} p(x_1^{n+m})^\gamma \right)^{1/\gamma} \le K \left(\sum_{x_1^n} p(x_1^n)^\gamma \right)^{1/\gamma} \left(\sum_{x_1^m} p(x_1^n)^\gamma \right)^{1/\gamma} \tag{9.34}$$

Hence, $a_{n+m} \ge a_n + a_m$ for $a_n := \frac{\gamma-1}{\gamma} H_\gamma(n) - \log K$ for $\gamma \in (0, \infty]$ by continuity and thus (9.33) holds by the Fekete lemma (Theorem 5.11). □

Similar statements hold also for rates h_γ^{cond}, which are equal to rates h_γ.

Theorem 9.8: *For a stationary process $(X_i)_{i\in\mathbb{Z}}$ over a finite alphabet which satisfies condition (9.31) and $\gamma \in (1,\infty]$, we have*

$$h_\gamma^{\mathrm{cond}} = \lim_{n\to\infty} \frac{H_\gamma^{\mathrm{cond}}(n)}{n} = h_\gamma \qquad (9.35)$$

Proof: It is sufficient to show that differences $[H_\gamma(n) - H_\gamma^{\mathrm{cond}}(n)]/n$ tend to zero. Write $p(x_1^n) := P(X_1^n = x_1^n)$ and $p(x_1^n|X_{n+1}^\infty) := P(X_1^n = x_1^n|X_{n+1}^\infty)$. For $\gamma \in (1,\infty)$, assuming (9.31), we obtain indeed

$$\frac{\gamma-1}{\gamma}[H_\gamma(n) - H_\gamma^{\mathrm{cond}}(n)] = \log \frac{\mathbf{E}\left(\sum_{x_1^n} p(x_1^n|X_{n+1}^\infty)^\gamma\right)^{1/\gamma}}{\left(\sum_{x_1^n} p(x_1^n)^\gamma\right)^{1/\gamma}}$$

$$\leq \log \operatorname{ess\,sup} \frac{P(X_1^n|X_{n+1}^\infty)}{P(X_1^n)} \leq \log K \qquad (9.36)$$

which remains true also for $\gamma = \infty$ by continuity. Hence, we obtain the claim. $\square$

It is an open question whether relationships (9.33) and (9.35) survive when we relax somehow condition (9.31). Answering this question is important since by the discussion in Chapter 8, we may suppose that condition (9.31) is too strong as for probabilistic modeling of natural language, for it implies that excess entropy is finite.

9.3 Recurrence Times

In this section, we will study some basic properties of recurrence times, i.e. the functions which return the first position in an infinite sequence on which the initial block of a given length reoccurs. Recurrence times can be related to both Rényi entropies (see Theorem 9.28) and maximal repetition (see Theorem 9.26). In this section, we will present a few more fundamental and more celebrated results. It should be noted that recurrence times are a classical topic in ergodic theory and information theory. Their fundamental links with probability and the Shannon entropy rate have been established by Kac (1947), Wyner and Ziv (1989), and Ornstein and Weiss (1993). Less recognized are their links with Rényi entropies (Ko, 2012). Recently, recurrence times have been also researched experimentally for natural language (Altmann et al., 2009).

The formal definition of the recurrence times is as follows.

Definition 9.4 *(Waiting and recurrence times):* The waiting time $W(y_1^k|x_1^\infty)$ is the first position in the infinite sequence $x_1^\infty \in \mathbb{X}^{\mathbb{N}}$ on which a copy of a string $y_1^k \in \mathbb{X}^*$ starts,

$$W(y_1^k|x_1^\infty) := \inf\{i \geq 1 : x_i^{i+k-1} = y_1^k\} \qquad (9.37)$$

where the infimum of the empty set equals infinity, $\inf \emptyset := \infty$. In contrast, the recurrence time $R(k|x_1^\infty)$ is the first position in the infinite sequence $x_1^\infty \in \mathbb{X}^{\mathbb{N}}$ on which a copy of the finite prefix $x_1^k \in \mathbb{X}^*$ restarts,

$$R(k|x_1^\infty) := W(x_1^k|x_2^\infty) = \inf\{i \geq 1 : x_{i+1}^{i+k} = x_1^k\} \qquad (9.38)$$

The first proposition concerning recurrence times is due to Kac (1947). It states that the conditional expectation of the recurrence time is equal to the inverse probability of the recurring event.

Theorem 9.9 *(Kac lemma):* *Consider a stationary ergodic process* $(Y_i)_{i\in\mathbb{Z}}$. *If* $P(Y_0 = y) > 0$ *then*

$$\mathbf{E}\,[W(y|Y_1^\infty)|Y_0 = y] = \frac{1}{P(Y_0 = y)} \qquad (9.39)$$

Proof: Denoting $Q_k(y) = P(Y_k = y, Y_n \neq y, 1 \leq n < k|Y_0 = y)$, we obtain

$$\mathbf{E}\,[W(y|Y_1^\infty)|Y_0 = y] = \sum_{k=1}^{\infty} k Q_k(y) = \sum_{j=0}^{\infty} \sum_{k=1}^{\infty} Q_{j+k}(y) \qquad (9.40)$$

Recall that $P(Y_0 = y) > 0$. Since $\sum_{n=1}^{\infty} \mathbf{1}\{Y_n = y\} = \sum_{n=-\infty}^{0} \mathbf{1}\{Y_n = y\} = \infty$ holds almost surely by the Birkhoff ergodic theorem (Theorem 4.6), then we obtain $P(A) = 1$ for

$$A := (Y_n = y \text{ for infinitely many } n \leq 0 \text{ and infinitely many } n \geq 1) \quad (9.41)$$

Subsequently, by stationarity we may write

$$1 = P(A) = \sum_{j=0}^{\infty} \sum_{k=1}^{\infty} P(Y_k = y, Y_j = y, Y_n \neq y, j < n < k)$$

$$= \sum_{j=0}^{\infty} \sum_{k=1}^{\infty} P(Y_j = y)P(Y_k = y, Y_n \neq y, j < n < k|Y_j = y)$$

$$= \sum_{j=0}^{\infty} \sum_{k=1}^{\infty} P(Y_j = y)Q_{j+k}(y)$$

$$= P(Y_0 = y)\mathbf{E}\,[W(y|Y_1^\infty)|Y_0 = y] \qquad (9.42)$$

which implies the claim. $\qquad\qquad\square$

Since for a stationary ergodic process $(X_i)_{i\in\mathbb{Z}}$, process $(Y_i)_{i\in\mathbb{Z}}$ consisting of blocks $Y_i = X_{i+1}^{i+k}$ is also stationary ergodic, as a consequence of the Kac lemma we obtain two probabilistic bounds for the recurrence time, established by Kontoyiannis (1998). The first Kontoyiannis bound applies unconditional probability.

Theorem 9.10 (*Kontoyiannis bound*): *For a stationary ergodic process $(X_i)_{i\in\mathbb{N}}$ over a countable alphabet, for any $C > 0$, we have*

$$P\left(R(k|X_1^\infty) \geq \frac{C}{P(X_1^k)}\right) \leq C^{-1} \tag{9.43}$$

Proof: By the conditional Markov inequality (see Theorem 3.11)

$$P(Y \geq \epsilon|Z) \leq \frac{\mathbf{E}[Y|Z]}{\epsilon}, \quad Y \geq 0 \tag{9.44}$$

and the Kac lemma (Theorem 9.9), we obtain

$$P\left(R(k|X_1^\infty) \geq \frac{C}{P(X_1^k)}\right) = \mathbf{E}\,P\left(R(k|X_1^\infty) \geq \frac{C}{P(X_1^k)}\Big|X_1^k\right)$$

$$\leq \mathbf{E}\,\mathbf{E}\left[\frac{P(X_1^k)\mathbf{E}\,(R(k|X_1^\infty)|X_1^k)}{C}\Big|X_1^k\right]$$

$$= \mathbf{E}\,\mathbf{E}\,[C^{-1}|X_1^k] = C^{-1} \tag{9.45}$$

which is the claim. □

The second Kontoyiannis bound applies conditional probability.

Theorem 9.11 (*Kontoyiannis bound*): *For a process $(X_i)_{i\in\mathbb{N}}$ over a finite alphabet of cardinality D, for any $C > 0$, we have*

$$P\left(R(k|X_1^\infty) \leq \frac{C}{P(X_1^k|X_{k+1}^\infty)}\Big|X_{k+1}^\infty\right) \leq C(1 + k\ln D) \tag{9.46}$$

Proof: By the conditional Markov inequality (9.44), we obtain

$$P\left(P(X_1^k|X_{k+1}^\infty) \leq \frac{C}{R(k|X_1^\infty)}\Big|X_{k+1}^\infty\right) \leq \mathbf{E}\left(\frac{C}{P(X_1^k|X_{k+1}^\infty)R(k|X_1^\infty)}\Big|X_{k+1}^\infty\right)$$

$$= \sum_{x_1^k}\frac{C}{W(x_1^k|x_2^kX_{k+1}^\infty)} \tag{9.47}$$

But for a fixed x_{k+1}^∞ and each $i \geq 1$ there exists at most one string x_1^k such that $W(x_1^k|x_2^\infty) = i$, so we have a uniform almost sure bound

$$\sum_{x_1^k} \frac{C}{W(x_1^k|x_2^\infty)} \leq \sum_{i=1}^{D^k} \frac{C}{i} \leq C\left(1 + \int_1^{D^k} \frac{1}{u} du\right) \leq C(1 + k \ln D) \qquad (9.48)$$

which implies the claim. □

A simple consequence of the Kontoyiannis bounds is the following theorem discovered by Wyner and Ziv (1989) and Ornstein and Weiss (1993), which is called the WZ/OW theorem. For a stationary ergodic process over a finite alphabet, the WZ/OW theorem equates the exponential growth rate of the recurrence time to the Shannon entropy rate.

Theorem 9.12 (*Almost sure WZ/OW theorem*): *For a stationary ergodic process* $(X_i)_{i \in \mathbb{N}}$ *over a finite alphabet, we have*

$$\lim_{k \to \infty} \frac{1}{k} \log R(k|X_1^\infty) = h \text{ almost surely} \qquad (9.49)$$

where h is the entropy rate of the process.

Proof: Putting $C = k^2$ in (9.43) and $C = k^{-3}$ in (9.46) we obtain by the Borel–Cantelli lemma (Theorem 3.12) that sandwich bound

$$\frac{1}{k}[-\log P(X_1^k|X_{k+1}^\infty)] - \frac{3}{k}\log k \leq \frac{1}{k}\log R(k|X_1^\infty) \leq \frac{1}{k}[-\log P(X_1^k)] + \frac{2}{k}\log k \qquad (9.50)$$

holds for all but finitely many k almost surely. But by the proof of the Shannon–McMillan–Breiman (SMB) theorem (Theorem 6.1), we have

$$\lim_{k \to \infty} \frac{1}{k}[-\log P(X_1^k)] = \lim_{k \to \infty} \frac{1}{k}[-\log P(X_1^k|X_{k+1}^\infty)] = h \text{ almost surely} \quad (9.51)$$

Hence the claim follows. □

A similar result holds in expectation for any stationary process over a finite alphabet. As it can be foreseen, the expected exponential growth rate of the recurrence time is also equal to the Shannon entropy rate.

Theorem 9.13 (*Expected WZ/OW theorem*): *For a stationary process* $(X_i)_{i \in \mathbb{N}}$ *over a finite alphabet, we have*

$$\lim_{k \to \infty} \frac{1}{k} \mathbf{E} \log R(k|X_1^\infty) = h \qquad (9.52)$$

where h is the Shannon entropy rate of the process.

Proof: Let $H(k) := \mathbf{E}\left[-\log P(X_1^k)\right]$ be the block entropy of process $(X_i)_{i\in\mathbb{Z}}$. Observe that $\mathbf{E}\,X \le \int_0^\infty P(X \ge p)dp$. Hence, applying Theorem 9.10, we also obtain

$$\mathbf{E}\log R(k|X_1^\infty) = \mathbf{E}\left[-\log P(X_1^k)\right] + \mathbf{E}\left[\log R(k|X_1^\infty)P(X_1^k)\right]$$

$$\le H(k) + \int_0^\infty P(R(k|X_1^\infty)P(X_1^k) \ge 2^p)dp$$

$$\le H(k) + \int_0^\infty 2^{-p}dp = H(k) + (\ln 2)^{-1} \qquad (9.53)$$

Conversely, we observe that given x_{k+1}^∞ there is a one-to-one correspondence between x_1^k and $R(k|x^\infty)$. Moreover, we observe that $\sum_{n=1}^\infty n^{-1}[\log n + 1]^{-2} < \infty$. Hence, $H(S) \le \mathbf{E}\log S + 2\log(\mathbf{E}\log S + 1) + C$ for a random variable S assuming values in natural numbers and a constant $C > 0$, see the proof of a similar bound in Theorem 8.17. Thus,

$$hk = H(X_1^k|X_{k+1}^\infty) = H(R(k|X_1^\infty)|X_{k+1}^\infty)$$

$$\le H(R(k|X_1^\infty))$$

$$\le \mathbf{E}\log R(k|X_1^\infty) + 2\log(\mathbf{E}\log R(k|X_1^\infty) + 1) + C$$

$$\le \mathbf{E}\log R(k|X_1^\infty) + 2\log(H(k) + (\ln 2)^{-1} + 1) + C \qquad (9.54)$$

As a result, we obtain a sandwich bound for expectations,

$$hk - 2\log(H(k) + (\ln 2)^{-1} + 1) - C \le \mathbf{E}\log R(k|X_1^\infty) \le H(k) + (\ln 2)^{-1}$$

$$(9.55)$$

Noticing that $\lim_{k\to\infty} H(k)/k = h$ completes the proof. □

Using the Kontoyiannis bounds, moments of the recurrence time $\mathbf{E}\,R(k|X_1^\infty)^\gamma$ can be also bounded by Rényi entropy rates of order γ. A related idea will be used in Theorem 9.28 to bound the maximal repetition, via recurrence times, with Rényi entropies.

9.4 Subword Complexity

In this section, we will investigate some properties of subword complexity, i.e. the function which counts how many different substrings of a given length appear in a given string. Subword complexity can be simply related to both recurrence times (see Theorem 9.15) and maximal repetition (see Theorem 9.20). Subword complexity has not been much researched in information theory but it is a simple characteristic of a string. It has been mostly studied by theoretical computer scientists (Ferenczi, 1999; de Luca, 1999; Janson et al., 2004; Gheorghiciuc and Ward, 2007; Ivanko, 2008). As we will show in this section, subword complexity can be used to estimate moments of recurrence times, so it should deserve more attention in stochastic processes.

The formal definition of the subword complexity is as follows.

Definition 9.5 *(Subword complexity)*: The subword complexity $f(k|x_1^n)$ is the number of distinct substrings of length k in string $x_1^n \in \mathbb{X}^*$. In symbols, we have

$$f(k|x_1^n) := \#\{x_{j+1}^{j+k} : 0 \le j \le n - k\}, \quad k \le n \tag{9.56}$$

whereas we put $f(k|x_1^n) := 0$ for $k > n$.

We note in passing that subword complexity $f(k|x_1^n)$ is the cardinality of the empirical vocabulary $V_k(x_1^n)$ defined in (6.57). Let us write

$$a \wedge b := \min\{a, b\}, \quad a \vee b := \max\{a, b\} \tag{9.57}$$

We note the well-known simple combinatorial bound for the subword complexity

$$f(k|x_1^n) \le D^k \wedge (n - k + 1) \tag{9.58}$$

where D is the cardinality of the alphabet of symbols x_i.

It is natural to ask whether subword complexity and recurrence times are related. Let us make the following simple observation, which holds in general.

Theorem 9.14: *Let x_1^∞ be an arbitrary infinite sequence. Then*

$$f(k|x_1^n) = \sum_{i=1}^{n+k-1} \mathbf{1}\{R(k|x_i^\infty) \ge n - k + 2 - i\} \tag{9.59}$$

Proof: Observe that for $1 \le i \le n - k + 1$ we have

$$f(k|x_i^n) - f(k|x_{i+1}^n) = \mathbf{1}\{R(k|x_i^\infty) \ge n - k + 2 - i\} \tag{9.60}$$

and since $f(k|x_{n-k+2}^n) = 0$ then we obtain the telescope sum

$$f(k|x_1^k) = \sum_{i=1}^{n-k+1} [f(k|x_i^n) - f(k|x_{i+1}^n)]$$

$$= \sum_{i=1}^{n+k-1} \mathbf{1}\{R(k|x_i^\infty) \ge n - k + 2 - i\} \tag{9.61}$$

Thus, we have derived the claim. □

Subsequently, let us define the subword complexity of an infinite sequence and the trimmed recurrence times of a string,

$$f(k|x_1^\infty) := \lim_{n \to \infty} f(k|x_1^n) \tag{9.62}$$

$$R(k|x_1^n) := R(k|x_1^\infty) \wedge (n - k + 1) \tag{9.63}$$

As a consequence of Theorem 9.14, an exact equality can be proved for the expectations of subword complexity and trimmed recurrence time, if we assume that the process which generates strings is stationary.

Theorem 9.15: Let $(X_i)_{i \in \mathbb{N}}$ be a stationary stochastic process over a countable alphabet. Then for $n \in \{k, k+1, \dots, \infty\}$, we have

$$\mathbf{E}\, f(k|X_1^n) = \mathbf{E}\, R(k|X_1^n) \tag{9.64}$$

Proof: By identity (9.59), we have

$$\mathbf{E}\, f(k|X_1^n) = \mathbf{E} \sum_{i=1}^{n-k+1} \mathbf{1}\{R(k|X_i^\infty) \geq n - k + 2 - i\}$$

$$= \sum_{i=1}^{n-k+1} P(R(k|X_i^\infty) \geq n - k + 2 - i) \tag{9.65}$$

Applying stationarity and inverting the order of summation, we obtain

$$\mathbf{E}\, f(k|X_1^n) = \sum_{i=1}^{n-k+1} P(R(k|X_1^\infty) \geq i) \tag{9.66}$$

Finally, using identity $\mathbf{E}\, Y = \sum_{i=1}^{\infty} P(Y \geq i)$ for a random variable Y taking values in natural numbers yields

$$\mathbf{E}\, f(k|X_1^n) = \sum_{i=1}^{\infty} P(R(k|X_1^\infty) \wedge (n - k + 1) \geq i)$$

$$= \mathbf{E}\, R(k|X_1^n) \tag{9.67}$$

which is the requested claim for finite n. The claim for infinite n follows hence by the monotone convergence (Theorem 3.8). $\qquad\square$

We note that the expectation of the recurrence time is $\mathbf{E}\, R(k|X_1^\infty) \leq D^k$ for a process $(X_i)_{i \in \mathbb{N}}$ over an alphabet of cardinality D – by the Kac lemma (Theorem 9.9). Thus, identity (9.64) is a neat complement of the simple bound (9.58). We find it intriguing that identity (9.64) holds for all stationary processes. Similar equalities do not seem to be true for other moments of subword complexity and recurrence times – and even the equality for the expectations may break down for nonstationary sources.

What is somehow surprising, the variance of recurrence time and, more generally, Jensen gaps of recurrence time can be expressed by a series of expected subword complexities. This is a consequence of a more general fact for Jensen gaps of random variables taking values in natural numbers. For a real random variable Y and a function $g : \mathbb{R} \to \mathbb{R}$, the Jensen gap is usually defined as difference

$\mathbf{E} g(Y) - g(\mathbf{E} Y)$. For random variable Y taking values in natural numbers and a function $g : \mathbb{N} \to \mathbb{R}$, we will define the Jensen gap as $\mathbf{E} g(Y) - g\{\mathbf{E} Y\}$, where formula

$$g\{r\} := (1 - r + \lfloor r \rfloor)g(\lfloor r \rfloor) + (r - \lfloor r \rfloor)g(\lfloor r \rfloor + 1), \quad r \in [1, \infty) \tag{9.68}$$

extends function $g : \mathbb{N} \to \mathbb{R}$ to a piecewise linear function $g : [1, \infty) \to \mathbb{R}$. For a convex function $g : \mathbb{N} \to \mathbb{R}$, so extended function $g : [1, \infty) \to \mathbb{R}$ is also convex and so defined Jensen gap is positive by the Jensen inequality (Theorem 3.14).

Now let us denote the central second difference operator

$$\delta^2 g(n) := g(n + 1) - 2g(n) + g(n - 1) \tag{9.69}$$

Let us observe that for a random variable Y taking values in natural numbers,

$$P(Y = n) = -\delta^2 \mathbf{E} (Y \wedge n) \tag{9.70}$$

since $\mathbf{E} Y = \sum_{i=1}^{\infty} P(Y \geq i)$ and thus $\mathbf{E} (Y \wedge n) - \mathbf{E} (Y \wedge (n - 1)) = P(Y \geq n)$. Hence, we obtain the following representation by summing by parts.

Theorem 9.16: *Let $Y : \Omega \to \{1, 2, \ldots, N\}$ be a bounded random variable assuming values in natural numbers. For an arbitrary function $g : \mathbb{N} \to \mathbb{R}$, we have*

$$\mathbf{E} g(Y) - g\{\mathbf{E} Y\} = \sum_{n=2}^{N-1} [(\mathbf{E} Y) \wedge n - \mathbf{E} (Y \wedge n)]\delta^2 g(n) \tag{9.71}$$

Proof: For functions $g, f : \mathbb{N} \to \mathbb{R}$, we observe the identity of summing by parts which takes form

$$\sum_{n=M+1}^{N-1} f(n)\delta^2 g(n) + f(M)g(M + 1) + f(N)g(N - 1)$$

$$= \sum_{n=M+1}^{N-1} g(n)\delta^2 f(n) + g(M)f(M + 1) + g(N)f(N - 1) \tag{9.72}$$

Now let us put $f(n) := (\mathbf{E} Y) \wedge n - \mathbf{E} (Y \wedge n)$. Since $1 \leq Y \leq N$, we obtain $f(n) = 0$ for $n \leq 1$ and $n \geq N$. This yields

$$\sum_{n=2}^{N-1} f(n)\delta^2 g(n) = \sum_{n=2}^{N-1} g(n)\delta^2 f(n) + g(1)f(2) + g(N)f(N - 1)$$

$$= \sum_{n=1}^{N} g(n)\delta^2 f(n) \tag{9.73}$$

Since

$$\delta^2(r \wedge n) = -(1 - r + \lfloor r \rfloor)\mathbf{1}\{n = \lfloor r \rfloor\} - (r - \lfloor r \rfloor)\mathbf{1}\{n = \lfloor r \rfloor + 1\} \tag{9.74}$$

and $P(Y = n) = -\delta^2 \mathbf{E}(Y \wedge n)$ by observation (9.70), we further derive

$$\sum_{n=1}^{N} g(n)\delta^2 f(n) = -g\{\mathbf{E}\,Y\} - \sum_{n=1}^{N} g(n)\delta^2 \mathbf{E}(Y \wedge n) = -g\{\mathbf{E}\,Y\} + \mathbf{E}\,g(Y)$$

$$(9.75)$$

Thus, the claim follows by conjunction of equalities (9.73) and (9.75). $\square$

Obviously, $(\mathbf{E}\,Y) \wedge n \geq \mathbf{E}(Y \wedge n)$. Thus, Theorem 9.16 has a simple corollary for convex functions, which follows by the monotone convergence.

Theorem 9.17: *Let* $Y : \Omega \to \mathbb{N}$ *be a random variable assuming values in natural numbers such that* $\mathbf{E}\,Y < \infty$*. Put* $Y_N := Y \wedge N$ *for* $N \in \mathbb{N}$*. For a convex function* $g : \mathbb{N} \to \mathbb{R}$ *we have*

$$\mathbf{E}\,g(Y_N) - g\{\mathbf{E}\,Y_N\} \leq \mathbf{E}\,g(Y) - g\{\mathbf{E}\,Y\} \tag{9.76}$$

$$\mathbf{E}\,g(Y) - g\{\mathbf{E}\,Y\} = \sum_{n=2}^{\infty} [(\mathbf{E}\,Y) \wedge n - \mathbf{E}(Y \wedge n)]\delta^2 g(n) \tag{9.77}$$

Proof: Consider $Y_N := Y \wedge N$ for $N \in \mathbb{N}$. We have

$$\mathbf{E}\,g(Y_N) - g\{\mathbf{E}\,Y_N\} = \sum_{n=2}^{N-1} [(\mathbf{E}\,Y_N) \wedge n - \mathbf{E}(Y \wedge n)]\delta^2 g(n) \tag{9.78}$$

Observe that for a convex function $g : \mathbb{N} \to \mathbb{R}$ we have $\delta^2 g(n) \geq 0$, whereas random variables $g(Y_N)$ are uniformly either increasing or decreasing to $g(Y)$ for sufficiently large N. Also terms $\mathbf{E}\,Y_N$ and $(\mathbf{E}\,Y_N) \wedge n$ increase with N to $\mathbf{E}\,Y$ and $(\mathbf{E}\,Y) \wedge n$ respectively. Hence the claims follow by the monotone convergence (Theorem 3.8). $\square$

It is an interesting problem whether inequality (9.76) can be further generalized. Let us note that there exist convex functions $g : \mathbb{R} \to \mathbb{R}$ and linear functions $f : \mathbb{R} \to \mathbb{R}$ such that the reverse inequality is true,

$$\mathbf{E}\,g(f(Y)) - g(\mathbf{E}\,f(Y)) > \mathbf{E}\,g(Y) - g(\mathbf{E}\,Y) \tag{9.79}$$

For example, we can take $g(y) = y^2$, $f(y) = ay + b$ and $P(Y = \pm 1) = 1/2$. Then inequality (9.79) is obtained for $a^2 > 1$. It may be possible, however, that the desired inequality

$$\mathbf{E}\,g(f(Y)) - g(\mathbf{E}\,f(Y)) \leq \mathbf{E}\,g(Y) - g(\mathbf{E}\,Y) \tag{9.80}$$

holds if function $g : \mathbb{R} \to \mathbb{R}$ is convex and function $f : \mathbb{R} \to \mathbb{R}$ satisfies a sort of contracting condition similar to $|f(x) - f(y)| \leq |x - y|$.

To make another observation, as a consequence of formula (9.77), we obtain a bound for the variance $\operatorname{Var}Y = \mathbf{E}\, Y^2 - (\mathbf{E}\, Y)^2$ of form

$$\mathbf{E}\, Y^2 - \{\mathbf{E}\, Y\}^2 = 2 \sum_{n=2}^{\infty} [(\mathbf{E}\, Y) \wedge n - \mathbf{E}\, (Y \wedge n)] \tag{9.81}$$

where $\{a\}^2 \in [a^2, (a+1)^2)$ by convexity of function $\mathbb{R} \ni x \mapsto x^2 \in \mathbb{R}$.

Let us come back to subword complexity and trimmed recurrence time in the case of a stationary process. Let us write their common expectation as

$$\mathcal{V}_k(n) := \mathbf{E}\, f(k|X_1^{n+k-1}) = \mathbf{E}\, R(k|X_1^{n+k-1}) \tag{9.82}$$

with $\mathcal{V}_k(0) := 0$. Since $P(R(k|X_1^{\infty}) = n) = -\delta^2 \mathcal{V}_k(n)$ by observation (9.70), then expectation $\mathcal{V}_k(n)$ is a positive, growing, and concave function of n – like the Shannon block entropy $H(n)$. Assume now that limit

$$\mathcal{V}_k := \lim_{n \to \infty} \mathcal{V}_k(n) = \mathbf{E}\, f(k|X_1^{\infty}) = \mathbf{E}\, R(k|X_1^{\infty}) \tag{9.83}$$

is finite and take a convex function $g : \mathbb{N} \to \mathbb{R}$. By Theorem 9.17 for $n \in \mathbb{N}$ we obtain

$$\mathbf{E}\, g(R(k|X_1^{n+k-1})) - g\{\mathcal{V}_k(n)\} \le \mathbf{E}\, g(R(k|X_1^{\infty})) - g\{\mathcal{V}_k\} \tag{9.84}$$

$$\mathbf{E}\, g(R(k|X_1^{\infty})) - g\{\mathcal{V}_k\} = \sum_{n=2}^{\infty} [\mathcal{V}_k \wedge n - \mathcal{V}_k(n)] \delta^2 g(n) \tag{9.85}$$

As some special cases of formula (9.85), we obtain for instance expressions

$$\mathbf{E}\, (R(k|X_1^{\infty}))^2 = \{\mathcal{V}_k\}^2 + 2 \sum_{n=2}^{\infty} [\mathcal{V}_k \wedge n - \mathcal{V}_k(n)] \tag{9.86}$$

$$\mathbf{E}\, \log R(k|X_1^{\infty}) = \log\{\mathcal{V}_k\} - \sum_{n=2}^{\infty} [\mathcal{V}_k \wedge n - \mathcal{V}_k(n)] \log \frac{n^2}{n^2 - 1} \tag{9.87}$$

$$\mathbf{E}\, (R(k|X_1^{\infty}))^{-1} = \{\mathcal{V}_k\}^{-1} + 2 \sum_{n=2}^{\infty} \frac{\mathcal{V}_k \wedge n - \mathcal{V}_k(n)}{n(n^2 - 1)} \tag{9.88}$$

We suppose that the variance of subword complexity is in general smaller than the variance of the respective trimmed recurrence time. Thus, potentially, plugging empirical averages of subword complexities instead of their expectations $\mathcal{V}_k(n)$ into the above formulas, we may estimate Jensen gaps of untrimmed recurrence time given a sample path of a stationary ergodic process.

In the following, we would like to provide some links between the subword complexity and certain entropy functions. Since subword complexity for arbitrary sequences can be quite large, it is convenient to consider the logarithm of the subword complexity. In this way, we obtain the Hartley entropy of a sequence, see Allouche and Shallit (2003).

Definition 9.6 *(Hartley entropy):* The Hartley entropy of a string or an infinite sequence is defined as

$$H_0(k|w) := \log f(k|w) \tag{9.89}$$

The Hartley entropy of a sequence can grow at most linearly for a finite alphabet of cardinality D since $H_0(k|w) \leq k \log D$ in this case. This entropy is naturally connected to the Hartley entropy of a generating stochastic process.

Theorem 9.18: *For a stationary process* $(X_i)_{i \in \mathbb{N}}$ *over a countable alphabet,*

$$H_0(k|X_1^\infty) \leq H_0(k) \text{ almost surely} \tag{9.90}$$

and if process $(X_i)_{i \in \mathbb{N}}$ *is ergodic, then we have equality in (9.90).*

Proof: Since the alphabet is countable, inequality $P(X_{i+1}^{i+k}) > 0$ holds almost surely for all $0 \leq i < \infty$, and hence $H_0(k|X_1^\infty) \leq H_0(k)$ holds almost surely. Additionally, if the process is ergodic then by the Birkhoff ergodic theorem (Theorem 4.6), if $P(X_{i+1}^{i+k} = x_1^k) > 0$ then the relative frequency of x_1^k in x_1^∞ is almost surely strictly greater than zero. Hence, $H_0(k|X_1^\infty) \geq H_0(k)$ holds almost surely. $\qquad\square$

Somewhat less trivial is a bound for the expected subword complexity in terms of the Shannon entropy. The following inequality is based on the results originally formulated by Dębowski (2016a,b, 2018b).

Theorem 9.19: *For a stationary process* $(X_i)_{i \in \mathbb{N}}$ *over a countable alphabet, if* $2^{H(k)} \leq n - k + 1$ *then*

$$\mathbf{E} f(k|X_1^n) \leq 2^{H(k)} + (\ln 2)H(k)[\mathrm{Li}(n - k + 1) - \mathrm{Li}(2^{H(k)})] \tag{9.91}$$

where we use the integral logarithm

$$\mathrm{Li}(x) := \int_2^x \frac{dp}{\ln p} \tag{9.92}$$

Proof: We will use the identity

$$f(k|X_1^n) = \sum_{w \in \mathbb{X}^k} \mathbf{1}\left\{ \sum_{i=0}^{n-k} \mathbf{1}\{X_{i+1}^{i+k} = w\} \geq 1 \right\} \tag{9.93}$$

Hence by the Markov inequality (Theorem 3.11),

$$\mathbf{E} f(k|X_1^n) = \sum_{w \in \mathbb{X}^k} P\left(\sum_{i=0}^{n-k} \mathbf{1}\{X_{i+1}^{i+k} = w\} \geq 1 \right)$$

$$\leq \sum_{w \in \mathbb{X}^k} \left[1 \wedge \mathbf{E} \sum_{i=0}^{n-k} \mathbf{1}\{X_{i+1}^{i+k} = w\} \right] = \sum_{w \in \mathbb{X}^k} [1 \wedge (n - k + 1)P(X_1^k = w)]$$

$$= \mathbf{E} \left(\frac{1}{P(X_1^k)} \wedge (n - k + 1) \right) = \int_0^{n+k-1} P\left(\frac{1}{P(X_1^k)} \geq p \right) dp \qquad (9.94)$$

Observe that $\mathbf{E}[-\log P(X_1^k)] = H(k)$. Therefore, by the Markov inequality (Theorem 3.11), we have

$$P\left(\frac{1}{P(X_1^k)} \geq p \right) = P(-\log P(X_1^k) \geq \log p) \leq \frac{H(k)}{\log p} \qquad (9.95)$$

for $p > 1$. Hence, for $2^{H(k)} \leq n - k + 1$, we further obtain from (9.94) that

$$\mathbf{E} f(k|X_1^n) \leq 1 + \int_1^{n-k+1} \left(1 \wedge \frac{H(k)}{\log p} \right) dp$$

$$\leq 2^{H(k)} + (\ln 2)H(k) \int_{2^{H(k)}}^{n-k+1} \frac{dp}{\ln p}$$

$$= 2^{H(k)} + (\ln 2)H(k)[\text{Li}(n - k + 1) - \text{Li}(2^{H(k)})] \qquad (9.96)$$

which is the requested claim. $\qquad \square$

We would like to stress that bound (9.91) is valid also for processes over an infinite alphabet. If the random symbols X_i are words of natural language (which are a priori countably infinitely many) then subword complexity $f(1|X_1^n)$ as a function of text length n is the empirical vocabulary size function, studied in quantitative linguistics (Khmaladze, 1988; Baayen, 2001). In this domain, also discussed is the type-token ratio $f(1|X_1^n)/n$. On page 9.4 we have observed that for a stationary process, the expected vocabulary size $V_1(n) = \mathbf{E} f(1|X_1^n)$ is a positive, growing, and concave function of text length n. In that case, by Theorem 9.19, the expected type-token ratio satisfies

$$\lim_{n \to \infty} \frac{V_1(n)}{n} \leq \lim_{n \to \infty} \frac{(\ln 2)H(1) \, \text{Li}(n)}{n} = 0 \qquad (9.97)$$

if additionally the Shannon entropy $H(1)$ is finite.

We note that for natural language, the empirical vocabulary size $f(1|X_1^n)$ grows approximately like $O(n^\beta)$ where $\beta < 1$ by the Herdan's law discussed in Section 1.3, whereas the unigram entropy of words was estimated cross-linguistically as $H(1) \approx 9$ bits (Bentz et al., 2017). Hence, Theorem 9.19 gives quite a loose bound in this case. It is interesting, however, to note that finiteness of word entropy is in general a sufficient cause to make the expected type-token ratio tend to zero, under the hypothesis that texts are generated by a stationary source.

9.5 Two Maximal Lengths

In this section, we will deal with two examples of maximal lengths: the longest match length and the maximal repetition. The longest match length has been researched mostly by information-theorists (Wyner and Ziv, 1989; Ornstein and Weiss, 1993; Szpankowski, 1993a; Kontoyiannis et al., 1998; Gao et al., 2008). In contrast, the maximal repetition has been studied by computer scientists (de Luca, 1999; Kolpakov and Kucherov, 1999b,a; Crochemore and Ilie, 2008), probabilists (Erdös and Rényi, 1970; Arratia and Waterman, 1989; Shields, 1992, 1997), and information theorists (Szpankowski, 1993b; Kontoyiannis and Suhov, 1994; Dębowski, 2011, 2017, 2018b).

The definition of the longest match length is as follows.

Definition 9.7 *(Longest match length)*: The longest match length $L(n|x_1^\infty)$ of an infinite sequence $x_1^\infty \in \mathbb{X}^*$ is the maximal length of its prefix that is repeated within the first n symbols. In symbols, we have

$$L(n|x_1^\infty) := \max\{k : x_{i+1}^{i+k} = x_1^k \text{ for some } 1 \le i \le n\} \tag{9.98}$$

Subsequently comes the definition of the maximal repetition.

Definition 9.8 *(Maximal repetition)*: The maximal repetition $L(x_1^n)$ of a string $x_1^n \in \mathbb{X}^*$ is the maximal length of a substring that is repeated in the string on possibly overlapping positions. In symbols, we have

$$L(x_1^n) := \max\{k : x_{i+1}^{i+k} = x_{j+1}^{j+k} \text{ for some } 0 \le i < j \le n - k\} \tag{9.99}$$

Maximal repetition can be computed in time which grows linearly with the length of string (Kolpakov and Kucherov, 1999a,b). Thus, we can easily compute the maximal repetition for empirical data.

The longest match length and the maximal repetition can be connected to recurrence times. To make the simple observation, we have

$$L(x_1^{n+k-1}) < k \implies L(n|x_1^\infty) < k \iff R(k|x_1^\infty) > n \tag{9.100}$$

so that we obtain

$$L(n|x_1^\infty) = \max\{k : R(k|x_1^\infty) \le n\} \le L(x_1^{n+L(n|x_1^\infty)-1}) \tag{9.101}$$

Hence by the almost sure WZ/OW theorem (Theorem 9.12) for a stationary ergodic process $(X_i)_{i\in\mathbb{Z}}$ with a Shannon entropy rate h, we obtain

$$\liminf_{n\to\infty} \frac{L(X_1^n)}{\log n} \ge \lim_{n\to\infty} \frac{L(n|X_1^\infty)}{\log n} = h^{-1} \text{ almost surely} \tag{9.102}$$

The exact proof is left as a simple exercise. As we can see, the rate of the longest match length is exactly equal to the Shannon entropy rate, whereas the rate of the maximal repetition is greater.

Indeed, the first inequality in (9.102) can be strict, as shown by Shields (1992). In particular, for any stationary ergodic process $(X_i)_{i\in\mathbb{Z}}$ and a function $f(n)$ such that $\lim_{n\to\infty} f(n)/n = 0$, there is a measurable function g of infinite sequences and a stationary ergodic process $(Y_i)_{i\in\mathbb{Z}}$, where $Y_i := g((X_{i+j})_{j\in\mathbb{Z}})$, such that

$$\limsup_{n\to\infty} \frac{L(Y_1^n)}{f(n)} \geq 1 \text{ almost surely} \tag{9.103}$$

Whereas the Shannon entropy rate of process $(Y_i)_{i\in\mathbb{Z}}$ is smaller than that of process $(X_i)_{i\in\mathbb{Z}}$ (Gray, 1990), a careful analysis of the proof of Shields (1992) shows that the difference between the two can be made arbitrarily small. Moreover, if we take $(X_i)_{i\in\mathbb{Z}}$ to be an IID process, then process $(Y_i)_{i\in\mathbb{Z}}$ is mixing.

Is there a general upper bound for the maximal repetition in terms of a different entropy function? According to Szpankowski (1993a,b), if process $(X_i)_{i\in\mathbb{Z}}$ satisfies both conditions (9.31) and (9.32) then not only the collision entropy rates satisfy

$$h_2 = \lim_{n\to\infty} \frac{H_2(n)}{n} = h_2^{\text{cond}} = \lim_{n\to\infty} \frac{H_2^{\text{cond}}(n)}{n} \tag{9.104}$$

which we have shown in Section 9.2 but also the maximal repetition obeys

$$\lim_{n\to\infty} \frac{L(X_1^n)}{\log n} = h_2^{-1} \text{ almost surely} \tag{9.105}$$

As it was mentioned in Section 1.13, it was shown by Dębowski (2012b, 2015b) that the maximal repetition for texts in natural language satisfies the power-law logarithmic growth and in this context we should rather contemplate condition

$$\limsup_{n\to\infty} \frac{L(X_1^n)}{(a\log n)^\alpha} = 1 \text{ almost surely} \tag{9.106}$$

for an $\alpha > 1$ and $a > 0$. Moreover, in the context of natural language modeling, probably we cannot assume condition (9.31) since it implies that the excess entropy is finite. In view of these empirical premises, it is valid to ask whether condition (9.106) implies in general that the collision entropy rate h_2 exists as a proper limit and is equal to zero. This question is more difficult than it seems and we will not give a definite answer. In the next section, we will only manage to show that if condition (9.106) holds then $h_\gamma^{\text{cond}} = 0$ for $\gamma > 1$.

It is easy to see that the growth rate of the maximal repetition should be somehow related to the growth rate of some sort of collision entropy. Namely, some simple idea of upper bounding the maximal repetition in terms of collision entropy is as follows. We have inequality

$$P(L(X_1^n) \geq k) = P(X_{i+1}^{i+k} = X_{j+1}^{j+k} \text{ for some } 0 \leq i < j \leq n-k)$$

$$\le \sum_{0 \le i < j \le n-k} P(X_{i+1}^{i+k} = X_{j+1}^{j+k})$$

$$\le n^2 \times \frac{1}{n} \sum_{i=1}^{n} P(X_{i+1}^{i+k} = X_1^k) \tag{9.107}$$

where as for a stationary ergodic process by identity (4.34) we obtain

$$\lim_{n \to \infty} \frac{1}{n} \sum_{i=1}^{n} P(X_{i+1}^{i+k} = X_1^k) = \sum_{x_1^k} P(X_1^k = x_1^k)^2 = 2^{-H_2(k)} \tag{9.108}$$

Hence we obtain $\lim_{n \to \infty} C(n, k) = 0$ for the difference

$$C(n, k) := \frac{1}{n} \sum_{i=1}^{n} P(X_{i+1}^{i+k} = X_1^k) - 2^{-H_2(k)} \tag{9.109}$$

and consequently

$$P(L(X_1^n) \ge k) \le n^2 (2^{-H_2(k)} + C(n, k)) \tag{9.110}$$

which seems to open way to using the Borel–Cantelli lemma. The ugly little problem, however, lies in upper bounding the difference $C(n, k)$ in terms of $H_2(k)$.

To show some fairly promising path that fails, consider the following. Let

$$\gamma_k := \{(X_1^k = x_1^k) : x_1^k \in \mathbb{X}^k\} \tag{9.111}$$

be the partitions of the probability space into cylinders of a fixed length and let $\gamma := \bigcup_{k \in \mathbb{N}} \gamma_k$ be their union. We note that we may bound uniformly

$$C(n, k) \le \sup_{k \in \mathbb{N}} \left| \frac{1}{n} \sum_{i=1}^{n} P(X_{i+1}^{i+k} = X_1^k) - 2^{-H_2(k)} \right|$$

$$= \sup_{k \in \mathbb{N}} \left| \int \sum_{A \in \gamma_k} I_A \left(\frac{1}{n} \sum_{i=1}^{n} T^{-i} I_A - P(A) \right) dp \right|$$

$$\le \sup_{k \in \mathbb{N}} \int \sup_{B \in \gamma_k} \left| \frac{1}{n} \sum_{i=1}^{n} T^{-i} I_B - P(B) \right| dp \le C(n) \tag{9.112}$$

where

$$C(n) := \int \sup_{B \in \gamma} \left| \frac{1}{n} \sum_{i=1}^{n} T^{-i} I_B - P(B) \right| dp \tag{9.113}$$

Subsequently, since class γ has a finite Vapnik–Chervonenkis dimension, we may apply the uniform ergodic theorem by Adams and Nobel (2010) and the dominated convergence (Theorem 3.10) to obtain $\lim_{n \to \infty} C(n) = 0$. But unfortunately, we have inequality $C(n) \ge 1/n$ if the distribution of process $(X_i)_{i \in \mathbb{Z}}$ is nonatomic, so bound (9.112) is mostly useless when combined with inequality (9.110), see also Problem 9.8.

Discussing bounds for the maximal repetition in terms of entropy functions, we should also take into account some basic algebraic links between maximal repetition, subword complexity, and recurrence times. The link between maximal repetition and subword complexity is captured by the following simple observation.

Theorem 9.20: *We have $f(k|x_1^n) < n - k + 1$ if and only if $L(x_1^n) \geq k$.*

Proof: String x_1^n contains $n - k + 1$ substrings of length k. Among them there can be at most $f(k|x_1^n)$ different substrings. If $f(k|x_1^n) < n - k + 1$, there must be a repeat of length k. Hence $L(x_1^n) \geq k$. Conversely, if $L(x_1^n) \geq k$, there is a repeat of length k and $f(k|x_1^n) < n - k + 1$. □

As a result we obtain these bounds.

Theorem 9.21: *We have*

$$P(L(X_1^n) < k) \leq \frac{\mathbf{E}\, f(k|X_1^n)}{n - k + 1} \tag{9.114}$$

$$P(L(X_1^n) \geq k) \leq n - k + 1 - \mathbf{E}\, f(k|X_1^n) \tag{9.115}$$

Proof: We have $f(k|X_1^n) = n - k + 1$ if $L(X_1^n) < k$ and $f(k|X_1^n) \leq n - k$ if $L(x_1^n) \geq k$. Hence

$$\mathbf{E}\, f(k|X_1^n) \geq (n - k + 1)P(L(X_1^n) < k) \tag{9.116}$$

$$\mathbf{E}\, f(k|X_1^n) \leq (n - k + 1)P(L(X_1^n) < k) + (n - k)P(L(X_1^n) \geq k) \tag{9.117}$$

from which the claims follow. □

Similarly, we can demonstrate the bound

$$P(L(X_1^n) < k) \leq \mathbf{E}\, \exp[f(k|X_1^n) - (n - k + 1)] \tag{9.118}$$

We suppose that the variance of subword complexity and also its higher moments $\mathbf{E}\,[f(k|X_1^n)]^p - [\mathbf{E}\, f(k|X_1^n)]^p$ are relatively small for stationary ergodic processes. If it is so, we may use the Taylor series

$$\mathbf{E}\, \exp f(k|X_1^n) = \exp \mathbf{E}\, f(k|X_1^n) + \sum_{p=2}^{\infty} \frac{\mathbf{E}\,[f(k|X_1^n)]^p - [\mathbf{E}\, f(k|X_1^n)]^p}{p!} \tag{9.119}$$

and consequently both probabilities $P(L(X_1^n) < k)$ and $P(L(X_1^n) \geq k)$ would be controlled by the difference

$$n - k + 1 - \mathbf{E}\, f(k|X_1^n) = \sum_{i=1}^{n-k+1} P(R(k|X_1^\infty) < i) \tag{9.120}$$

This difference can get sufficiently large if probabilities of short recurrence times are large enough. We recall from the Kontoyiannis bounds in Section 9.3 that this may be roughly the case when the collision entropy $H_2(k)$ is small. This sketches some idea of bounding probability $P(L(X_1^n) < k)$ with entropy $H_2(k)$. We will not pursue it, however, since there are a few gaps on the way such as bounding the higher moments in equality (9.119), see also Problem 9.10.

9.6 Logarithmic Power Laws

Eventually, we would like to be constructive and to present a few approaches to bounding the maximal repetition that do work in the general stationary ergodic case. Let us notice that formally the power-law logarithmic growth of maximal repetition can be linked to a power-law growth of some generalized block entropies and vanishing of some generalized entropy rates. We observe that condition (9.106) can be written as

$$\text{hilb}_{n\to\infty} \mathcal{P}_\bullet(L(X_1^n)) = 1 \text{ almost surely} \tag{9.121}$$

for some generalized block entropy function $H_\bullet(k) = a^{-1} k^{1/\alpha}$ and its related entropy power

$$\mathcal{P}_\bullet(k) := 2^{H_\bullet(k)} \tag{9.122}$$

Let us observe that if $H_\bullet(k)$ is an increasing function of block length k then the Hilberg exponent of random variable $\mathcal{P}_\bullet(L(X_1^n))$ is almost surely constant for a stationary ergodic process by Theorem 8.3. In fact, in this case we can often state a stronger proposition that the Hilberg exponent of function $\mathcal{P}_\bullet(L(X_1^n))$ for some entropy function $H_\bullet(k)$ is bounded away from zero and infinity. Having such a statement, we can then use this linking result.

Theorem 9.22:   *Let $(X_i)_{i\in\mathbb{N}}$ be a stationary ergodic process. Suppose that*

$$\text{hilb}_{n\to\infty} \mathcal{P}_\bullet(L(X_1^n)) \in [C_1, C_2] \text{ almost surely} \tag{9.123}$$

where $\mathcal{P}_\bullet(k) := 2^{H_\bullet(k)}$ and $H_\bullet(k)$ is an increasing function of k. Then for $\alpha \geq 1$ we have implications

$$\liminf_{n\to\infty} \frac{L(X_1^n)}{(a \log n)^\alpha} \geq 1 \text{ almost surely} \implies \limsup_{k\to\infty} \frac{H_\bullet(k)}{k^{1/\alpha}} \leq \frac{C_2}{a} \tag{9.124}$$

$$\limsup_{n\to\infty} \frac{L(X_1^n)}{(a \log n)^\alpha} \leq 1 \text{ almost surely} \implies \limsup_{k\to\infty} \frac{H_\bullet(k)}{k^{1/\alpha}} \geq \frac{C_1}{a} \tag{9.125}$$

Proof: We notice that for a stationary ergodic process $(X_i)_{i\in\mathbb{N}}$, the upper and the lower limit of $\dfrac{L(X_1^n)}{(a\log n)^\alpha}$ are almost surely constant by a reasoning similar to the proof of Theorem 8.3. Hence, first, we have

$$\liminf_{n\to\infty}\frac{L(X_1^n)}{(a\log n)^\alpha}\geq 1 \text{ almost surely} \Rightarrow \limsup_{n\to\infty}\frac{H_\bullet(\lfloor(a\log n)^\alpha\rfloor)}{\log n}\leq C_2$$

$$\Rightarrow \limsup_{k\to\infty}\frac{H_\bullet(k)}{k^{1/\alpha}}\leq\frac{C_2}{a} \qquad (9.126)$$

Second, we have

$$\limsup_{k\to\infty}\frac{H_\bullet(k)}{k^{1/\alpha}}<\frac{C_1}{a} \Rightarrow \limsup_{n\to\infty}\frac{a^{-1}L(X_1^n)^{1/\alpha}}{\log n}>1 \text{ almost surely}$$

$$\Rightarrow \limsup_{n\to\infty}\frac{L(X_1^n)}{(a\log n)^\alpha}>1 \text{ almost surely} \qquad (9.127)$$

Hence, by contradiction, we obtain the second implication. □

Resuming, by Theorem 9.22, if the Hilberg exponent of function $\mathcal{P}_\bullet(L(X_1^n))$ for some entropy $H_\bullet(k)$ is bounded away from zero and infinity then the power-law logarithmic growth of maximal repetition is roughly equivalent to the power-law growth of the generalized entropy function $H_\bullet(k)$ and implies in particular that the respective entropy rate satisfies $h_\bullet=\lim_{n\to\infty}H_\bullet(k)/k=0$. In the following, we will produce a few bounds of form (9.123) using Rényi–Arimoto entropies.

Let us observe that the result of Szpankowski (1993a, b) for processes $(X_i)_{i\in\mathbb{N}}$ that satisfy both conditions (9.31) and (9.32) can be written in a slightly weaker form as

$$\text{hilb }\mathcal{P}_2(L(X_1^n))=1 \text{ almost surely} \qquad (9.128)$$

where $\mathcal{P}_2(k):=2^{H_2(k)}$ is the power of the collision entropy of the process. Unfortunately, this statement rests on the strong assumptions (9.31) and (9.32) which do not seem to hold for natural language. Thus we need some other bounds that hold more generally.

A simple application of Theorem 9.20 is the following lower bound applying the concept of Hartley entropy of a sequence, the largest of all Rényi entropies.

Theorem 9.23: *For a sequence x_1^∞ over a finite alphabet, let $\mathcal{P}_0(k|x_1^\infty):=2^{H_0(k|x_1^\infty)}$ be the power of the Hartley entropy of the sequence. Assume that $\lim_{n\to\infty}L(x_1^n)/n=0$. Then we have*

$$\text{hilb }\mathcal{P}_0(L(x_1^n)|x_1^\infty)\geq 1 \qquad (9.129)$$

Proof: By Theorem 9.20, $H_0(L(x_1^n)|x_1^\infty) + \log D \geq H_0(L(x_1^n) + 1|x_1^\infty) \geq H_0$ $(L(x_1^n) + 1|x_1^n) = \log(n - L(x_1^n))$, where D is cardinality of the alphabet. Consequently, since $\lim_{n\to\infty} L(x_1^n)/n = 0$, hence the claim follows. □

This result has a stochastic counterpart.

Theorem 9.24: *For a stationary process* $(X_i)_{i\in\mathbb{N}}$ *over a finite alphabet, let* $\mathcal{P}_0(k) :=$ $2^{H_0(k)}$ *be the power of the block Hartley entropy of the process. Assume that* $\lim_{n\to\infty} L(x_1^n)/n = 0$ *almost surely. Then we have*

$$\text{hilb}_{n\to\infty} \mathcal{P}_0(L(X_1^n)) \geq 1 \text{ almost surely} \tag{9.130}$$

Proof: By Theorem 9.18, $H_0(k|X_1^\infty) \leq H_0(k)$ almost surely. Hence the claim is a corollary of Theorem 9.23. □

Abound similar to Theorem 9.24 was observed by Dębowski (2015b).

Now we will derive an upper bound of a similar simplicity. The upper bound applies the concept of the conditional min-entropy, which is the smallest of all Rényi–Arimoto entropies.

Theorem 9.25: *For a stationary process* $(X_i)_{i\in\mathbb{N}}$ *over a finite alphabet, let* $\mathcal{P}_\infty^{\text{cond}}(k) := 2^{H_\infty^{\text{cond}}(k)}$ *be the power of the block conditional min-entropy. We have*

$$\text{hilb}_{n\to\infty} \mathcal{P}_\infty^{\text{cond}}(L(X_1^n)) \leq 2 \text{ almost surely} \tag{9.131}$$

Proof: For $0 \leq i < j < \infty$ we have

$$P(X_{i+1}^{i+k} = X_{j+1}^{j+k}) = \mathbf{E}\, P(X_{i+1}^{i+k} = X_{j+1}^{j+k}|X_{i+k+1}^\infty)$$
$$\leq \mathbf{E}\, \max_{x_1^k} P(X_{i+1}^{i+k} = x_1^k|X_{i+k+1}^\infty) = 2^{-H_\infty^{\text{cond}}(k)} \tag{9.132}$$

Thus by the Bonferroni inequality (2.31),

$$P(L(X_1^n) \geq k) = P(X_{i+1}^{i+k} = X_{j+1}^{j+k} \text{ for some } 0 \leq i < j \leq n-k)$$
$$\leq \sum_{0\leq i<j\leq n-k} P(X_{i+1}^{i+k} = X_{j+1}^{j+k}) \leq n^2 2^{-H_\infty^{\text{cond}}(k)} \tag{9.133}$$

In particular, it follows that

$$P([\mathcal{P}_\infty^{\text{cond}}(L(X_1^n))]^\delta \geq l) \leq \frac{n^2}{l^{1/\delta}}, \quad l \geq 1 \tag{9.134}$$

Since $\mathbf{E}\, Y = \int_0^\infty P(Y \geq y)dy$ for a positive random variable Y, using substitution $u = l/n^{2\delta}$, we obtain that

$$\mathbf{E}\,[\mathcal{P}_\infty^{\text{cond}}(L(X_1^n))]^\delta \leq n^{2\delta} + \int_{n^{2\delta}}^\infty \frac{n^2 dl}{l^{1/\delta}} = n^{2\delta} + n^{2\delta}\int_1^\infty \frac{du}{u^{1/\delta}} \tag{9.135}$$

where the integral is finite for $0 < \delta < 1$. Hence by Theorem 8.4, we infer in this case

$$\underset{n\to\infty}{\text{hilb}}\ \mathcal{P}_\infty^{\text{cond}}(L(X_1^n)) \leq \frac{1}{\delta}\underset{n\to\infty}{\text{hilb}}\ \mathbf{E}\left[\mathcal{P}_\infty^{\text{cond}}(L(X_1^n))\right]^\delta \leq 2 \text{ almost surely} \qquad (9.136)$$

$\square$

Bounds similar to Theorem 9.25 were observed in part by Erdös and Rényi (1970), Shields (1997), and Kontoyiannis and Suhov (1994).

Subsequently, we will strengthen Theorems 9.24 and 9.25, narrowing the sandwich bound for the maximal repetition. We will show that we may replace the block Hartley entropy $H_0(n)$ with the block Shannon entropy $H(n)$ and the conditional block min-entropy $H_\infty^{\text{cond}}(n)$ with the conditional block Rényi entropies $H_\gamma^{\text{cond}}(n)$, where $\gamma \in (1, \infty)$, slightly improving the constants. For this goal we will apply recurrence times and the Kontoyiannis bounds from Section 9.3.

Theorem 9.26: *For a stationary process $(X_i)_{i\in\mathbb{N}}$ over a countable alphabet, we have*

$$P(L(X_1^{n+k-1}) \geq k) \in [P(R(k|x_1^\infty) \leq n), nP(R(k|x_1^\infty) \leq n)] \qquad (9.137)$$

Proof: By relationships (9.101) and (9.59) respectively, we have

$$(L(X_1^{n+k-1}) \geq k) \subseteq (L(n|X_1^\infty) \geq k) = (R(k|x_1^\infty) \leq n) \qquad (9.138)$$

$$(L(X_1^{n+k-1}) \geq k) = (f(k|X_1^{n+k-1}) < n) = \bigcup_{i=0}^{n-1}(R(k|X_i^\infty) < n - i) \qquad (9.139)$$

Hence, claim follows by stationarity and the Bonferroni inequality (2.31). $\square$

Now we will state the strengthening of Theorem 9.24. A bound similar to the following proposition, albeit slightly weaker, was established by Dębowski (2018b).

Theorem 9.27: *For a stationary process $(X_i)_{i\in\mathbb{N}}$ over a countable alphabet, let $\mathcal{P}(k) := 2^{H(k)}$ be the power of the block Shannon entropy. Assume that*

$$\beta := \inf_{k\in\mathbb{N}} \frac{\log H(k)}{\log k} > 0 \qquad (9.140)$$

Then we have

$$\underset{n\to\infty}{\text{hilb}}\ \mathcal{P}(L(X_1^n)) \geq 1 \text{ almost surely} \qquad (9.141)$$

Proof: By inequality (9.137), the Markov inequality (Theorem 3.11), and inequality (9.55), we obtain

$$P(L(X_1^n) < k) \leq P(R(k|X_1^\infty) > n - k + 1) \leq \frac{\mathbf{E}\log R(k|X_1^\infty)}{\log(n - k + 1)}$$

$$\leq \frac{H(k) + (\ln 2)^{-1}}{\log(n - k + 1)} \leq \frac{H(k) + (\ln 2)^{-1}}{\log(n - H(k)^{1/\beta} + 1)} \tag{9.142}$$

Hence,

$$P(H(L(X_1^n)) < l) \leq \frac{l + (\ln 2)^{-1}}{\log(n - l^{1/\beta} + 1)} \tag{9.143}$$

and in particular, for $p \in (0, 1)$, the respective quantile of the entropy power satisfies

$$\mathbf{Q}_p H(L(X_1^n)) \geq \sup\left\{l : \frac{l + (\ln 2)^{-1}}{\log(n - l^{1/\beta} + 1)} \leq p\right\} \tag{9.144}$$

It follows hence that $\mathrm{hilb}_{n \to \infty} \mathbf{Q}_p \mathcal{P}(L(X_1^n)) \geq p$ for any $p \in (0, 1)$ and we obtain the claim by Theorem 8.5. $\qquad\square$

An alternative proof of Theorem 9.27 can make use of inequalities (9.91) and (9.114). See Dębowski (2018b).

Finally, we will state the strengthening of Theorem 9.25. A bound similar to the following proposition was shown by Dębowski (2018b).

Theorem 9.28: *For a stationary process* $(X_i)_{i \in \mathbb{N}}$ *over a finite alphabet, let* $\gamma \in (1, \infty)$ *and* $\mathcal{P}_\gamma^{\mathrm{cond}}(k) := 2^{H_\gamma^{\mathrm{cond}}(k)}$ *be the power of the block conditional Rényi–Arimoto entropy. Assume that*

$$\beta := \inf_{k \in \mathbb{N}} \frac{\log H_\gamma^{\mathrm{cond}}(k)}{\log k} > 0 \tag{9.145}$$

Then we have

$$\mathrm{hilb}_{n \to \infty} \mathcal{P}_\gamma^{\mathrm{cond}}(L(X_1^n)) \leq 2 + \frac{1}{\gamma - 1} \text{ almost surely} \tag{9.146}$$

Proof: Let D be the cardinality of the alphabet of process $(X_i)_{i \in \mathbb{N}}$. By inequality (9.137), the second Kontoyiannis bound (9.46), and the Markov inequality (Theorem 3.11), we obtain

$$P(L(X_1^{n+k-1}) \geq k) \leq nP(R(k|X_1^\infty) \leq n)$$

$$\leq n \, \mathbf{E} \min_{C > 0}\left[P\left(\frac{C}{P(X_1^k|X_{k+1}^\infty)} \leq n \Big| X_{k+1}^\infty\right) + P\left(R(k|X_1^\infty) \leq \frac{C}{P(X_1^k|X_{k+1}^\infty)} \Big| X_{k+1}^\infty\right)\right]$$

$$\leq n\mathbf{E} \min_{C > 0}\left[P\left(\frac{C}{P(X_1^k|X_{k+1}^\infty)} \leq n \Big| X_{k+1}^\infty\right) + Ck(1 + \ln D)\right]$$

$$\leq n \, \mathbf{E} \min_{C > 0}\left[\frac{n^{\gamma - 1}\sum_{x_1^k} P(X_1^k = x_1^k|X_{k+1}^\infty)^\gamma}{C^{\gamma - 1}} + Ck(1 + \ln D)\right]$$

$$\leq n\mathbf{E}\,2\left(n^{\gamma-1}\sum_{x_1^k}P(X_1^k = x_1^k|X_{k+1}^\infty)^\gamma\right)^{1/\gamma}[k(1+\ln D)]^{1-1/\gamma}$$

$$= 2n^{2-1/\gamma}[k(1+\ln D)]^{1-1/\gamma}(2^{-H_\gamma^{\mathrm{cond}}(k)})^{1-1/\gamma} \tag{9.147}$$

Hence in particular for $\gamma > 1$ we obtain

$$P([\mathcal{P}_\gamma^{\mathrm{cond}}(L(X_1^n))]^\delta \geq l) \leq \frac{2n^{2-1/\gamma}(\log l)^{1/\beta}(1+\ln D)}{(l^{1/\delta})^{1-1/\gamma}}, \quad l \geq 1 \tag{9.148}$$

Let us write $\zeta := 2 + 1/(\gamma-1)$. Since $\mathbf{E}\,Y = \int_0^\infty P(Y \geq y)dy$ for a positive random variable Y, we obtain

$$\mathbf{E}\,[\mathcal{P}_\gamma^{\mathrm{cond}}(L(X_1^n))]^\delta \leq n^{\delta\zeta} + \int_{n^{\delta\zeta}}^\infty \frac{2n^{2-1/\gamma}(\log l)^{1/\beta}(1+\ln D)}{(l^{1/\delta})^{1-1/\gamma}}dl \tag{9.149}$$

Substituting $u = l/n^{\delta\zeta}$, we further derive

$$\mathbf{E}\,[\mathcal{P}_\gamma^{\mathrm{cond}}(L(X_1^n))]^\delta \leq n^{\delta\zeta} + n^{\delta\zeta}\int_1^\infty \frac{2(\log un^{\delta\zeta})^{1/\beta}(1+\ln D)}{(u^{1/\delta})^{1-1/\gamma}}du$$

$$\leq 2(1+\ln D)n^{\delta\zeta}[1 + A + (\log n^{\delta\zeta})^{1/\beta}B] \tag{9.150}$$

where integrals

$$A := \int_1^\infty \frac{(2\log u)^{1/\beta}}{(u^{1/\delta})^{1-1/\gamma}}du, \quad B := \int_1^\infty \frac{2^{1/\beta}}{(u^{1/\delta})^{1-1/\gamma}}du \tag{9.151}$$

are finite for $0 < \delta < 1 - 1/\gamma$. Hence by Theorem 8.4, we infer in this case

$$\underset{n\to\infty}{\mathrm{hilb}}\,\mathcal{P}_\gamma^{\mathrm{cond}}(L(X_1^n)) \leq \frac{1}{\delta}\underset{n\to\infty}{\mathrm{hilb}}\,\mathbf{E}\,[\mathcal{P}_\gamma^{\mathrm{cond}}(L(X_1^n))]^\delta \leq \frac{2\gamma-1}{\gamma-1} \text{ almost surely}$$

$$\tag{9.152}$$

$\square$

It remains an open question whether Theorems 9.27 and 9.28 can be sharpened.

Problems

9.1 Let random variables $X_1, \dots, X_n$ be independent. Show that equality $H_\gamma(X_1^n) = \sum_{i=1}^n H_\gamma(X_i)$ holds for the Rényi entropies for any $\gamma \geq 0$.

9.2 Show that Definitions 9.1–9.2 and the theorems from Section 9.1 can be generalized to random variables with a countably infinite alphabet.

9.3 Prove inequality $H_\gamma(X, Y|Z) \leq H_0(X|Z) + H_\gamma(Y|X, Z)$. Compare your solution with Fehr and Berens (2014) afterwards.

9.4 For $\gamma \in (0, 1)$ and (9.32) show that

$$h_\gamma^{\text{cond}} = \lim_{n \to \infty} \frac{H_\gamma^{\text{cond}}(n)}{n} = h_\gamma = \lim_{n \to \infty} \frac{H_\gamma(n)}{n} \tag{9.153}$$

9.5 Let $(X_i)_{i \in \mathbb{N}}$ be an arbitrary process over a countable alphabet. For any $\gamma > 1$ show that

$$-\log P(X_1^n) \geq H_\gamma(X_1^n) - \frac{2}{\gamma - 1} \log n \tag{9.154}$$

for all but finitely many n almost surely.

9.6 Prove relationships (9.102) for a stationary ergodic process over a finite alphabet.

9.7 Show that subword complexity satisfies $f(k + 1|x_1^n) \geq f(k|x_1^n)$ for $k \leq L(x_1^n)$, whereas $f(k|X_1^n) = n - k + 1$ for $k > L(x_1^n)$, where $L(x_1^n)$ is the maximal repetition of x_1^n. Having completed the task, compare your solution with de Luca (1999).

9.8 Let $D < \infty$ be the cardinality of the alphabet of process $(X_i)_{i \in \mathbb{N}}$. Show that

$$\sum_{i=1}^{D^k} P(X_{i+1}^{i+k} = X_1^k) \geq 1 \tag{9.155}$$

9.9 Let $(X_i)_{i \in \mathbb{N}}$ be an arbitrary process over a countable alphabet. Let us define the empirical distribution

$$P_{kn}(w) := \frac{1}{n - k + 1} \sum_{i=0}^{n-k} \mathbb{1}\{X_{i+1}^{i+k} = w\} \tag{9.156}$$

and let $H(p) := -\sum_x p(x) \log p(x)$ be the Shannon entropy of distribution p. Argue that $H(P_{kn}) \leq \log(n - k + 1)$, whereas $\mathbf{E} H(P_{kn}) \leq H(k)$ if process $(X_i)_{i \in \mathbb{N}}$ is stationary. Finally, show that for $k \leq n$ we have

$$0 \leq f(k|X_1^n) - 2^{H(P_{kn})} \leq (\ln 2)H(P_{kn})[\text{Li}(n - k + 1) - \text{Li}(2^{H(P_{kn})})] \tag{9.157}$$

9.10 Consider a bounded random variable $0 \leq X \leq m < \infty$. Show that

$$\mathbf{E} \exp(X - m) \leq \exp(\mathbf{E} X - m) + \sqrt{\text{Var } X} \tag{9.158}$$

Hint: Use $a^n - b^n = (a - b)(a^{n-1} + a^{n-2}b + \cdots + b^{n-1})$.

10

AMS Processes

An important property of natural language – and of programming languages, as well – is the double articulation: written texts consist of words, whereas words consist of letters. Whereas the number of distinct letters is bounded and rather small, the number of distinct words is difficult to bound and we can imagine it as countably infinite. The theorems about facts and words discussed in Section 8.4 suggest that the double articulation arises from a mere semantic constraint that texts describe infinitely many independent elementary facts in a repetitive way. From this perspective, it is natural to switch between two alternative views of a text: one in which the text is a string of letters and another in which the text is a string of words.

By an analogy, if we consider discrete stochastic processes, we can also switch between viewing them as sequences of random letters and sequences of random words. It pays off to study this operation on stochastic processes and their probability measures in its own right. The operation does not preserve stationarity of the investigated process, but it preserves a more general property, called asymptotic mean stationarity (AMS). Curiously, this natural operation has not been much investigated in the literature. We notice that the theory of AMS measures and respective operations on them is quite technical, and probably underdeveloped, but we will need it for construction of some linguistically motivated examples of stochastic processes in Chapter 11.

Consequently, in this chapter, we will come back to the general setting of ergodic theory, and we will consider AMS measures and some useful operations on them. It turns out that AMS probability measures are the largest class of measures for which the relative frequencies of all events converge almost surely. This property makes AMS measures attractive from a theoretical point of view. The ugly little problem is that it is usually not trivial to demonstrate that a particular measure is AMS since we need to verify convergence of relative frequencies for all measurable events rather than just cylinder sets. There exist simple counterexamples of measures which satisfy the condition for all cylinder sets but are not AMS. We call

Information Theory Meets Power Laws: Stochastic Processes and Language Models, First Edition.
Łukasz Dębowski.
© 2021 John Wiley & Sons, Inc. Published 2021 by John Wiley & Sons, Inc.

these measures pseudo AMS. A nontrivial example of pseudo AMS measures may be random hierarchical association (RHA) processes to be discussed in Section 11.4. We have not managed to prove that they are AMS.

The detailed organization of this chapter is following. Section 10.1 introduces the formal definitions of AMS and pseudo AMS probability measures. The defining property of these measures is that they have a stationary mean, which is an appropriately taken average of the probability measure with respect to a measurable transformation of the probability space. It turns out that AMS processes are the largest class of processes for which the relative frequency function exists almost surely, i.e. they are the largest class of processes that satisfy the claim of the Birkhoff ergodic theorem. Unfortunately, there seems to be no analogue of Kolmogorov process theorem for general AMS measures, so these measures must be constructed differently.

In Section 10.2, we present an operation which can be used to construct nontrivial AMS measures, called quasiperiodic coding. This operation generalizes the aforementioned switching between the process consisting of letters and the process consisting of words, where the quasiperiod is the length of the first word of the process measured in letters. We show that quasiperiodic coding preserves the AMS property if the expectation of the quasiperiod for the input measure is strictly positive and finite almost surely. In some cases, there exists also a formula that links the stationary means of an AMS measure and its image under quasiperiodic coding.

Quasiperiodic coding is a convenient operation to encode a stochastic process over one alphabet into a process over another alphabet. We can subsequently ask whether this operation preserves not only the AMS property but also ergodicity. It turns out that the answer is positive for some special quasiperiodic codings called synchronizable codings. In the special case of switching between the process consisting of letters and the process consisting of words, a synchronizable coding corresponds to the intuitive requirement that words are separated by some delimiters such as spaces or commas. These topics will be discussed in Section 10.3.

In Section 10.4, we will deal with entropy rate of AMS and pseudo AMS processes. We show that the entropy rate of a pseudo AMS process is less than the entropy rate of its stationary mean. In contrast, for two-sided AMS processes, not only the respective entropy rates are equal but also the process satisfies the claim of the Shannon–McMillan–Breiman (SMB) theorem discussed in Chapter 6. This result can be generalized also for one-sided AMS processes but we do not exhibit the respective proof.

10.1 AMS and Pseudo AMS Measures

In this section, we will define AMS and pseudo AMS probability measures. These measures enjoy the property of having a stationary mean for all or for some events, respectively. To define the stationary mean, we have to come back to the setting of Chapter 4, concerning relative frequencies and ergodic properties. That is, the investigated probability measures will live on some abstract dynamical system $(\Omega, \mathcal{J}, T)$, being a measurable space $(\Omega, \mathcal{J})$ with a measurable operation $T : \Omega \to \Omega$.

Let us consider this definition of the stationary mean $\bar{P}(A)$ for an event A, which is analogous to Definition 4.2 of relative frequencies $\Phi_n(A)$ and $\Phi(A)$.

Definition 10.1 *(Stationary mean):* For a probability measure P on a dynamical system $(\Omega, \mathcal{J}, T)$, the stationary mean $\bar{P} : \mathcal{J} \mapsto \mathbb{U}_\perp$ is defined through

$$\bar{P}_n(A) := \frac{1}{n} \sum_{i=0}^{n-1} P(T^{-i}(A)) \tag{10.1}$$

$$\bar{P}(A) := \begin{cases} \lim_{n\to\infty} \bar{P}_n(A) & \text{if the limit exists} \\ \perp & \text{else} \end{cases} \tag{10.2}$$

We will say that $\bar{P}(A)$ exists if $\bar{P}(A) \neq \perp$.

It is obvious that stationary mean is a stationary set function and $\bar{P} = \bar{P}_n = P$ if measure P is stationary itself. Moreover, we have two simple results. The first one says that the stationary mean of a probability measure equals the probability measure for invariant events.

Theorem 10.1: $\bar{P}(A) = P(A)$ *for* $A \in \mathcal{J}$.

Proof: Obvious by the definition of the invariant σ-field $\mathcal{J}$. □

The second result links the stationary mean with relative frequency Φ:

Theorem 10.2: *For each* $A \in \mathcal{J}$, *if* $\Phi(A) \neq \perp$ *holds P-almost surely, then*

$$\bar{P}(A) = \int \Phi(A) dP \neq \perp \tag{10.3}$$

Proof: Observe that $\bar{P}_n(A) = \int \Phi_n(A) dP$. Hence, if $\Phi(A) \neq \perp$ holds P-almost surely, then the claim follows by the dominated convergence (Theorem 3.10). □

Consequently, probability measures which admit the stationary mean for all events will be called AMS (asymptotically mean stationary). Probability measures for which the stationary mean exists only for events in a certain countable generating field will be called pseudo AMS.

Definition 10.2 *(AMS and pseudo AMS measures)*: A probability measure P on a dynamical system $(\Omega, \mathcal{J}, T)$ is called AMS if stationary means $\bar{P}(A)$ exist for all $A \in \mathcal{J}$. When $\mathcal{J} = \sigma(\mathcal{K})$ for a certain countable field $\mathcal{K}$, then a probability measure P on $(\Omega, \mathcal{J}, T)$ is called pseudo AMS if stationary means $\bar{P}(A)$ exist for all $A \in \mathcal{K}$.

Some accounts of AMS measures were given by Gray and Kieffer (1980), Fontana et al. (1981), Gray (2009), and Kakihara (1999), whereas the definition of pseudo AMS measures is our invention motivated by the forthcoming application in Chapter 11. In particular, a nontrivial example of pseudo AMS measures are RHA processes to be discussed in Section 11.4. So far, we have not managed to show that they are AMS.

The following observation applies the Vitali–Hahn–Saks theorem.

Theorem 10.3: *We have:*

1. *If P is AMS then $\bar{P}$ is a stationary measure.*
2. *If P is pseudo AMS then $\bar{P}|_{\mathcal{K}}$ can be extended to a stationary measure on $\mathcal{J}$.*

Proof:

1. If limits $\bar{P}(A)$ exist for all $A \in \mathcal{J}$, then by the Vitali–Hahn–Saks theorem (Theorem 4.2) they define a stationary measure on $\mathcal{J}$.
2. If limits $\bar{P}(A)$ exist for all $A \in \mathcal{K}$, then by the Vitali–Hahn–Saks theorem (Theorem 4.2) they define a stationary measure on field $\mathcal{K}$, which can be uniquely extended to a stationary measure on $\mathcal{J}$ by Theorem 2.4. □

We recall from Theorem 4.4 that relative frequencies $\Phi(A)$ usually cannot exist for all events $A \in \mathcal{J}$ simultaneously. According to the next proposition, AMS measures are exactly those probability measures for which the relative frequencies $\Phi(A)$ exist almost surely for each event $A \in \mathcal{J}$ separately.

Theorem 10.4: *Measure P is AMS if and only if $\Phi(A) \neq \perp$ holds P-almost surely for all $A \in \mathcal{J}$.*

Proof: By Theorem 10.2, it suffices to show that if P is AMS, then $\Phi(A) \neq \perp$ holds P-almost surely for all $A \in \mathcal{J}$. For this aim, observe that sets $B = (\Phi(A) \neq \perp)$, where $A \in \mathcal{J}$, are invariant. Hence, by the Birkhoff ergodic theorem (Theorem 4.6) for measure $\bar{P}$ and Theorem 10.1, we obtain that $1 = \bar{P}(B) = P(B)$. □

There are some other exact characterizations of AMS measures, see Gray and Kieffer (1980), Fontana et al. (1981), and Problem 10.1. These characterizations are mostly of a theoretical importance since they require checking conditions for events not being cylinder sets. However, they can be useful sometimes, see Theorem 10.7.

For pseudo AMS measures, there are no such equivalences. Let $\Omega_\mathbb{S}$ and $\Omega_\mathbb{E}$ be the sets of stationary and stationary ergodic points, defined in Problem 4.7.

Theorem 10.5: *We have the following implications:*

$$P \text{ is AMS} \implies P(\Omega_\mathbb{E}) = 1 \implies P(\Omega_\mathbb{S}) = 1 \implies P \text{ is pseudo AMS} \qquad (10.4)$$

Proof: If P is AMS, then we obtain $P(\Omega_\mathbb{E}) = \bar{P}(\Omega_\mathbb{E}) = 1$ by the Birkhoff ergodic theorem (Theorem 4.6) and the first ergodic decomposition theorem (Theorem 4.19). Hence, $P(\Omega_\mathbb{S}) = 1$ by $\Omega_\mathbb{E} \subseteq \Omega_\mathbb{S}$. Next, if $P(\Omega_\mathbb{S}) = 1$ then $\Phi_n(A)$ converge P-almost surely for all $A \in \mathcal{K}$ by the definition of $\Omega_\mathbb{S}$. Consequently, by Theorem 10.2, we obtain that P is pseudo AMS. □

Pseudo AMS measures seem to be a kind of an ugly duckling but they arise naturally when we consider measures concentrated on a single point. Let P_ω stand for the measure concentrated on a point $\omega \in \Omega$, i.e. $P_\omega(\{\omega\}) = 1$. By the definition of the set of stationary points $\Omega_\mathbb{S}$, for each $\omega \in \Omega_\mathbb{S}$, measure P_ω is pseudo AMS with the stationary mean $\bar{P}_\omega(A) = \Phi(A)(\omega)$ for all $A \in \mathcal{K}$. In contrast, by Theorem 4.4, we know that measure P_ω is AMS if and only if point ω is periodic. Thus ensuring existence of stationary mean for all $A \in \mathcal{J}$ can be difficult.

The question arises how we can construct AMS measures which are not stationary. There is no analogue of the Kolmogorov process theorems (Theorems 2.7 and 4.5) for AMS measures. Nonetheless, as we will see in the next section, AMS measures arise naturally when we apply an operation which we call quasiperiodic coding to other AMS measures, e.g. to stationary measures in particular.

10.2 Quasiperiodic Coding

Suppose we have a stochastic process over some alphabet $\mathbb{X}$ and we want to transform it into a process over another alphabet $\mathbb{Y}$. A typical situation which we will consider in this book is $\mathbb{X}$ being a countably infinite set, such as the set of natural numbers, and $\mathbb{Y}$ being a finite set, such as the set of digits. Let us stress that the respective coding operation does not preserve stationarity in general, but given reasonable conditions it preserves the AMS property, as we will inspect it in this section.

To begin the formal construction, we may have two spaces of infinite sequences $(\mathbb{X}^\mathbb{T}, \mathcal{X}^\mathbb{T})$ and $(\mathbb{Y}^\mathbb{T}, \mathcal{Y}^\mathbb{T})$, where $\mathbb{T} = \mathbb{N}$ or $\mathbb{T} = \mathbb{Z}$, with shift operations

$$T_1 : \mathbb{X}^\mathbb{T} \ni (x_i)_{i\in\mathbb{T}} \mapsto (x_{i+1})_{i\in\mathbb{T}} \in \mathbb{X}^\mathbb{T} \tag{10.5}$$

$$T_2 : \mathbb{Y}^\mathbb{T} \ni (y_i)_{i\in\mathbb{T}} \mapsto (y_{i+1})_{i\in\mathbb{T}} \in \mathbb{Y}^\mathbb{T} \tag{10.6}$$

These two spaces may be related by means of a code $f : \mathbb{X} \to \mathbb{Y}^*$ (Definition 7.1) and its infinite extensions $f^\mathbb{T} : \mathbb{X}^\mathbb{T} \to \mathbb{Y}^\mathbb{T}$ (Definition 7.3).

In fact, it is possible to consider a more general setting of a quasiperiodic mapping between two dynamical systems.

Definition 10.3 *(Quasiperiodic coding)*: Consider two dynamical systems $(\Omega_1, \mathcal{J}_1, T_1)$ and $(\Omega_2, \mathcal{J}_2, T_2)$. A function $g : \Omega_1 \to \Omega_2$ measurable from $\mathcal{J}_1$ to $\mathcal{J}_2$ is called a quasiperiodic coding with respect to T_1 and T_2 if for all $\omega_1 \in \Omega_1$ there exists an $l \in \{0, 1, 2, ...\}$ such that

$$T_2^l(g(\omega_1)) = g(T_1(\omega_1)) \tag{10.7}$$

The smallest l for which condition (10.7) holds is called the quasiperiod of ω and is denoted $L(\omega_1)$.

In this setting, we can cast extended codes $f^\mathbb{T} : \mathbb{X}^\mathbb{T} \to \mathbb{Y}^\mathbb{T}$ with $\mathbb{T} = \mathbb{N}$ or $\mathbb{T} = \mathbb{Z}$.

Theorem 10.6: *Let us put $\Omega_1 = \mathbb{X}^\mathbb{T}$ and $\Omega_2 = \mathbb{Y}^\mathbb{T}$. Then, the extended code $f^\mathbb{T} : \mathbb{X}^\mathbb{T} \to \mathbb{Y}^\mathbb{T}$ is a quasiperiodic coding for the shift operations (10.5) and (10.6), whereas the quasiperiod is*

$$L((x_i)_{i\in\mathbb{T}}) = |f(x_1)| \tag{10.8}$$

Proof: Left as an easy exercise. □

Thus, a quasiperiodic coding arises naturally when we want to transform a stochastic process from one alphabet into another alphabet. The general concept of a quasiperiodic coding provides a convenient layer of abstraction which allows to simultaneously discuss the product spaces $(\mathbb{X}^\mathbb{T}, \mathcal{X}^\mathbb{T})$, where $\mathbb{T} = \mathbb{N}$ or $\mathbb{T} = \mathbb{Z}$. Possibly, we could imagine other kinds of quasiperiodic codings.

Consequently, let introduce some probability measures into the general setting of a quasiperiodic coding. If we have a function $f : \Omega_1 \to \Omega_2$ measurable from $\mathcal{J}_1$ to $\mathcal{J}_2$ and a probability measure P_1 on $\mathcal{J}_1$, we may define measure a probability measure $P_2 = P_1 \circ f^{-1}$ on $\mathcal{J}_2$ as the transported measure, i.e. $P_2(A) = P_1(f^{-1}(A))$ for all $A \in \mathcal{J}_2$. It is quite a general fact that P_2 is AMS if P_1 is AMS and f is a quasiperiodic coding with a strictly positive and finite expected quasiperiod.

Theorem 10.7: *Consider two dynamical systems* $(\Omega_1, \mathcal{J}_1, T_1)$ *and* $(\Omega_2, \mathcal{J}_2, T_2)$ *linked by a quasiperiodic coding* $g : \Omega_1 \to \Omega_2$. *Let* P_1 *be a probability measure on* $\mathcal{J}_1$ *and let* $P_2 = P_1 \circ g^{-1}$ *be the transported measure on* $\mathcal{J}_2$. *If* P_1 *is AMS and the quasiperiod satisfies*

$$\mathbf{E}_{\bar{P}_1}(L|\mathcal{J}_1) \in (0, \infty) \tag{10.9}$$

$\bar{P}_1$-*almost surely, where* $\mathcal{J}_1$ *is the invariant σ-field of* $(\Omega_1, \mathcal{J}_1, T_1)$, *then* P_2 *is also AMS with the stationary mean*

$$\bar{P}_2(A) = \int \frac{\mathbf{E}_{\bar{P}_1}(G(A)|\mathcal{J}_1)}{\mathbf{E}_{\bar{P}_1}(L|\mathcal{J}_1)} d\bar{P}_1 \tag{10.10}$$

where

$$G(A) := \sum_{k=0}^{L-1} I_A \circ T^k \circ g \tag{10.11}$$

Proof: By Theorem 10.4, to demonstrate that P_2 is AMS, it is sufficient to show that relative frequencies $\Phi(A)$ exist P_2-almost surely for all $A \in \mathcal{J}_2$. In fact, by the Birkhoff ergodic theorem (Theorem 4.6), we obtain that limit

$$\lim_{m\to\infty} \frac{1}{m} \sum_{j=0}^{m} I_A \circ T^j \circ g = \lim_{n\to\infty} \frac{\sum_{i=0}^{n} G(A) \circ T^i}{\sum_{i=0}^{n} L \circ T^i} = \frac{\mathbf{E}_{\bar{P}_1}(G(A)|\mathcal{J}_1)}{\mathbf{E}_{\bar{P}_1}(L|\mathcal{J}_1)} \tag{10.12}$$

exists $\bar{P}_1$-almost surely. By Theorem 10.1, this equality holds also P_1-almost surely, so we obtain that limit

$$\Phi(A) = \lim_{m\to\infty} \frac{1}{m} \sum_{j=0}^{m} I_A \circ T^j \tag{10.13}$$

exists P_2-almost surely. Hence, formula (10.10) follows by Theorem 10.2. □

Under a stronger assumption, there exists a simpler expression for the stationary mean $\bar{P}_2$ in terms of $\bar{P}_1$, originally noticed for $\bar{P}_1 = P_1$ by Gray and Kieffer (1980, example 6). Namely, if additionally $\mathbf{E}_{\bar{P}_1}(L|\mathcal{J}_1)$ is $\bar{P}_1$-almost surely constant then

$$\bar{P}_2(A) = \frac{\mathbf{E}_{\bar{P}_1} G(A)}{\mathbf{E}_{\bar{P}_1} L} \tag{10.14}$$

What is somewhat surprising, even when $\mathbf{E}_{\bar{P}_1}(L|\mathcal{I}_1)$ is not constant then the right-hand side of expression (10.14) still defines a stationary measure. In other words, we may always construct a stationary measure by averaging the transported stationary mean over a randomized shift within the quasiperiod.

Theorem 10.8: *Consider two dynamical systems* $(\Omega_1, \mathcal{I}_1, T_1)$ *and* $(\Omega_2, \mathcal{I}_2, T_2)$ *linked by a quasiperiodic coding* $g : \Omega_1 \rightarrow \Omega_2$. *Let* P_1 *be a probability measure on* $\mathcal{I}_1$. *If* P_1 *is AMS and the quasiperiod satisfies*

$$\mathbf{E}_{\bar{P}_1} L \in (0, \infty) \tag{10.15}$$

then function

$$R(A) = \frac{\mathbf{E}_{\bar{P}_1} G(A)}{\mathbf{E}_{\bar{P}_1} L} \tag{10.16}$$

is a stationary measure on $\mathcal{I}_2$.

Proof: First of all, we have $R(\Omega_2) = 1$, whereas the countable additivity follows by the dominated convergence theorem (Theorem 3.10). As for the stationarity, we obtain

$$R(T_2^{-1}A) - R(A) = (\mathbf{E}_{\bar{P}_1} L)^{-1} \mathbf{E}_{\bar{P}_1}(I_A \circ T_2^L \circ g - I_A \circ T_2^0 \circ g)$$
$$= (\mathbf{E}_{\bar{P}_1} L)^{-1} \mathbf{E}_{\bar{P}_1}(I_A \circ g \circ T_1 - I_A \circ g) = 0 \tag{10.17}$$

from $\bar{P}_1 \circ T^{-1} = \bar{P}_1$. $\qquad\square$

It is an interesting open question how the above operation interacts with the ergodic decomposition of AMS measures – a topic which we do not develop here, see Gray (2009) and Schönhuth (2008).

In this section, we have presented constructions originally described by Dębowski (2010). Some similar constructions were considered by Cariolaro and Pierobon (1977), Gray and Kieffer (1980), and Timo et al. (2007).

10.3 Synchronizable Coding

This section also reports some ideas of Dębowski (2010) in a slightly generalized fashion. Whereas quasiperiodic coding is a convenient construction to encode a stochastic process into another alphabet, a question arises whether this transformation preserves some important properties of the process such as ergodicity. In general, the answer is negative, but we may restrict ourselves to some special quasiperiodic coding which does preserve ergodicity. This special case will be called a synchronizable coding.

Definition 10.4 *(Synchronizable coding)*: Consider two dynamical systems $(\Omega_1, \mathcal{J}_1, T_1)$ and $(\Omega_2, \mathcal{J}_2, T_2)$ where both T_1 and T_2 are invertible. An injection $g : \Omega_1 \rightarrow \Omega_2$ measurable from $\mathcal{J}_1$ to $\mathcal{J}_2$ is called a synchronizable coding with respect to T_1 and T_2 if:

1. $T_1^j(\omega_1) = \omega_1'$ for a $j \in \mathbb{Z}$ implies $T_2^i(g(\omega_1)) = g(\omega_1')$ for an $i \in \mathbb{Z}$.
2. $T_2^i(g(\omega_1)) = g(\omega_1')$ for an $i \in \mathbb{Z}$ implies $T_1^j(\omega_1) = \omega_1'$ for a $j \in \mathbb{Z}$.

Let us recall the definition of a comma-separated code (Definition 7.5). It turns out that if $f : \mathbb{X} \rightarrow \mathbb{Y}^*$ is a comma-separated code then the respective code extension is a synchronizable coding.

Theorem 10.9: *Let* $\Omega_1 = \mathbb{X}^{\mathbb{Z}}$ *and* $\Omega_2 = \mathbb{Y}^{\mathbb{Z}}$, *whereas* $f : \mathbb{X} \rightarrow \mathbb{Y}^*$ *be a comma-separated code. Then, the infinite extension* $f^{\mathbb{Z}} : \mathbb{X}^{\mathbb{Z}} \rightarrow \mathbb{Y}^{\mathbb{Z}}$ *is a synchronizable coding for the shift operations (10.5) and (10.6).*

Proof: Left as an easy exercise. □

In general, for a measurable injection $g : \Omega_1 \rightarrow \Omega_2$, the transported operation

$$T_g := g \circ T_1 \circ g^{-1} \tag{10.18}$$

considered by Gray and Kieffer (1980, example 6), constitutes a measurable injection $g(\Omega_1) \rightarrow g(\Omega_1)$. If we have a stationary measure P_1 on dynamical system $(\Omega_1, \mathcal{J}_1, T_1)$, then transported measure $P_2 = P_1 \circ g^{-1}$ is stationary on dynamical system $(g(\Omega_1), \mathcal{J}_2 \cap g(\Omega_1), T_g)$ but not necessarily on dynamical system $(\Omega_2, \mathcal{J}_2, T_2)$.

In the following, we will show an intuitive fact that the invariant σ-fields of T_2 and T_g are almost equal for a synchronizable coding g. Some apparent technical difficulty is that set $g(\Omega_1)$ usually does not belong to the invariant σ-field of T_2. However, this can be overcome relatively easily given a certain care. Define first the proper invariant σ-fields

$$\mathcal{J}_1 := \{A_1 \in \mathcal{J}_1 : T_1^{-1}(A_1) = A_1\} \tag{10.19}$$

$$\mathcal{J}_2 := \{A_2 \in \mathcal{J}_2 : T_2^{-1}(A_2) = A_2\} \tag{10.20}$$

Next, we define σ-fields

$$\mathcal{Q}_1 := g(\mathcal{J}_1) \tag{10.21}$$

$$\mathcal{Q}_2 := \mathcal{J}_2 \cap g(\Omega_1) \tag{10.22}$$

Now we have the aforementioned result.

Theorem 10.10: *For a synchronizable coding g* $: \Omega_1 \to \Omega_2$,

$$\mathcal{Q}_2 = \mathcal{Q}_1 \tag{10.23}$$

Proof: First, we will prove $\mathcal{Q}_2 \subseteq \mathcal{Q}_g$. Let $B \in \mathcal{Q}_2$. We have $B = A \cap g(\Omega_1)$ for a certain $A \in \mathcal{I}_2$. Since g is a synchronizable coding, by condition (1), we obtain

$$B \subseteq \bigcup_{i \in \mathbb{Z}} T_g^i(B) \subseteq \bigcup_{i \in \mathbb{Z}} T_2^i(B) \cap g(\Omega_1) \subseteq \bigcup_{i \in \mathbb{Z}} T_2^i(A) \cap g(\Omega_1) = A \cap g(\Omega_1) = B \tag{10.24}$$

Hence, $B = \bigcup_{i \in \mathbb{Z}} T_g^i(B) \in \mathcal{Q}_1$.

Now we will show that $\mathcal{Q}_1 \subseteq \mathcal{Q}_2$. Let $A \in \mathcal{Q}_1$. Since g is a synchronizable coding, by condition (2), we obtain

$$A \subseteq \bigcup_{i \in \mathbb{Z}} T_2^i(A) \cap g(\Omega_1) \subseteq \bigcup_{i \in \mathbb{Z}} T_g^i(A) = A \tag{10.25}$$

Hence, $A = \bigcup_{i \in \mathbb{Z}} T_2^i(A) \cap g(\Omega_1) \in \mathcal{Q}_2$. □

To complete the result, we add some probability measures to the considerations. The established equality of σ-fields implies equality of transported measures.

Theorem 10.11: *Consider a synchronizable coding g* $: \Omega_1 \to \Omega_2$, *a measure* P_1 *on* $(\Omega_1, \mathcal{I}_1)$ *and a measure* $P_2 = P_1 \circ g^{-1}$ *on* $(\Omega_2, \mathcal{I}_2)$. *For each* $B \in \mathcal{I}_2$, *there exists an* $A \in \mathcal{I}_1$, *and for each* $A \in \mathcal{I}_1$, *there exists an* $B \in \mathcal{I}_2$ *such that*

$$P_2(B) = P_2(B \cap g(\Omega_1)) = P_1(A) \tag{10.26}$$

Proof: For $B \in \mathcal{I}_2$, take $A = g^{-1}(B)$. For $A \in \mathcal{I}_1$, take $B = \bigcup_{i \in \mathbb{Z}} T_2^i g(A)$. Then, $B \cap g(\Omega_1) = g(A)$ and (10.26) follows from Theorem 10.10. □

Let us trace some consequences of the above proposition. First, for a synchronizable coding g, both measures P_1 and P_2 are ergodic or neither of them has this property. Moreover, let us recall that $\bar{P}_1(A) = P_1(A)$ for $A \in \mathcal{I}_1$ and an AMS measure P_1. The analogical equality $\bar{P}_2(B) = P_2(B)$ holds for $B \in \mathcal{I}_2$ and an AMS measure P_2 – which is guaranteed by Theorem 10.7 for an AMS P_1 if g is also a quasiperiodic coding. Some oddness of equalities (10.26) is buried in the fact that $\bar{P}_2(B)$ does not necessarily equal $\bar{P}_2(B \cap g(\Omega_1))$. It is only the support of P_2 that is confined to $g(\Omega_1)$, whereas $g(\Omega_1)$ need not be T_2-invariant, as it has been remarked.

10.4   Entropy Rate in the AMS Case

So far we wanted to state our results in the most general way which we could imagine. Now we will scale down to our standard application of processes over a countable alphabet, and we will investigate Shannon entropies for AMS and pseudo AMS processes. First, we will consider AMS processes defined as follows.

Definition 10.5   *(AMS process):*   A stochastic process $(X_i)_{i \in \mathbb{T}}$, where $\mathbb{T} = \mathbb{N}$ or $\mathbb{T} = \mathbb{Z}$, on a probability space $(\Omega, \mathcal{J}, P)$, is called AMS when the distribution $\mu(A) = P((X_i)_{i \in \mathbb{T}} \in A)$ of the process is an AMS measure with respect to the shift operation (4.1).

Similarly, we will define pseudo AMS processes. Some particular twist in the respective definition is that we choose a particular countable generating field, being the field generated by cylinder sets.

Definition 10.6   *(Pseudo AMS process):*   A stochastic process $(X_i)_{i \in \mathbb{T}}$, where $\mathbb{T} = \mathbb{N}$ or $\mathbb{T} = \mathbb{Z}$ and $X_i : \Omega \to \mathbb{X}$ for a countable alphabet $\mathbb{X}$, on a probability space $(\Omega, \mathcal{J}, P)$, is called pseudo AMS when the distribution $\mu(A) = P((X_i)_{i \in \mathbb{T}} \in A)$ of the process is a pseudo AMS measure with respect to the shift operation (4.1) and the countable field $\sigma_0(\{[x]_k^{\mathbb{T}} : x \in \mathbb{X}, k \in \mathbb{T}\})$.

For AMS and pseudo AMS processes, we are interested whether we can generalize the concept of the entropy rate defined as limit (5.30). As we will see, the answer is partly positive. In the following, we will assume that $(X_i)_{i \in \mathbb{N}}$ is a process on a probability space $(\mathbb{X}^{\mathbb{N}}, \mathcal{X}^{\mathbb{N}}, P)$, where $X_j : \mathbb{X}^{\mathbb{N}} \ni (x_i)_{i \in \mathbb{Z}} \mapsto x_j \in \mathbb{X}$ and the alphabet $\mathbb{X}$ is countable. Then, $\mu = P$ and the stationary mean of P will be denoted $\bar{P}$. First, we will prove an auxiliary proposition. Denote conditional entropies

$$H_P(X|A) := - \sum_{x \in \mathbb{X}} P(X = x|A) \log P(X = x|A) \tag{10.27}$$

$$H_P(X|Y) := \sum_{y \in \mathbb{Y}} P(Y = y) H_P(X|Y = y) \tag{10.28}$$

We have the following bound for conditional entropies.

Theorem 10.12:   *For a pseudo AMS process $(X_i)_{i \in \mathbb{N}}$ and a cylinder set A,*

$$\bar{P}(A) H_{\bar{P}}(X_1|A) \geq \limsup_{n \to \infty} \frac{1}{n} \sum_{i=0}^{n-1} P(T^{-i}A) H_P(X_{i+1}|T^{-i}A) \tag{10.29}$$

Proof: Recall measures $\bar{P}_n := n^{-1} \sum_{i=0}^{n-1} P \circ T^{-i}$. We have

$$\bar{P}_n(X_1 = x|A) = \frac{1}{n} \sum_{i=0}^{n-1} \frac{P(T^{-i}A)}{\bar{P}_n(A)} P(X_{i+1} = x|T^{-i}A) \tag{10.30}$$

Hence, by the Jensen inequality (Theorem 3.14), for function $p \mapsto -p \log p$,

$$H_{\bar{P}_n}(X_1|A) \geq \frac{1}{n} \sum_{i=0}^{n-1} \frac{P(T^{-i}A)}{\bar{P}_n(A)} H_P(X_{i+1}|T^{-i}A) \tag{10.31}$$

Since probabilities $\bar{P}_n(B)$ tend to $\bar{P}(B)$ for every cylinder set B then

$$H_{\bar{P}}(X_1|A) \geq \limsup_{n \to \infty} \frac{1}{n} \sum_{i=0}^{n-1} \frac{P(T^{-i}A)}{\bar{P}_n(A)} H_P(X_{i+1}|T^{-i}A)$$

$$= \frac{1}{\bar{P}(A)} \limsup_{n \to \infty} \frac{1}{n} \sum_{i=0}^{n-1} P(T^{-i}A) H_P(X_{i+1}|T^{-i}A) \tag{10.32}$$

In view of this, we obtain the requested claim. $\qquad\square$

Now, we will show a bound for the block entropy of a pseudo AMS process. It turns out that the rate of entropy of a pseudo AMS process is less than the entropy rate of its stationary mean.

Theorem 10.13: *For a pseudo AMS process* $(X_i)_{i \in \mathbb{N}}$ *with* $H_P(X_1^n) < \infty$,

$$\limsup_{n \to \infty} \frac{H_P(X_1^n)}{n} \leq h_{\bar{P}} \tag{10.33}$$

Proof: In view of Theorem 10.12, we obtain

$$H_{\bar{P}}(X_j|X_1^{j-1}) \geq \limsup_{n \to \infty} \frac{1}{n} \sum_{i=0}^{n-1} H_P(X_{i+j}|X_{i+1}^{i+j-1})$$

$$\geq \limsup_{n \to \infty} \frac{H_P(X_j^{n+j-1}|X_1^{j-1})}{n}$$

$$= \limsup_{n \to \infty} \frac{H_P(X_1^n)}{n} \tag{10.34}$$

where the last equality follows from the bound $H_P(X_1^{j-1}) < \infty$. Recalling that $h_{\bar{P}} = \inf_{j \in \mathbb{N}} H_{\bar{P}}(X_j|X_1^{j-1})$ yields the claim. $\qquad\square$

Inequality (10.33) can be sharp as the following example asserts. Let P_ω stand for the measure concentrated on a single point $\omega \in \mathbb{X}^{\mathbb{N}}$, i.e. $P_\omega(\{\omega\}) = 1$. As we have written in Section 10.1, measure P_ω is pseudo AMS for each stationary point

$\omega \in \Omega_{\mathbb{S}}$ and the stationary mean satisfies $\bar{P}_{\omega}(A) = \Phi(A)(\omega)$ for all cylinder sets A. It can be easily verified that $H_{P_{\omega}}(X_1^n) = 0$ regardless of the choice of ω, whereas choosing a suitable point $\omega \in \Omega_{\mathbb{S}}$, we can achieve an arbitrary entropy rate $h_{\bar{P}_{\omega}}$. Thus, even when we have the equality in (10.33), the measure P need not be AMS.

In contrast, the existence of the entropy rate for an AMS process is guaranteed by a generalization of the SMB theorem. According to this generalization, the entropy rate of an AMS process is equal to the entropy rate of its stationary mean. The following version of the SMB theorem for two-sided AMS processes rests on the SMB theorem for the stationary case (Theorem 6.9).

Theorem 10.14 *(SMB theorem): Let $(X_i)_{i\in\mathbb{Z}}$ be an AMS process on $(\Omega, \mathcal{J}, P)$, where $X_i : \Omega \to \mathbb{X}$ and the alphabet $\mathbb{X}$ is finite. Let $\bar{P}$ be the stationary mean of measure P, let $\mathcal{J}$ be the shift invariant σ-field of the process $(X_i)_{i\in\mathbb{Z}}$, let $\bar{F}(A) = \bar{P}(A|\mathcal{J})$ be the random ergodic measure of the process, and let $h_{\bar{F}}$ be the entropy rate of $(X_i)_{i\in\mathbb{Z}}$ with respect to measure $\bar{F}$. For any universal semidistribution $Q : \mathbb{X}^* \to \mathbb{R}$, we have*

$$\lim_{n\to\infty} \frac{1}{n} \mathbf{E}_P[-\log Q(X_1^n)] = \lim_{n\to\infty} \frac{1}{n} \mathbf{E}_P[-\log P(X_1^n)] = h_{\bar{P}} = \mathbf{E}_P h_{\bar{F}} \qquad (10.35)$$

$$\lim_{n\to\infty} \frac{1}{n}[-\log Q(X_1^n)] = \lim_{n\to\infty} \frac{1}{n}[-\log P(X_1^n)] = h_{\bar{F}} \ P\text{-almost surely} \qquad (10.36)$$

Proof: By the Barron lemma (Theorem 3.20), when $\pi(n)Q(X_1^n)$ with $\pi(n) := 6\pi^{-2}n^{-2}$ is a semidistribution and $P(X_1^n)$ is the probability distribution, we obtain that

$$\sup_{i\in\mathbb{Z}} \limsup_{n\to\infty} \frac{1}{n}[-\log Q(X_{i+1}^{i+n})] \geq \sup_{i\in\mathbb{Z}} \limsup_{n\to\infty} \frac{1}{n}[-\log P(X_{i+1}^{i+n})] \qquad (10.37)$$

holds P-almost surely and the above event is $\mathcal{J}$-measurable. In contrast, when $\pi(n)P(X_1^n)$ with $\pi(n) := 6\pi^{-2}n^{-2}$ is a semidistribution and $\bar{P}(X_1^n)$ is the probability distribution, we obtain that

$$\inf_{i\in\mathbb{Z}} \liminf_{n\to\infty} \frac{1}{n}[-\log P(X_{i+1}^{i+n})] \geq \inf_{i\in\mathbb{Z}} \liminf_{n\to\infty} \frac{1}{n}[-\log \bar{P}(X_{i+1}^{i+n})] \qquad (10.38)$$

holds $\bar{P}$-almost surely and the above event is $\mathcal{J}$-measurable. Moreover, by equality (6.62), we have that

$$\sup_{i\in\mathbb{Z}} \limsup_{n\to\infty} \frac{1}{n}[-\log Q(X_{i+1}^{i+n})] = \inf_{i\in\mathbb{Z}} \liminf_{n\to\infty} \frac{1}{n}[-\log \bar{P}(X_{i+1}^{i+n})] = h_{\bar{F}} \qquad (10.39)$$

holds $\bar{P}$-almost surely and the above event is $\mathcal{J}$-measurable. Recall now that $P(A) = \bar{P}(A)$ for $A \in \mathcal{J}$ (Theorem 10.1). Hence, we may combine the above three displayed formulas to obtain (10.36).

Now, we will prove (10.35). By the Gibbs inequality (Theorem 3.18),

$$\mathbf{E}_P[-\log Q(X_1^n)] \geq \mathbf{E}_P[-\log P(X_1^n)] \qquad (10.40)$$

Since we have (6.21) then by the dominated convergence (Theorem 3.10) we obtain from (10.36) that

$$\lim_{n\to\infty} \frac{1}{n} \mathbf{E}_P[-\log Q(X_1^n)] = \mathbf{E}_P h_{\bar{F}} = \mathbf{E}_{\bar{P}} h_{\bar{F}} = h_{\bar{P}} \tag{10.41}$$

In contrast, by the Fatou lemma (Theorem 3.9), we obtain

$$\liminf_{n\to\infty} \frac{1}{n} \mathbf{E}_P[-\log P(X_1^n)] \geq \mathbf{E}_P \liminf_{n\to\infty} \frac{1}{n}[-\log P(X_1^n)] = h_{\bar{P}} \tag{10.42}$$

From the last three displayed formulas, we derive (10.35). $\qquad\square$

It is known that the SMB theorem is true also for a one-sided AMS process $(X_i)_{i\in\mathbb{N}}$, see Gray and Kieffer (1980). The respective proof is more complicated.

Problems

10.1 For the probability space $(\mathbb{X}^{\mathbb{Z}}, \mathcal{X}^{\mathbb{Z}}, P)$ with the shift operation (4.1), show that if measure P is AMS, then there holds measure domination $P \ll \bar{P}$, i.e. $\bar{P}(G) = 0$ implies $P(G) = 0$ for all events $G \in \mathcal{X}^{\mathbb{Z}}$. Subsequently, show that the measure P is AMS if and only if there exists a stationary measure R such that $P \ll R$.
Hint: Consider events $\bigcup_{i\in\mathbb{Z}} T^i(G)$. This result was shown by Fontana et al. (1981).

10.2 Prove Theorems 10.6 and 10.9.

10.3 Construct examples of quasiperiodic or synchronizable codings which are different from code extensions $f^{\mathbb{Z}} : \mathbb{X}^{\mathbb{Z}} \to \mathbb{Y}^{\mathbb{Z}}$.

10.4 Let us recall that a subset of strings $A \subseteq \mathbb{Y}^*$ is called:
(a) prefix-free if $w \neq zs$ for $w, z \in A$ and $s \in \mathbb{Y}^+$;
(b) suffix-free if $w \neq sz$ for $w, z \in A$ and $s \in \mathbb{Y}^+$;
(c) fix-free if it is both prefix-free and suffix-free;
(d) complete if it satisfies the Kraft equality $\sum_{w\in A} (\#\mathbb{Y})^{-|w|} = 1$.

For instance, set $\{01,000, 100,110, 111, 0010, 0011, 1010, 1011\}$ is complete fix-free with respect to $\mathbb{Y} = \{0,1\}$ (Gillman and Rivest, 1995; Ahlswede et al., 1996). Subsequently, a function $f : \mathbb{X} \to \mathbb{Y}^*$ is called a (complete) prefix/suffix/fix-free code if f is an injection and the image $f(\mathbb{X})$ is respectively (complete) prefix/suffix/fix-free. For finite $f(\mathbb{X})$, function f is called finite.
Now, suppose that $\mathbb{X}$ is finite and $f : \mathbb{X} \to \mathbb{Y}^*$ is a complete fix-free code. Show that:

(a) Code extension f^Z is a bijection $\mathbb{X}^Z \to \mathbb{Y}^Z$.
(b) Measure $\mu = \nu \circ f^Z$ is a stationary probability measure on $(\mathbb{X}^Z, \mathcal{X}^Z)$ if ν is a stationary probability measure on $(\mathbb{Y}^Z, \mathcal{Y}^Z)$.

The above partial converse to quasiperiodic coding was noticed by Dębowski (2010).

10.5 Consider a code $f : \mathbb{X} \to \mathbb{Y}^*$ and its extension $f^Z : \mathbb{X}^Z \to \mathbb{Y}^Z$. For a AMS process $(X_i)_{i \in \mathbb{Z}}$ on $(\Omega, \mathcal{J}, P)$, where $X_i : \Omega \to \mathbb{X}$, we introduce process

$$(Y_i)_{i \in \mathbb{Z}} := f^Z((X_i)_{i \in \mathbb{Z}}) \tag{10.43}$$

with random variables $Y_i : \Omega \to \mathbb{Y}$. Let $L_i := |f(X_i)|$.

(a) Show that limit

$$L := \lim_{n \to \infty} \frac{1}{n} \sum_{i=0}^{n-1} L_i \in [0, \infty] \tag{10.44}$$

exists almost surely and process $(Y_i)_{i \in \mathbb{Z}}$ is also AMS.

(b) Let process $(\bar{Y}_i)_{i \in \mathbb{Z}}$ be the stationary mean of process $(Y_i)_{i \in \mathbb{Z}}$, i.e. let it have the distribution

$$P((\bar{Y}_i)_{i \in \mathbb{Z}} \in A) = \lim_{n \to \infty} \frac{1}{n} \sum_{j=0}^{n-1} P((Y_{j+i})_{i \in \mathbb{Z}} \in A) \tag{10.45}$$

Suppose further that process $(X_i)_{i \in \mathbb{Z}}$ is stationary and the average length of a symbol encoding tends to its expectation,

$$\lim_{n \to \infty} \frac{1}{n} \sum_{i=0}^{n-1} L_i = L := \mathbf{E} L_i \tag{10.46}$$

Show that process $(\bar{Y}_i)_{i \in \mathbb{Z}}$ can be constructed in the following way. First, we construct a random variable $N : \Omega \to \mathbb{N} \cup \{0\}$, called a random shift, and a nonstationary process $(\bar{X}_i)_{i \in \mathbb{Z}}$ where $\bar{X}_i : \Omega \to \mathbb{X}$. Precisely, random variable N and process $(\bar{X}_i)_{i \in \mathbb{Z}}$ are conditionally independent given variable $\bar{X}_0$ and their joint distribution is

$$P(\bar{X}_k^l = x_k^l) = P(X_k^l = x_k^l) \times \frac{|f(x_0)|}{L}, \qquad k \le 0 \le l \tag{10.47}$$

$$P(N = n | \bar{X}_0 = x_0) = \frac{\mathbf{1}\{0 \le n \le |f(x_0)| - 1\}}{|f(x_0)|}, \quad n \in \mathbb{N} \cup \{0\} \tag{10.48}$$

In the sequel, process $(\bar{Y}_i)_{i \in \mathbb{Z}}$ can be obtained as

$$(\bar{Y}_i)_{i \in \mathbb{Z}} = T^{-N} f^Z((\bar{X}_i)_{i \in \mathbb{Z}}) \tag{10.49}$$

where $T((y_i)_{i \in \mathbb{Z}}) := (y_{i+1})_{i \in \mathbb{Z}}$ is the shift operation.

Remark: The above construction of process $(\bar{Y}_i)_{i\in\mathbb{Z}}$ will be considered in Section 11.3 and called a stationary coding of process $(X_i)_{i\in\mathbb{Z}}$ in Definition 11.6. A different concept of a stationary coding has been considered by Gray (1990). He calls process $(Y_i)_{i\in\mathbb{Z}}$ a stationary coding of a process $(X_i)_{i\in\mathbb{Z}}$ when $Y_i := g((X_{i+j})_{j\in\mathbb{Z}})$ for g being a measurable function of infinite sequences. We have encountered this construction in Section 9.5. It is a remarkable fact that the Shannon entropy rate of stationary process $(Y_i)_{i\in\mathbb{Z}}$ is in general less than that of stationary process $(X_i)_{i\in\mathbb{Z}}$.

11

Toy Examples

In this chapter, we will discuss some toy examples of stochastic processes that can or cannot model some particular statistical phenomena possibly exhibited by natural texts, i.e. texts in natural language. As mentioned in Section 1.15, there are four phenomena pertaining to natural texts which we may consider first:

- The strictly positive Shannon entropy rate:

$$h = \lim_{n\to\infty} \frac{H(X_1^n)}{n} > 0 \tag{11.1}$$

- The power-law growth of block mutual information:

$$I(X_1^n; X_{n+1}^{2n}) \propto n^\beta, \quad 0 < \beta < 1 \tag{11.2}$$

- The power-law logarithmic growth of maximal repetition:

$$L(X_1^n) \propto (\log n)^\alpha, \quad \alpha > 1 \tag{11.3}$$

- The power-law decay of mutual information between individual symbols:

$$I(X_0; X_n) \propto n^{-2\gamma}, \quad \gamma > 0 \tag{11.4}$$

Respectively, the above laws have been proposed by Shannon (1951), Hilberg (1990), Dębowski (2012b), and Lin and Tegmark (2017).

Intentionally, we have mentioned above only those statistical phenomena that can be expressed using purely information-theoretic concepts, such as entropy, mutual information, or maximal repetition. It should be noted that quantitative linguists have identified a few other laws, see Köhler et al. (2005). Potentially, for the approach developed in our book, the most interesting of these omitted laws is the Menzerath–Altmann law, which states that the longer a given linguistic construction is, the shorter its parts are on average (Menzerath, 1928; Altmann, 1980). Dębowski (2007b) tried to relate the Menzerath–Altmann law to minimal grammar-based codes, which also exhibit hierarchical constructions. Thus, in principle, minimal grammar-based codes could allow to discuss the

Information Theory Meets Power Laws: Stochastic Processes and Language Models, First Edition. Łukasz Dębowski.
© 2021 John Wiley & Sons, Inc. Published 2021 by John Wiley & Sons, Inc.

Menzerath–Altmann law for arbitrary stochastic processes, but Dębowski (2007b) did not succeed in observing any particular pattern for an independent identically distributed (IID) process in a simple numerical simulation. In consequence, we have concentrated our efforts on providing mathematical examples for statistical laws that are simpler to analyze and model.

The present chapter is divided into five sections discussing progressively more complicated examples of processes and relating them to laws (11.1)–(11.4). In Section 11.1, we will discuss finite and ultrafinite energy processes. Finite and ultrafinite energy processes are defined by a condition that the probability of a block of random variables is an exponentially decaying function of the block length. Finite and ultrafinite energy processes have a strictly positive Shannon entropy rate (11.1) but they do not satisfy either a power-law growth of mutual information (11.2) or a power-law logarithmic growth of maximal repetition (11.3). Since typical hidden Markov processes are finite energy then we should treat these results as an indication that natural texts cannot be accurately modeled by hidden Markov processes, although these processes were quite successfully used in computational linguistics.

In Section 11.2, we will exhibit a few variations of Santa Fe processes, introduced informally in Section 1.11 and defined formally in Section 8.3. As motivated in Section 1.11, a Santa Fe process is a series of abstract random propositions making assertions about an infinite random reality. In particular, we may split a Santa Fe process into a static description of a time-independent reality and a narration process which chooses which facts about the reality are described by texts. Tampering further, we may generalize the definition of a Santa Fe process so that the reality described texts becomes slowly evolving rather then time-independent.

Whereas the original definition of a Santa Fe process makes use of a countably infinite alphabet, the process can be successfully encoded into a finite alphabet using the quasiperiodic coding introduced in Chapter 10. This construction will be performed in Section 11.3, and we will show that it preserves all vital properties of a Santa Fe process. Given simple conditions, the encoded Santa Fe processes are strongly nonergodic, perigraphic, or none of them, which provides simple examples and counterexamples for theorems about facts and words demonstrated in Chapter 8. The examples of Santa Fe processes considered in Sections 11.2 and 11.3 enjoy both a strictly positive Shannon entropy rate (11.1) and power-law growing block mutual information (11.2). However, the considered encodings of Santa Fe processes into a finite alphabet seem to be finite energy and fail to satisfy power-law logarithmic growth of maximal repetition (11.3). Maybe there exist other encodings of Santa Fe processes which do not have this drawback.

In Section 11.4, we will present a class of random hierarchical association (RHA) processes, which were mentioned without definition in Section 1.13 and which will be analyzed to a lesser extent. RHA processes are constructed by forming an infinite pool of random texts by random concatenation, subsequently sampling progressively longer and longer texts from the pool and taking the stationary mean of the resulting pseudo AMS process – as defined in Chapter 10. We can show that for a simple choice of parameters, called perplexities, which control the number of random texts in the pool, RHA processes satisfy both a power-law growth of mutual information resembling (11.2) and a power-law logarithmic bound for maximal repetition resembling (11.3). Unfortunately, these RHA processes do not have a strictly positive Shannon entropy rate (11.1). However, as mentioned in Section 1.14, we suppose that exactly these RHA processes satisfy the power-law decay of mutual information between individual symbols (11.4). The exact calculation is left as an open problem.

In Section 11.5, we will recapitulate the above results and put them in context of statistical language modeling. We will discuss advantages and disadvantages of the toy examples of processes presented in the preceding sections and we will sketch some further research ideas. In our opinion, there is an important open problem whether the power-law logarithmic growth of maximal repetition (11.3) and the power-law decay of mutual information between individual symbols (11.4) imply the power-law growth of block mutual information (11.2). The motivating idea inspired by our thinking about natural language is that the complex reality repeatedly described in natural texts may emerge as a result of our tendency to repeat previously encountered text chunks.

11.1   Finite and Ultrafinite Energy

In this section we will study finite and ultrafinite energy processes. The finite energy condition is more general than that of ultrafinite energy. Some examples of ultrafinite energy processes are hidden Markov processes with bounded emission probabilities and a finite number of hidden states, whereas some examples of finite energy processes are also Zipfian Santa Fe processes. We will show that ultrafinite energy processes have a finite excess entropy and a strictly positive Shannon entropy rate (11.1) whereas finite energy processes have a strictly positive conditional min-entropy rate. Therefore, all ultrafinite energy processes fail to satisfy the power-law growth of block mutual information (11.2), whereas all finite energy processes fail to satisfy the power-law logarithmic growth of maximal repetition (11.3).

To begin, we consider these definitions.

Definition 11.1 *(Finite energy process)*:   A discrete process $(X_i)_{i \in \mathbb{T}}$, where $\mathbb{T} = \mathbb{N}$ or $\mathbb{T} = \mathbb{Z}$ and $X_i : \Omega \to \mathbb{X}$, is called a finite energy process if

$$P(X_{i+1}^{i+m+n} = x_1^{m+n}) \leq Kc^n P(X_{i+1}^{i+m} = x_1^m) \tag{11.5}$$

for all $m, n = 0, 1, ...$, all $i \in \mathbb{T} \cup \{0\}$, all $x_i \in \mathbb{X}$, and certain constants $K < \infty$ and $0 < c < 1$.

Definition 11.2 *(Ultrafinite energy process):* A discrete process $(X_i)_{i \in \mathbb{T}}$, where $\mathbb{T} = \mathbb{N}$ or $\mathbb{T} = \mathbb{Z}$ and $X_i : \Omega \to \mathbb{X}$, is called an ultrafinite energy process if

$$P(X_{i+1}^{i+m} = x_1^m) \le Kc^m \tag{11.6}$$

$$P(X_{i+1}^{i+m+n} = x_1^{m+n}) \le KP(X_{i+1}^{i+m} = x_1^m)P(X_{i+m+1}^{i+m+n} = x_{m+1}^{m+n}) \tag{11.7}$$

for all $m, n = 0, 1, ...$, all $i \in \mathbb{T} \cup \{0\}$, all $x_i \in \mathbb{X}$, and certain constants $K < \infty$ and $0 < c < 1$.

We have already encountered condition (11.7) written as condition (9.31). The following fact can be easily seen.

Theorem 11.1: *Any ultrafinite energy process is a finite energy process.*

Before we discuss some general properties of finite energy and ultrafinite energy processes, let us present some examples of these processes. In fact, a large subclass of IID, Markov, and hidden Markov processes are both finite energy and ultrafinite energy processes.

Theorem 11.2: *For a stationary hidden Markov process* $(X_i)_{i \in \mathbb{T}}$ *denote the underlying Markov process* $(Y_i)_{i \in \mathbb{T}}$. *Process* $(X_i)_{i \in \mathbb{T}}$ *is finite energy if*

$$c := \sup_{x \in \mathbb{X}, y \in \mathbb{Y}} P(X_i = x | Y_i = y) < 1 \tag{11.8}$$

Proof: We observe that random variables X_{i+m+1} and X_{i+1}^{i+m} are conditionally independent given Y_{i+m+1}. Hence,

$$P\left(X_{i+m+1} = x_{m+1} | X_{i+1}^{i+m} = x_1^m\right)$$

$$= \sum_{y \in \mathbb{Y}} P\left(X_{i+m+1} = x_{m+1} | Y_{i+m+1} = y\right) P\left(Y_{i+m+1} = y | X_{i+1}^{i+m} = x_1^m\right)$$

$$\le \sum_{y \in \mathbb{Y}} cP\left(Y_{i+m+1} = y | X_{i+1}^{i+m} = x_1^m\right) < c \tag{11.9}$$

In consequence, we obtain

$$P(X_{i+1}^{i+m+n} = x_1^{m+n}) \le c^n P(X_{i+1}^{i+m} = x_1^m) \tag{11.10}$$

Hence, process $(X_i)_{i \in \mathbb{T}}$ is finite energy. $\square$

Theorem 11.3: *For a stationary hidden Markov process* $(X_i)_{i\in\mathbb{T}}$ *denote the underlying Markov process* $(Y_i)_{i\in\mathbb{T}}$. *Process* $(X_i)_{i\in\mathbb{T}}$ *is ultrafinite energy if we have (11.8) and*

$$K := \sup_{y,y'\in\mathbb{Y}} \frac{P(Y_{i-1}=y, Y_i=y')}{P(Y_{i-1}=y)P(Y_i=y')} < \infty \tag{11.11}$$

Condition (11.11) is satisfied if the hidden alphabet $\mathbb{Y}$ is finite.

Proof: By Theorem 11.2, conditions (11.5) and (11.6) are satisfied if (11.8) holds true. It remains to prove (11.7). We observe that random variables X_{i+1}^{i+m} and X_{i+m+1}^{i+m+n} are conditionally independent given (Y_{i+m}, Y_{i+m+1}). Thus

$$P\left(X_{i+1}^{i+m+n} = x_1^{m+n}\right)$$
$$= \sum_{y,y'\in\mathbb{Y}} P\left(X_{i+m+1}^{i+m+n} = x_{m+1}^{m+n}|Y_{i+m+1}=y'\right)$$
$$\times P(Y_{i+m+1}=y'|Y_{i+m}=y)P(Y_{i+m}=y|X_{i+1}^{i+m}=x_1^m)P\left(X_{i+1}^{i+m}=x_1^m\right)$$
$$\leq \sum_{y,y'\in\mathbb{Y}} P\left(X_{i+m+1}^{i+m+n} = x_{m+1}^{m+n}|Y_{i+m+1}=y'\right)$$
$$\times KP(Y_{i+m+1}=y')P\left(Y_{i+m}=y|X_{i+1}^{i+m}=x_1^m\right)P(X_{i+1}^{i+m}=x_1^m)$$
$$= KP\left(X_{i+1}^{i+m}=x_1^m\right)P\left(X_{i+m+1}^{i+m+n}=x_{m+1}^{m+n}\right) \tag{11.12}$$

Hence, process $(X_i)_{i\in\mathbb{T}}$ is ultrafinite energy. □

Since all IID processes and all Markov processes are instances of hidden Markov processes, they are also finite energy and ultrafinite energy given some versions of conditions (11.8) and (11.11). Moreover, nonergodic processes made by averaging the distributions of such processes are also finite energy processes, as it can be asserted in the following result.

Theorem 11.4: *A stationary process is finite energy if the process is finite energy with fixed constants K and c with respect to the random ergodic measure* $F(A) = P(A|\mathcal{J})$ *almost surely, where* $\mathcal{J}$ *is the invariant* σ-*field.*

Proof: If $F(X_{i+1}^{i+m+n} = x_1^{m+n}) \leq Kc^n F(X_{i+1}^{i+m} = x_1^m)$ holds almost surely then $P(X_{i+1}^{i+m+n} = x_1^{m+n}) \leq Kc^n P(X_{i+1}^{i+m} = x_1^m)$ holds for $P(A) = \mathbf{E}F(A)$. □

Recall now Zipfian Santa Fe processes defined in Definitions 8.4 and 8.8.

Theorem 11.5: *The mixture and individual Zipfian Santa Fe processes are finite energy processes.*

Proof: We have $F(X_{i+1}^{i+m+n} = x_1^{m+n}) = P(K_{i+1}^{i+m+n} = k_1^{m+n})$ almost surely where $(K_i)_{i \in \mathbb{Z}}$ is a finite energy IID process. Hence, process $(X_i)_{i \in \mathbb{Z}}$ is finite energy by Theorem 11.4. □

Another subclass of finite energy processes are uniformly dithered processes.

Definition 11.3 *(Uniformly dithered process)*: Let $(\mathbb{X}, *)$ be a group, i.e. a pair of a set $\mathbb{X}$ and a binary operation $*: \mathbb{X} \times \mathbb{X} \to \mathbb{X}$ which is associative, that is, $(a * b) * c = a * (b * c)$, has an identity element $e \in \mathbb{X}$, that is, $e * a = a * e = e$, and for each $a \in \mathbb{X}$ there exists the inverse element a^{-1}, that is, $a^{-1} * a = a * a^{-1} = e$. A discrete stochastic process $(X_i)_{i \in \mathbb{Z}}$ is called uniformly dithered if it satisfies

$$X_i = W_i * Z_i \tag{11.13}$$

where $(W_i)_{i \in \mathbb{Z}}$ is an arbitrary discrete process and $(Z_i)_{i \in \mathbb{Z}}$ is an independent IID process with

$$c := \max_{a \in \mathbb{X}} P(Z_i = a) < 1 \tag{11.14}$$

Information theorists may regard uniformly dithered processes as stochastic processes passed through a special noisy channel which distorts symbols of the process in quite a symmetric way. We have this result.

Theorem 11.6: *Any uniformly dithered process is a finite energy process.*

Proof: Consider process $X_i = W_i * Z_i$. Then,

$$P\left(Z_{i+m+1}^{i+m+n} = z_{m+1}^{m+n}\right) \le c^n \tag{11.15}$$

Let us denote $w_i^k * z_i^k = (w_i * z_i, w_{i+1} * z_{i+1}, \ldots, w_k * z_k)$ and $(z_i^k)^{-1} = (z_i^{-1}, z_{i+1}^{-1}, \ldots, z_k^{-1})$. Using the fact that $(W_i)_{i \in \mathbb{Z}}$ and $(Z_i)_{i \in \mathbb{Z}}$ are independent, we can further write

$$P\left(X_{i+m+1}^{i+m+n} = x_{m+1}^{m+n} | X_{i+1}^{i+m} = x_1^m\right)$$

$$= \sum_{z_{m+1}^{m+n} \in \mathbb{X}^n} P\left(W_{i+m+1}^{i+m+n} = x_{m+1}^{m+n} * (z_{m+1}^{m+n})^{-1}, Z_{i+m+1}^{i+m+n} = z_{m+1}^{m+n} | X_{i+1}^{i+m} = x_1^m\right)$$

$$= \sum_{z_{m+1}^{m+n} \in \mathbb{X}^n} P\left(W_{i+m+1}^{i+m+n} = x_{m+1}^{m+n} * (z_{m+1}^{m+n})^{-1} | X_{i+1}^{i+m} = x_1^m\right) P\left(Z_{i+m+1}^{i+m+n} = z_{m+1}^{m+n}\right)$$

$$\le \sum_{z_{m+1}^{m+n} \in \mathbb{X}^n} P\left(W_{i+m+1}^{i+m+n} = x_{m+1}^{m+n} * (z_{m+1}^{m+n})^{-1} | X_{i+1}^{i+m} = x_1^m\right) c^n \le c^n \tag{11.16}$$

since

$$\sum_{z^{m+n}_{m+1} \in \mathbb{X}^n} P\left(W^{i+m+n}_{i+m+1} = x^{m+n}_{m+1} * (z^{m+n}_{m+1})^{-1} | X^{i+m}_{i+1} = x^m_1\right)$$

$$= \sum_{w^{m+n}_{m+1} \in \mathbb{X}^n} P\left(W^{i+m+n}_{i+m+1} = w^{m+n}_{m+1} | X^{i+m}_{i+1} = x^m_1\right) = 1 \qquad (11.17)$$

Hence, $(X_i)_{i \in \mathbb{T}}$ is a finite energy process. □

We suppose that uniformly dithered processes need not be ultrafinite energy.

Let us observe that for a stationary process $(X_i)_{i \in \mathbb{Z}}$, the finite energy condition (11.5) is equivalent to

$$P(X^n_1 = x^n_1 | X^0_{-\infty}) \le Kc^n \text{ almost surely} \qquad (11.18)$$

by the Levy law (3.68). There are two related Doeblin conditions

$$P(X_r = x_r | X^0_{-\infty}) \ge d \text{ almost surely} \qquad (11.19)$$

$$P(X_r = x_r | X^0_{-\infty}) \le D \text{ almost surely} \qquad (11.20)$$

for some $r \ge 1$ and $0 < d, D < 1$, see Gao et al. (2008) and Kontoyiannis et al. (1998). The first condition can be satisfied for a finite alphabet only. It turns out that both Doeblin conditions are stronger than finite energy.

Theorem 11.7: *If a process assuming more than one value satisfies condition (11.19) then it satisfies condition (11.20). Moreover, if a stationary process satisfies condition (11.20) then it is finite energy.*

Proof: First, assume condition (11.19). Then obviously

$$P(X_r = x_r | X^0_{-\infty}) = 1 - \sum_{x'_r \ne x_r} P(X_r = x'_r | X^0_{-\infty}) \le 1 - d =: D \qquad (11.21)$$

so we obtain condition (11.20). Next, assume condition (11.20). Then

$$P(X^r_1 = x'_1 | X^0_{-\infty}) \le P(X_r = x_r | X^0_{-\infty}) \le D \qquad (11.22)$$

and, by stationarity, $P(X^n_1 = x^n_1 | X^0_{-\infty}) \le D^{\lfloor n/r \rfloor} \le D^{n/r-1}$, so (11.18) follows. □

Now let us review some general properties of finite energy and ultrafinite energy processes. We have three basic facts.

Theorem 11.8: *No stationary finite energy process is asymptotically deterministic.*

Proof: For a stationary finite energy process $(X_i)_{i \in \mathbb{Z}}$, we have $H(n) \ge -\log K + n \log c^{-1}$. Hence, the Shannon entropy rate is $h > 0$. □

Theorem 11.9: *No stationary ultrafinite energy process is infinitary.*

Proof: For a stationary ultrafinite energy process $(X_i)_{i \in \mathbb{Z}}$, we have $I(X_1^n; X_{n+1}^{2n}) \leq \log K$. Hence, the excess entropy is $E \leq \log K$. □

Theorem 11.10: *For a finite energy process $(X_i)_{i \in \mathbb{N}}$, we have*

$$L(X_1^n) < A \log n \tag{11.23}$$

for sufficiently large n almost surely, for some $A > 0$.

Proof: We have $H_\infty^{\text{cond}}(n) \geq -\log K + n \log c^{-1}$. Thus the claim follows by Theorems 9.22 and 9.25. □

The results discussed in this section come mostly from Dębowski (2015b, 2018b). Finite energy processes were introduced by Shields (1997) for a study of the maximal repetition. The condition of ultrafinite energy is our invention for the purpose of this book. Let us note that Ko (2012) introduced a slightly more general class of processes, which he called simple mixing processes. The condition for a simple mixing process is just (11.7), in contrast to mixing processes defined in Definition 4.8. Related concepts of mixing have been known in the literature, see Szpankowski (1993b) and Bradley (2005). Uniformly dithered processes were introduced by Dębowski (2015b). These processes generalize a construction by Shields (1997), who only considered binary processes and addition modulo 2.

Finite and ultrafinite energy processes have a strictly positive Shannon entropy rate by Theorem 11.8, probably like the natural language. Yet in view of Theorems 11.9 and 11.10, finite and ultrafinite energy processes do not satisfy either a power-law logarithmic growth of maximal repetition or a power-law growth of mutual information, in contrast to the natural language. Since a large subclass of hidden Markov processes are finite energy processes then we should treat the above facts as an indication that texts in natural language cannot be faithfully modeled by hidden Markov processes. – In fact, hidden Markov processes were widely used in natural language engineering until they were replaced by artificial neural networks. – The same remark applies to Zipfian Santa Fe processes discussed at length in the next two sections. It was also supposed by Kontoyiannis (1997) that natural language satisfies the Doeblin condition, but we have shown that this condition is stronger than the finite energy condition, whereas natural language does not seem to be finite energy.

11.2 Santa Fe Processes and Alike

In this section, we will study some generalizations of Zipfian Santa Fe processes, introduced half-formally in Section 1.11 and formally defined in Section 8.3. A general Santa Fe process can be interpreted a series of random propositions making assertions about an infinite random reality. In particular, we may split a Santa Fe process into a complete description of the reality and a narration process which selects which facts about the reality are described in a given text. As we will see, we can choose the described reality to be static or to evolve in time. Respectively, we obtain strongly nonergodic or mixing processes with a power-law growth of mutual information (11.2). These constructions, combined with a stationary coding into a finite alphabet – to be discussed in Section 11.3, provide conceptually simple examples for theorems about facts and words proved in Chapter 8.

First, we will study such a generalization of Definition 8.4.

Definition 11.4 (*Mixture Santa Fe process*): Process $(X_i)_{i \in \mathbb{Z}}$ of form

$$X_i = (K_i, Z_{K_i}) \tag{11.24}$$

will be called a mixture Santa Fe process when:

1. Process $(Z_k)_{k \in \mathbb{N}}$ is a fair-coin process.
2. Process $(K_i)_{i \in \mathbb{Z}}$ is a stationary process with a probability distribution $P(K_i = k | \mathcal{J}) > 0$ almost surely for any $k \in \mathbb{N}$, where $\mathcal{J}$ is the invariant σ-field.

For so defined mixture Santa Fe processes, we can generalize Theorem 8.13.

Theorem 11.11: *Mixture Santa Fe processes are strongly nonergodic.*

Proof: Consider functions $g : \mathbb{N} \times \mathbb{X}^* \to \{0, 1, 2\}$ of form (8.79). Then,

$$P\left(g(k, X_{j+1}^{j+n}) = Z_k\right) = P((K_i = k \text{ for some } j + 1 \le i \le j + n)$$

$$= \mathbf{E1} \left\{ \sum_{i=j+1}^{j+n} \mathbf{1}\{K_i = k\} > 0 \right\} \tag{11.25}$$

Now, the process $(K_i)_{i \in \mathbb{Z}}$ is assumed to be stationary with $P(K_i = k | \mathcal{J}) > 0$. Hence, by the ergodic theorem (Theorem 4.6),

$$\lim_{n \to \infty} \frac{1}{n} \sum_{i=j+1}^{j+n} \mathbf{1}\{K_i = k\} = P(K_i = k | \mathcal{J}) > 0 \tag{11.26}$$

holds almost surely. Hence, by the monotone convergence (Theorem 3.8),

$$
\lim_{n\to\infty} P\left(g(k,X_{j+1}^{j+n}) = Z_k\right) = \lim_{n\to\infty} \mathbf{E}\mathbf{1}\left\{\sum_{i=j+1}^{j+n} \mathbf{1}\{K_i = k\} > 0\right\}
$$

$$
= \mathbf{E}\left[\lim_{n\to\infty} \mathbf{1}\left\{\sum_{i=j+1}^{j+n} \mathbf{1}\{K_i = k\} > 0\right\}\right]
$$

$$
= \mathbf{E}1 = 1 \tag{11.27}
$$

To complete the proof, we apply Theorem 8.12. □

The above construction of the mixture Santa Fe processes can be motivated linguistically. If $(X_i)_{i\in\mathbb{Z}}$ is a random infinite text in natural language, then we can imagine that it consists from some propositions $X_i = (k,z)$ which assert in particular that the kth fact equals z. Consequently, we may split the process $(X_i)_{i\in\mathbb{Z}}$ into a static description of some time-independent facts $(Z_k)_{k\in\mathbb{N}}$ and a narration process $(K_i)_{i\in\mathbb{Z}}$. Theorem 11.11 asserts that under mild condition on the narration process $(K_i)_{i\in\mathbb{Z}}$, the resulting process $(X_i)_{i\in\mathbb{Z}}$ is strongly nonergodic.

In particular, if we choose the process $(K_i)_{i\in\mathbb{Z}}$ to be an IID process with the power-law distribution (8.77), we obtain the mixture Zipfian Santa Fe process with an exponent α. A salient feature of the mixture Zipfian Santa Fe processes is a power-law growth of the number of inferrable facts (8.14). Now, if we could apply the theorem about facts and mutual information (Theorem 8.18), then we would obtain that the mutual information between blocks for the mixture Zipfian Santa Fe process must grow at least as fast as the expected number of inferrable facts,

$$
\operatorname*{hilb}_{n\to\infty} I(X_1^n; X_{n+1}^{2n}) \geq \operatorname*{hilb}_{n\to\infty} \mathbf{E}\#U(X_1^n) = 1/\alpha \tag{11.28}
$$

In fact, we should be cautious with such a statement since Santa Fe processes in general are not processes over a finite alphabet, so they do not satisfy an important assumption of Theorem 8.18.

Nevertheless, using a different method, we are able to show that the block mutual information for a Zipfian Santa Fe process follows an asymptotic power-law growth with the same exponent as the number of inferrable facts.

Theorem 11.12: *For the mixture Zipfian Santa Fe process with an exponent α,*

$$
\lim_{n\to\infty} \frac{I(X_1^n; X_{n+1}^{2n})}{n^\beta} = \frac{(2 - 2^\beta)\Gamma(1 - \beta)}{[\zeta(\beta^{-1})]^\beta}, \quad \beta = 1/\alpha \tag{11.29}
$$

Proof: In the following, two simple corollaries of the chain rule

$$I(X;Y,Z|W) = I(X;Y|W) + I(X;Z|Y,W) \tag{11.30}$$

will be used:

1. $I(X;Y) = I(X;Y;Z) + I(X;Y|Z)$ for the entropies $H(X), H(Y) < \infty$, where we define the interaction information

$$I(X;Y;Z) := I(X;Z) + I(Y;Z) - I((X,Y);Z) \tag{11.31}$$

2. $I(X;Z|Y) = I(X;Z)$ for X and Y independent and conditionally independent given Z.

The second identity follows from

$$I(X;(Y,Z)) = I(X;Y) + I(X;Z|Y) = I(X;Z) + I(X;Y|Z) \tag{11.32}$$

where both $I(X;Y) = 0$ and $I(X;Y|Z) = 0$.

Now, notice that random variables $Z_k, k \in \mathbb{N}$, are independent and conditionally independent given any finite block X_n^m. Hence,

$$I(X_1^n;(Z_k)_{k\in\mathbb{N}}) = \sum_{k=1}^{\infty} I(X_1^n;Z_k|Z_1^{k-1}) = \sum_{k=1}^{\infty} I(X_1^n;Z_k) \tag{11.33}$$

Also blocks X_1^n and X_{n+1}^{2n} are conditionally independent given $(Z_k)_{k\in\mathbb{N}}$. Hence, $I(X_1^n;X_{n+1}^{2n}|(Z_k)_{k\in\mathbb{N}}) = 0$. Both results yield

$$I(X_1^n;X_{n+1}^{2n}) = I\left(X_1^n;X_{n+1}^{2n};(Z_k)_{k\in\mathbb{N}}\right) + I\left(X_1^n;X_{n+1}^{2n}|(Z_k)_{k\in\mathbb{N}}\right)$$

$$= I\left(X_1^n;X_{n+1}^{2n};(Z_k)_{k\in\mathbb{N}}\right)$$

$$= 2I\left(X_1^n;(Z_k)_{k\in\mathbb{N}}\right) - I(X_1^{2n};(Z_k)_{k\in\mathbb{N}})$$

$$= \sum_{k=1}^{\infty} [2I(X_1^n;Z_k) - I(X_1^{2n};Z_k)]$$

$$= \sum_{k=1}^{\infty} I(X_1^n;X_{n+1}^{2n};Z_k) \tag{11.34}$$

Computing simple expressions

$$H(Z_k|X_1^n) = 1 \times P(K_i \neq k \text{ for all } i \in \{1,\dots,n\})$$

$$+ 0 \times P(K_i = k \text{ for some } i \in \{1,\dots,n\}) \tag{11.35}$$

$$I(X_1^n;Z_k) = P(K_i = k \text{ for some } i \in \{1,\dots,n\})$$

$$= (1 - [1 - P(K_i = k)]^n) \tag{11.36}$$

we obtain the interaction information

$$I(X_1^n; X_{n+1}^{2n}; Z_k) = (1 - [1 - P(K_i = k)]^n)^2 \tag{11.37}$$

and the block mutual information

$$I\left(X_1^n; X_{n+1}^{2n}\right) = \sum_{k=1}^{\infty} \left(1 - \left(1 - \frac{A}{k^{1/\beta}}\right)^n\right)^2 \tag{11.38}$$

where $A := 1/\zeta(\beta^{-1})$.

The summation in the right-hand side of (11.38) equals up to an additive constant ≤ 1 to the integral

$$\int_1^{\infty} \left(1 - \left(1 - \frac{A}{k^{1/\beta}}\right)^n\right)^2 dk = \beta(An)^{\beta} \int_{(1-A)^n}^1 f_n(u)du \tag{11.39}$$

where we use substitution

$$u := (1 - Ak^{-1/\beta})^n \tag{11.40}$$

and functions

$$f_n(u) := \frac{(1-u)^2}{u^{1-1/n}[n(1-u^{1/n})]^{\beta+1}} \tag{11.41}$$

We have the limit

$$\lim_{n \to \infty} f_n(u) = f(u) := \frac{(1-u)^2}{u(-\log u)^{\beta+1}} \tag{11.42}$$

with the upper bound

$$\frac{f_n(u)}{f(u)} \leq \sup_{0<x<1} \frac{x(-\log x)^{\beta+1}}{(1-x)^{\beta+1}} = 1, \quad u, \beta \in (0,1) \tag{11.43}$$

Moreover, function $f(u)$ is integrable on $u \in (0,1)$. Hence,

$$\lim_{n \to \infty} \frac{I\left(X_1^n; X_{n+1}^{2n}\right)}{n^{\beta}} = \beta A^{\beta} \int_0^1 f(u)du \tag{11.44}$$

follows by the dominated convergence (Theorem 3.10).

It remains to compute $\int f(u)du$. Putting $t := -\log u$ yields

$$\int_0^1 f(u)du = \int_0^{\infty} (1 - e^{-t})^2 \, t^{-\beta-1}dt$$

$$= \int_0^{\infty} [e^{-2t} - 1 - 2(e^{-t} - 1)] \, t^{-\beta-1}dt$$

$$= (2 - 2^{\beta})\beta^{-1}\Gamma(1 - \beta) \tag{11.45}$$

where integral

$$\int_0^{\infty} (e^{-kt} - 1)t^{-\beta-1}dt = (e^{-kt} - 1)(-\beta^{-1})t^{-\beta}|_0^{\infty}$$

$$-\int_0^\infty (-ke^{-kt})(-\beta^{-1})t^{-\beta}dt = -k^\beta\beta^{-1}\Gamma(1-\beta) \tag{11.46}$$

can be integrated by parts for the considered β. □

Theorem 11.12 was established by Dębowski (2012a). As we have seen in the above proof, the calculation of limit (11.29) is eased by a decomposition of mutual information between blocks X_1^n and X_{n+1}^{2n} into a series of interaction information among blocks X_1^n and X_{n+1}^{2n} and variables Z_k. This decomposition is a particular property of Zipfian Santa Fe processes.

By the construction of individual Zipfian Santa Fe processes, we may suppose that a power-law growth of algorithmic mutual information between blocks can occur also for ergodic processes. We may also ask whether strong nonergodicity of a process is a necessary condition for a power-law growth of the Shannon mutual information between blocks. To answer this question, we will tamper with the definition of Zipfian Santa Fe processes to obtain some mixing processes that satisfy a power-law growth of the block Shannon mutual information. Since mixing processes are ergodic then the answer to our question is negative.

The idea is quite simple. We will consider a modified Santa Fe process, where the probabilistic facts Z_k are replaced by some Markov processes $(Z_{ik})_{i\in\mathbb{Z}}$. In the forthcoming definition of a pseudo Santa Fe process, processes $(Z_{ik})_{i\in\mathbb{Z}}$ describe the time evolution of the kth probabilistic fact, where Z_{ik} is the state of the kth fact at instant i and the probability that the kth bit flips at a given instant equals some cross-over probability p_k. For vanishing cross-over probabilities, $p_k = 0$, the pseudo Santa Fe processes collapse to the mixture Santa Fe processes.

Definition 11.5 *(Zipfian pseudo Santa Fe process)*: A process $(X_i)_{i\in\mathbb{Z}}$ consisting of random variables

$$X_i = (K_i, Z_{i,K_i}) \tag{11.47}$$

is called a Zipfian pseudo Santa Fe process with an exponent α and cross-over probabilities $(p_k)_{k\in\mathbb{N}}$, when processes $(K_i)_{i\in\mathbb{Z}}$ and $(Z_{ik})_{i\in\mathbb{Z}}$, where $k \in \mathbb{N}$, are probabilistically independent, $(K_i)_{i\in\mathbb{Z}}$ is an IID process with distribution (8.77), and each process $(Z_{ik})_{i\in\mathbb{Z}}$ is a Markov process with the marginal distribution

$$P(Z_{ik} = 0) = P(Z_{ik} = 1) = 1/2 \tag{11.48}$$

and the cross-over probabilities

$$P(Z_{ik} = 0|Z_{i-1,k} = 1) = P(Z_{ik} = 1|Z_{i-1,k} = 0) = p_k \tag{11.49}$$

The Zipfian pseudo Santa Fe processes are not strongly nonergodic if all cross-over probabilities are different from 0 or 1. In fact, we can show that these processes are mixing, and hence, they are ergodic by Theorem 4.11.

Theorem 11.13: *Zipfian pseudo Santa Fe processes are mixing for $p_k \in (0,1)$.*

Proof: Introduce an auxiliary process $(W_i)_{i\in\mathbb{Z}}$, where $W_i = (K_i, (Z_{ik})_{k\in\mathbb{N}})$. Process $(W_i)_{i\in\mathbb{Z}}$ is a product of independent processes $(K_i)_{i\in\mathbb{Z}}$, $(Z_{i1})_{i\in\mathbb{Z}}$, $(Z_{i2})_{i\in\mathbb{Z}}$, ..., which are all mixing for $p_k \in (0,1)$ by Theorem 4.15. Hence, process $(W_i)_{i\in\mathbb{Z}}$ is mixing by Theorem 4.17. Having established the mixing property for process $(W_i)_{i\in\mathbb{Z}}$, we notice that $X_i = f(W_i)$ for some measurable function f. Hence, process $(X_i)_{i\in\mathbb{Z}}$ is mixing by Theorem 4.16. $\qquad\square$

Although the Zipfian pseudo Santa Fe processes are not strongly nonergodic, we can show that the block mutual information grows like a power law if the cross-over probabilities are sufficiently small, e.g. $p_k \leq P(K_i = k)$. Thus, we may generalize Theorem 11.12.

Theorem 11.14: *The block mutual information $I(X_1^n; X_{n+1}^{2n})$ for a Zipfian pseudo Santa Fe process obeys*

$$\limsup_{n\to\infty} \frac{I(X_1^n; X_{n+1}^{2n})}{n^\beta} \leq \frac{(2 - 2^\beta)\Gamma(1 - \beta)}{[\zeta(\beta-1)]^\beta} \tag{11.50}$$

The lower limits in particular cases are as follows:

1. *If $p_k \leq P(K_i = k)$, then*

$$\liminf_{n\to\infty} \frac{I(X_1^n; X_{n+1}^{2n})}{n^\beta} \geq \frac{A(\beta)}{[\zeta(\beta-1)]^\beta} \tag{11.51}$$

   where

$$A(\beta) := \sup_{\delta\in(1/2,1)} (1 - \mathcal{H}(\delta))^\beta \int_{\sqrt{\delta}}^1 \frac{(1 - u)^2 \, du}{u \, (-\log u)^{\beta+1}} \tag{11.52}$$

   Above, we apply the entropy $\mathcal{H}(p)$ of a two-point distribution, defined in (5.106).
2. *If $\lim_{k\to\infty} p_k/P(K_i = k) = 0$ then $I(X_1^n; X_{n+1}^{2n})$ obeys (11.29).*

Proof: Observe that processes $\tilde{Z}_k := (Z_{ik})_{i\in\mathbb{Z}}$, where $k \in \mathbb{N}$, are independent and conditionally independent given any finite block X_n^m. Also, blocks X_1^n and X_{n+1}^{2n} are conditionally independent given $(\tilde{Z}_k)_{k\in\mathbb{N}}$. Thus, we obtain

$$I(X_1^n; X_{n+1}^{2n}) = \sum_{k=1}^\infty I(X_1^n; X_{n+1}^{2n}; \tilde{Z}_k) \tag{11.53}$$

by replacing Z_k with $\tilde{Z}_k$ in derivation (11.34) from the proof of Theorem 11.12.

By the assumed Markov property, process $\tilde{Z}_k = (Z_{ik})_{i\in\mathbb{Z}}$ is independent from X_1^n given $(Z_{ik})_{1\leq i\leq n}$. This yields

$$I(X_1^n; X_{n+1}^{2n}; \tilde{Z}_k) = 2I(X_1^n; (Z_{ik})_{1\leq i\leq n}) - I(X_1^{2n}; (Z_{ik})_{1\leq i\leq 2n}) \tag{11.54}$$

The expressions on the right-hand side can be analyzed as

$$I(X_1^n; (Z_{ik})_{1 \le i \le n}) = \sum_{i=1}^{n} I(X_i; Z_{ik} | X_1^{i-1}) \tag{11.55}$$

because $(Z_{ik})_{1 \le i \le n}$ is independent from X_i given Z_{ik} and X_1^{i-1}. Moreover,

$$I(X_i; Z_{ik} | X_1^{i-1}) = H(Z_{ik} | X_1^{i-1}) - H(Z_{ik} | X_1^i) \tag{11.56}$$

To evaluate the conditional entropies, put $a_{nk} := \mathcal{H}(P(Z_{ik} = z | Z_{i-n,k} = z))$ and $b_k := P(K_i = k)$. Notice that by the Markovianity of $(Z_{ik})_{i \in \mathbb{Z}}$, we have

$$H(Z_{ik} | X_1^{i-1}) = \sum_{n=1}^{i-1} a_{nk} P(K_j \ne k \text{ for } i - n < j \le i - 1) P(K_{i-n} = k)$$
$$+ \mathcal{H}(P(Z_{ik} = z)) P(K_j \ne k \text{ for } 1 \le j \le i - 1)$$
$$= \sum_{n=1}^{i-1} a_{nk} b_k (1 - b_k)^{n-1} + (1 - b_k)^{i-1} \tag{11.57}$$

Similarly, since $a_{0k} = 0$, we obtain

$$H(Z_{ik} | X_1^i) = \sum_{n=0}^{i-1} a_{nk} P(K_j \ne k \text{ for } i - n < j \le i) P(K_{i-n} = k)$$
$$+ \mathcal{H}(P(Z_{ik} = z)) P(K_j \ne k \text{ for } 1 \le j \le i)$$
$$= \sum_{n=1}^{i-1} a_{nk} b_k (1 - b_k)^n + (1 - b_k)^i \tag{11.58}$$

Thus, we may reconstruct

$$I(X_i; Z_{ik} | X_1^{i-1}) = \sum_{n=1}^{i-1} a_{nk} b_k^2 (1 - b_k)^{n-1} + [(1 - b_k)^{i-1} - (1 - b_k)^i] \tag{11.59}$$

$$I(X_1^n; (Z_{ik})_{1 \le i \le n}) = \sum_{m=1}^{n-1} (n - m) a_{mk} b_k^2 (1 - b_k)^{m-1} + [1 - (1 - b_k)^n] \tag{11.60}$$

and

$$I(X_1^n; X_{n+1}^{2n}; \tilde{Z}_k) = - \sum_{m=1}^{n-1} m a_{mk} b_k^2 (1 - b_k)^{m-1}$$
$$- \sum_{m=n}^{2n-1} (2n - m) a_{mk} b_k^2 (1 - b_k)^{m-1} + [1 - (1 - b_k)^n]^2 \tag{11.61}$$

For a fixed b_k, we see that $I(X_1^n; X_{n+1}^{2n}; \tilde{Z}_k)$ is minimized for $a_{mk} = 1$. This case arises when $p_k = 1/2$ and $(Z_{ik})_{i \in \mathbb{Z}}$ are sequences of independent random variables. A direct evaluation yields then $H(Z_{ik}|X_1^{i-1}) = 1$, $H(Z_{ik}|X_1^i) = 0$, $I(X_1^n;(Z_{ik})_{1 \le i \le n}) = n$, and $I(X_1^n;X_{n+1}^{2n};\tilde{Z}_k) = 0$. In this way, we have proved that

$$\sum_{m=1}^{n-1} mb_k^2(1 - b_k)^{m-1} + \sum_{m=n}^{2n-1} (2n - m)b_k^2(1 - b_k)^{m-1} = [1 - (1 - b_k)^n]^2$$

$$(11.62)$$

On the other hand, $I(X_1^n;X_{n+1}^{2n};\tilde{Z}_k)$ is maximized for $a_{mk} = 0$. This holds if $p_k = 0$ or $p_k = 1$. We recall that for $p_k = 0$, the process $(X_i)_{i \in \mathbb{Z}}$ collapses to a Zipfian Santa Fe process.

Continuing the calculations, by equality (11.62), we obtain

$$I(X_1^n;X_{n+1}^{2n};\tilde{Z}_k) \in [(1 - \epsilon)[1 - (1 - b_k)^n]^2, [1 - (1 - b_k)^n]^2] \qquad (11.63)$$

if $a_{mk} \le \epsilon$ for $m \le 2n - 1$. To bound coefficients a_{mk}, observe

$$P(Z_{ik} = z|Z_{i-n,k} = z) \ge (1 - p_k)^n \qquad (11.64)$$

Hence, $a_{mk} \le \mathcal{H}(\delta)$ for $m \le 2n - 1$ if $(1 - p_k)^{2n} \ge \delta \ge 1/2$. Thus we obtain

$$I(X_1^n;X_{n+1}^{2n}) \in$$

$$\left[(1 - \mathcal{H}(\delta)) \sum_{k \in \mathbb{N}:(1-p_k)^{2n} \ge \delta} [1 - (1 - b_k)^n]^2, \sum_{k \in \mathbb{N}} [1 - (1 - b_k)^n]^2 \right] \qquad (11.65)$$

The most tedious part of the proof is completed.

The limiting behavior of the upper bound in (11.65) has been analyzed in the proof of Theorem 11.12, and by that reasoning (11.50) holds. Now we will consider the limit of the lower bound in (11.65). As in the previous proof, we will approximate the respective sum with an integral. Recall that $b_k = Ak^{-1/\beta}$ with $A = 1/\zeta(\beta^{-1})$. Let us define b_k for real k in the same way.

1. For $p_k \le b_k$: notice that $(1 - b_k)^{2n} \ge \delta$ implies $(1 - p_k)^{2n} \ge \delta$. Thus $I(X_1^n;X_{n+1}^{2n})/(1 - \mathcal{H}(\delta)) + 1$ is greater than

$$\int_{(1-b_k)^n \ge \sqrt{\delta}}^{\infty} (1 - (1 - b_k)^n)^2 dk = \beta(An)^\beta \int_{\sqrt{\delta}}^1 f_n(u)du \qquad (11.66)$$

where we use substitution (11.40) and functions (11.41). This yields (11.51) by the dominated convergence (Theorem 3.10).

2. For $\lim_k p_k/b_k = 0$: let $k(n)$ be the largest number k such that $(1 - p_k)^{2n} < \delta$ or put $k(n) = 1$ if there is no such number. Then, $I(X_1^n;X_{n+1}^{2n})/(1 - \mathcal{H}(\delta)) + 1$ is greater than

$$\int_{k(n)}^{\infty} (1 - (1 - b_k)^n)^2 dk = \beta(An)^\beta \int_{u(n)}^1 f_n(u)du \qquad (11.67)$$

where $u(n) := (1 - b_{k(n)})^n$. We have $\lim_n u(n) = 0$ if $\lim_n k(n) < \infty$. On the other hand, if $\lim_n k(n) = \infty$ then we use $\lim\inf_n n p_{k(n)} > -\log\sqrt{\delta}$ and $\lim_k p_k/b_k = 0$ to infer $\lim_n n b_{k(n)} = \infty$ and hence, $\lim_n u(n) = 0$. Thus, the dominated convergence (Theorem 3.10) in both cases yields

$$\liminf_{n\to\infty} \frac{I(X_1^n; X_{n+1}^{2n})}{n^\beta} \geq (1 - \mathcal{H}(\delta))\beta A^\beta \int_0^1 f(u)du \qquad (11.68)$$

Taking $\delta \to 1$ yields (11.29). □

Theorem 11.14 was established by Dębowski (2012a).

Theorem 11.14 shows that there exist mixing processes which obey a power-law growth of the block mutual information. Since these processes are ergodic in particular, this demonstrates that in general we do not have equality in inequality (8.107), which links the power-law growth of the block mutual information and the number of inferrable facts. To make a historical remark, the first example of a mixing process with infinite excess entropy was exhibited by Bradley (1980).

The Zipfian Santa Fe and pseudo Santa Fe processes are obviously a huge idealization with respect to natural language. They were proposed as very simple models of the power-law growth of the block mutual information and its possible links to nonergodicity. Yet these processes seem to capture, in an idealized form, some phenomena that may actually take place in natural language. Confronted with a text in natural language we have an impression that it consists of a description of some unchanging reality $(Z_k)_{k\in\mathbb{N}}$ or evolving reality $(Z_{ik})_{i\in\mathbb{Z},k\in\mathbb{N}}$ which is controlled by some narration process $(K_i)_{i\in\mathbb{Z}}$. The unchanging reality referred to in the texts may concern objects that do not change in time, like mathematical or physical constants. The evolving reality referred to in the texts may concern objects that evolve with a varied speed, like various facts about culture, language, or geography. It is important to notice that the distinction between the unchanging reality and the evolving reality may be a matter of some modeling approximation since some of the referred objects may evolve so slowly that practically they may be considered constant.

11.3 Encoding into a Finite Alphabet

The topic which we would like to address in this section concerns finite alphabet versions of Santa Fe processes and their generalizations. This topic, albeit technical, is important, since a few interesting theorems, such as the theorems about facts and words (Theorems 8.22 and 8.23), assume that the considered stochastic process is over a finite alphabet. A natural solution of the problem is to encode a process over a countably infinite alphabet, such as a Santa Fe process, into a process over a finite alphabet, using an operation to be called stationary coding. In the stationary coding, we will replace individual symbols from the larger alphabet by sequences over the smaller alphabet and we will concatenate the result, subsequently taking a stationary mean of the process distribution. In fact, this operation is a special case of quasiperiodic coding introduced in Section 10.2, see also Problem 10.5.

Formally, let a code $f : \mathbb{X} \to \mathbb{Y}^*$, called here a variable length code, map symbols from alphabet $\mathbb{X}$, say countably infinite, into sequences over another alphabet $\mathbb{Y}$, say finite. We define the extension $f^{\mathbb{Z}} : \mathbb{X}^{\mathbb{Z}} \to \mathbb{Y}^{\mathbb{Z}}$ of a variable length code f to doubly infinite sequences as

$$f^{\mathbb{Z}}((x_i)_{i\in\mathbb{Z}}) := ... f(x_{-1})f(x_0).f(x_1)f(x_2)... \tag{11.69}$$

where $x_i \in \mathbb{X}$ and the bold-face dot separates the zeroth and the first symbol. Then, for a stochastic process $(X_i)_{i\in\mathbb{Z}}$ on $(\Omega, \mathcal{J}, P)$, where $X_i : \Omega \to \mathbb{X}$, we introduce process

$$(Y_i)_{i\in\mathbb{Z}} := f^{\mathbb{Z}}((X_i)_{i\in\mathbb{Z}}) \tag{11.70}$$

with random variables $Y_i : \Omega \to \mathbb{Y}$.

Some drawback of this simple construction is that, even when process $(X_i)_{i\in\mathbb{Z}}$ is stationary, the process $(Y_i)_{i\in\mathbb{Z}}$ need not be stationary. In our applications, however, it is often assumed that the process of interest is stationary. To solve this problem, we notice that transformation $f^{\mathbb{Z}}$ preserves a weaker property. Let

$$L_i := |f(X_i)| \tag{11.71}$$

be the length of the encoding of symbol X_i. Suppose that the average length of a symbol encoding tends to some real random variable L, i.e. limit

$$L := \lim_{n\to\infty} \frac{1}{n} \sum_{i=0}^{n-1} L_i \tag{11.72}$$

exists almost surely. By the generalized ergodic theorem (see Theorem 10.4), this condition is satisfied with $L \in [0, \infty]$ for any asymptotically mean stationary (AMS) process $(X_i)_{i\in\mathbb{Z}}$. Now, by Theorem 10.7, if L is almost surely strictly positive and finite, that is $L \in (0, \infty)$, and process $(X_i)_{i\in\mathbb{Z}}$ is AMS then process $(Y_i)_{i\in\mathbb{Z}}$ is also AMS. By definition, process $(Y_i)_{i\in\mathbb{Z}}$ is AMS if there exists a stationary process $(\bar{Y}_i)_{i\in\mathbb{Z}}$, called the stationary mean, with distribution

$$P((\bar{Y}_i)_{i\in\mathbb{Z}} \in A) = \lim_{n\to\infty} \frac{1}{n} \sum_{j=0}^{n-1} P((Y_{j+i})_{i\in\mathbb{Z}} \in A) \tag{11.73}$$

for all events A measurable with respect to process $(Y_i)_{i\in\mathbb{Z}}$. In particular, all stationary processes $(Y_i)_{i\in\mathbb{Z}}$ are AMS and their stationary means $(\bar{Y}_i)_{i\in\mathbb{Z}}$ have the same distributions as processes $(Y_i)_{i\in\mathbb{Z}}$.

Thus, if we need a stationary process over a finite alphabet we can do as follows. First, we take a suitable stationary process $(X_i)_{i\in\mathbb{Z}}$ over a countably infinite alphabet and we encode it into an AMS process $(Y_i)_{i\in\mathbb{Z}}$ over a finite alphabet via construction (11.70). In the second step, we apply transformation (11.73) to construct a stationary process $(\bar{Y}_i)_{i\in\mathbb{Z}}$.

Definition 11.6 *(Stationary coding)*: The above construction of process $(\bar{Y}_i)_{i\in\mathbb{Z}}$, if it exists, will be called the stationary coding of process $(X_i)_{i\in\mathbb{Z}}$ using the variable length code f.

We are interested in a stationary coding which preserves some interesting properties of the encoded process. The first such property is (non)-ergodicity. It has been shown in Chapter 10 that (non)-ergodicity of the encoded process is preserved for some nice variable length codes, called synchronizable coding, see Definition 10.4 and Theorem 10.11. A particular application of Theorem 10.11 is as follows.

Theorem 11.15: *Put $\mathbb{Y} = \{0, 1, 2\}$ and let $(\bar{Y}_i)_{i\in\mathbb{Z}}$ be the stationary coding of the Zipfian pseudo Santa Fe process (11.47) using the variable length code*

$$f(k, z) = B(k)z2 \tag{11.74}$$

where $B(k) \in \{0, 1\}^*$ is the binary representation of a natural number k, i.e. $b(1) = 1$, $b(2) = 10$, $b(3) = 11$, $b(4) = 100$, and so on. Then, process $(\bar{Y}_i)_{i\in\mathbb{Z}}$ exists and we have:

1. *Process $(\bar{Y}_i)_{i\in\mathbb{Z}}$ is ergodic if and only if process $(X_i)_{i\in\mathbb{Z}}$ is ergodic.*
2. *Process $(\bar{Y}_i)_{i\in\mathbb{Z}}$ is strongly nonergodic if and only if process $(X_i)_{i\in\mathbb{Z}}$ is strongly nonergodic.*

Proof: Process $(L_i)_{i\in\mathbb{Z}}$ is stationary and ergodic since

$$L_i = \lfloor \log K_i \rfloor + 3 \tag{11.75}$$

Hence, by the Birkhoff ergodic theorem (Theorem 4.6), we have almost surely that

$$\lim_{n\to\infty} \frac{1}{n} \sum_{i=0}^{n-1} L_i = \mathbf{E}L_i = \sum_{k=1}^{\infty} (\lfloor \log k \rfloor + 3)\frac{k^{-\alpha}}{\zeta(\alpha)} \in (0, \infty) \tag{11.76}$$

Hence, process $(\bar{Y}_i)_{i\in\mathbb{Z}}$ exists by Theorem 10.7. Moreover, the variable length code f is a comma-separated code (see Definition 7.5) and thus it constitutes a synchronizable coding. Hence, claims (1) and (2) follow by Theorem 10.11. □

Computing the distribution of process $(\bar{Y}_i)_{i \in \mathbb{Z}}$ is in general nontrivial, but it can be done in some special cases. Suppose that process $(X_i)_{i \in \mathbb{Z}}$ is stationary and the average length of a symbol encoding tends to its expectation,

$$\lim_{n \to \infty} \frac{1}{n} \sum_{i=0}^{n-1} L_i = L := \mathbf{E} L_i \tag{11.77}$$

This condition is strictly weaker than ergodicity of the process $(X_i)_{i \in \mathbb{Z}}$. As we have observed in the proof of Theorem 11.15, for the variable length code (11.74) and a Santa Fe process $(X_i)_{i \in \mathbb{Z}}$, condition (11.77) holds if the narration process $(K_i)_{i \in \mathbb{Z}}$ is ergodic. It turns out that it is exactly the case when we can do some computations for process $(\bar{Y}_i)_{i \in \mathbb{Z}}$.

Formally, let process $(X_i)_{i \in \mathbb{Z}}$ be stationary and let condition (11.77) hold true. Then, we may apply Theorem 10.8. According to this proposition, as it was shown in Problem 10.5, we may construct process $(\bar{Y}_i)_{i \in \mathbb{Z}}$ in the following way. First, we construct a random variable $N : \Omega \to \mathbb{N} \cup \{0\}$, called a random shift, and a non-stationary process $(\breve{X}_i)_{i \in \mathbb{Z}}$ where $\breve{X}_i : \Omega \to \mathbb{X}$. Precisely, random variable N and process $(\breve{X}_i)_{i \in \mathbb{Z}}$ are conditionally independent given variable $\breve{X}_0$ and their joint distribution is

$$P(\breve{X}_k^l = x_k^l) = P(X_k^l = x_k^l) \times \frac{|f(x_0)|}{L}, \quad k \le 0 \le l \tag{11.78}$$

$$P(N = n | \breve{X}_0 = x_0) = \frac{\mathbf{1}\{0 \le n \le |f(x_0)| - 1\}}{|f(x_0)|}, \quad n \in \mathbb{N} \cup \{0\} \tag{11.79}$$

In the sequel, process $(\bar{Y}_i)_{i \in \mathbb{Z}}$, being the stationary coding of process $(X_i)_{i \in \mathbb{Z}}$ using function f, can be obtained as

$$(\bar{Y}_i)_{i \in \mathbb{Z}} = T^{-N} f^{\mathbb{Z}}((\breve{X}_i)_{i \in \mathbb{Z}}) \tag{11.80}$$

where $T((y_i)_{i \in \mathbb{Z}}) := (y_{i+1})_{i \in \mathbb{Z}}$ is the shift operation. Construction (11.80) does not yield the distribution of process $(\bar{Y}_i)_{i \in \mathbb{Z}}$ explicitly, but it is sufficiently simple to provide interesting bounds for some statistics of process $(\bar{Y}_i)_{i \in \mathbb{Z}}$.

In case of the Zipfian pseudo Santa Fe processes, there are some further simplifications. For this goal, we define first

$$\bar{L}_i := |f(\breve{X}_i)| \tag{11.81}$$

$$\bar{L} := \mathbf{E} \bar{L}_0 \tag{11.82}$$

and we observe the following.

Theorem 11.16: *Denote blocks $\breve{X}_k^l$ with $\breve{X}_0$ removed as $\breve{X}_k^{l \backslash 0}$. For a Zipfian pseudo Santa Fe process $(X_i)_{i \in \mathbb{Z}}$ and the variable length code (11.74), variables $\breve{X}_k^{l \backslash 0}$ and $X_k^{l \backslash 0}$ have the same distribution, whereas variables $\breve{X}_k^{l \backslash 0}$ and $\bar{L}_0$ are probabilistically independent.*

Proof: Notice that $|f(X_0)|$ does not depend on Z_{0,K_0} and variable K_0 is independent of block $X_k^{l\backslash 0}$. Hence,

$$P(\bar{X}_k^{l\backslash 0} = w)$$

$$= P(X_k^{l\backslash 0} = w) \sum_{x_0} \frac{|f(x_0)|}{L} P(X_0 = x_0 | X_k^{l\backslash 0} = w)$$

$$= P(X_k^{l\backslash 0} = w) \sum_{k_0, z_0} \frac{|f(k_0, 1)|}{L} P(K_0 = k_0) P(Z_{0,k_0} = z_0 | X_k^{l\backslash 0} = w)$$

$$= P(X_k^{l\backslash 0} = w) \tag{11.83}$$

Similarly,

$$P(\bar{X}_k^{l\backslash 0} = w, \bar{L}_0 = l)$$

$$= P(X_k^{l\backslash 0} = w) \sum_{x_0} \frac{|f(x_0)|}{L} P(X_0 = x_0 | X_k^{l\backslash 0} = w)\mathbf{1}\{|f(x_0)| = l\}$$

$$= P(X_k^{l\backslash 0} = w) \sum_{k_0, z_0} \frac{|f(k_0, 1)|}{L} P(K_0 = k_0) P(Z_{0,k_0} = z_0 | X_k^{l\backslash 0} = w)\mathbf{1}\{|f(k_0, 1)| = l\}$$

$$= P(X_k^{l\backslash 0} = w) \sum_{k_0} \frac{l}{L} P(K_0 = k_0)\mathbf{1}\{|f(k_0, 1)| = l\}$$

$$= P(\bar{X}_k^{l\backslash 0} = w)P(\bar{L}_0 = l) \tag{11.84}$$

Let us continue writing $L_i := |f(X_i)|$ and $\bar{L}_i := |f(\bar{X}_i)|$. In the discussed case, variables L_i are IID. For these variables, we define indices

$$L_t^+ := \frac{1}{t} \ln \mathbf{E}\, e^{tL_i} \tag{11.85}$$

$$L_t^- := -\frac{1}{t} \ln \mathbf{E}\, e^{-tL_i} \tag{11.86}$$

where $t > 0$. For the given distribution of L_i, we have $0 < L_t^-, L_t^+ < \infty$ for sufficiently small t. By the Jensen Inequality, L_t^+ is a growing function of t and L_t^- is a decreasing function of t. Jensen inequality implies also $L_t^- \leq L \leq L_t^+$.

Theorem 11.17: *Consider a Zipfian pseudo Santa Fe process and the variable length code (11.74). We have*

$$\lim_{t \to 0} L_t^+ = \lim_{t \to 0} L_t^- = L \tag{11.87}$$

Proof: Consider function $g(t, x) = t^{-2}(e^{tx} - 1 - tx)$. For $x > 0$, it is a growing function of t. Consider next such a t_0 that $\mathbf{E}\,e^{t_0 L_i} < \infty$. For $0 < t \leq t_0$, we obtain

$$\mathbf{E}\,e^{tL_i} = 1 + t\mathbf{E}\,L_i + t^2 \mathbf{E}\,g(t, L_i) \leq 1 + t\mathbf{E}\,L_i + t^2 \mathbf{E}\,g(t_0, L_i) \tag{11.88}$$

This yields

$$L \leq L_t^+ \leq \frac{1}{t} \ln(1 + t\mathbf{E}L_i + t^2\mathbf{E}g(t_0, L_i)) \xrightarrow{t \to 0} L \tag{11.89}$$

On the other hand, for $t > 0$, we have

$$\mathbf{E}e^{-tL_i} \leq 1 - t\mathbf{E}L_i + \frac{t^2\mathbf{E}L_i^2}{2} \tag{11.90}$$

Hence,

$$L \geq L_t^- \geq -\frac{1}{t} \ln\left(1 - t\mathbf{E}L_i + \frac{t^2\mathbf{E}L_i^2}{2}\right) \xrightarrow{t \to 0} L \tag{11.91}$$

$\square$

Subsequently, we will derive a special case of the Chernoff bounds.

Theorem 11.18 *(Chernoff bounds):* *Consider a Zipfian pseudo Santa Fe process and the variable length code (11.74). For $t > 0$, we have*

$$P\left(\sum_{i=1}^{n} L_i \geq m\right) \leq \exp(-t(m - L_t^+ n)) \tag{11.92}$$

$$P\left(\sum_{i=1}^{n} L_i \leq m\right) \leq \exp(-t(L_t^- n - m)) \tag{11.93}$$

Proof: Because variables L_i are IID, using Markov inequality (Theorem 3.11) we observe

$$P\left(e^{t\sum_{i=1}^{n} L_i} \geq e^{tm}\right) \leq \frac{\mathbf{E}e^{t\sum_{i=1}^{n} L_i}}{e^{tm}} = \exp(-t(m - L_t^+ n)) \tag{11.94}$$

$$P\left(e^{-t\sum_{i=1}^{n} L_i} \geq e^{-tm}\right) \leq \frac{\mathbf{E}e^{-t\sum_{i=1}^{n} L_i}}{e^{-tm}} = \exp(-t(L_t^- n - m)) \tag{11.95}$$

$\square$

Now we can easily check that the number of inferrable facts for the stationary coding of the mixture Zipfian Santa Fe process (i.e. a strongly nonergodic Zipfian pseudo Santa Fe process) is bounded below in the similar fashion as for the unencoded mixture Zipfian Santa Fe process (see Theorem 8.14). In the following, we write $a \sqsubseteq b$ when string a is a substring of sequence b.

Theorem 11.19: *Put $\mathbb{Y} = \{0, 1, 2\}$ and let $(\bar{Y}_i)_{i\in\mathbb{Z}}$ be the stationary coding of the mixture Zipfian Santa Fe process with an exponent α using the variable length code (11.74). Consider functions*

$$\bar{g}(k, w) := \begin{cases} 0, & \text{if } 2B(k)02 \sqsubseteq w \text{ and } 2b(k)12 \not\sqsubseteq w \\ 1, & \text{if } 2B(k)12 \sqsubseteq w \text{ and } 2b(k)02 \not\sqsubseteq w \\ 2, & \text{else} \end{cases} \tag{11.96}$$

for all strings and infinite sequences $w \in \mathbb{Y}^{\mathbb{Z}} \cup \mathbb{Y}^*$.

1. *Define random variables $\bar{Z}_k = \bar{g}(k, (\bar{Y}_i)_{i \in \mathbb{Z}})$. Process $(\bar{Z}_k)_{k \in \mathbb{N}}$ is a fair-coin process measurable with respect to the invariant σ-field of process $(\bar{Y}_i)_{i \in \mathbb{Z}}$ and it satisfies condition*

$$\lim_{m \to \infty} P(\bar{g}(k, \bar{Y}_{j+1}^{j+m}) = \bar{Z}_k) = 1 \qquad (11.97)$$

   for all $j \in \mathbb{Z}$ and $k \in \mathbb{N}$.

2. *Define the sets of inferrable facts*

$$\bar{U}(\bar{Y}_1^m) := \{l \in \mathbb{N} : \bar{g}(k, \bar{Y}_1^m) = \bar{Z}_k \text{ for all } k \le l\} \qquad (11.98)$$

   We have

$$\underset{m \to \infty}{\text{hilb}} \ \#\bar{U}(\bar{Y}_1^m) = 1/\alpha \text{ almost surely} \qquad (11.99)$$

$$\underset{m \to \infty}{\text{hilb}} \ \mathbf{E}\#\bar{U}(\bar{Y}_1^m) = 1/\alpha \qquad (11.100)$$

Proof:

1. Let $(X_i)_{i \in \mathbb{Z}}$ be the mixture Zipfian Santa Fe process. For the process $(Y_i)_{i \in \mathbb{Z}}$ given by (11.70), we have $Z_k = \bar{g}(k, (Y_i)_{i \in \mathbb{Z}})$, where Z_k are independent random bits measurable with respect to the invariant σ-field of process $(Y_i)_{i \in \mathbb{Z}}$. Since process $(Y_i)_{i \in \mathbb{Z}}$ is AMS, its stationary mean $(\bar{Y}_i)_{i \in \mathbb{Z}}$ has the same distribution on the invariant σ-field, by Theorem 10.1. Hence, process $\bar{Z}_k = \bar{g}(k, (\bar{Y}_i)_{i \in \mathbb{Z}})$ is a fair-coin process measurable with respect to the invariant σ-field of process $(\bar{Y}_i)_{i \in \mathbb{Z}}$. Using the technique from the proof of Theorem 8.10, we can show that variables $\bar{Z}_k$ satisfy condition (11.97).

2. First, we will prove (11.99). Denote $M_n := \sum_{i=1}^{n} \bar{L}_i$. Observe that

$$|\#\bar{U}(\bar{Y}_1^m) - \#\bar{U}(\bar{Y}_{1-N}^{m-N})| \le 2N \qquad (11.101)$$

   whereas

$$\#U(\bar{X}_1^{n-1}) \le \#\bar{U}(\bar{Y}_{1-N}^{M_n-N}) \le \#U(\bar{X}_1^n) \qquad (11.102)$$

   where

$$U(\bar{X}_1^n) := \{l \in \mathbb{N} : g(k, \bar{X}_1^m) = Z_k \text{ for all } k \le l\} \qquad (11.103)$$

   for the function g given in (8.79). Hence, we obtain

$$\underset{m \to \infty}{\text{hilb}} \ \#\bar{U}(\bar{Y}_1^m) = (\underset{n \to \infty}{\text{hilb}} \ \#U(\bar{X}_1^n)) \left(\lim_{n \to \infty} \frac{\log M_n}{\log n} \right) \qquad (11.104)$$

   almost surely if the last limit exists. But by Theorem 11.16 and the Birkhoff ergodic theorem (Theorem 4.6), we have

$$\lim_{n \to \infty} \frac{\log M_n}{\log n} = 1 \text{ almost surely} \qquad (11.105)$$

Thus, applying Theorem 11.16 and (8.82), we obtain (11.99).

To obtain equality (11.100), we notice that $\#\bar{U}(\bar{Y}_1^m)$ is a positive growing function of m and we invoke Theorem 8.4 for the lower bound, whereas for the upper bound, we observe the following. First, by (11.101) and (11.102), we obtain

$$\#\bar{U}(\bar{Y}_1^m) \leq \#U(\bar{X}_1^n) + 2N \text{ for } M_n \geq m \tag{11.106}$$

Hence, since $\#U(\bar{X}_1^m)$ is positive and $\#\bar{U}(\bar{Y}_1^m) \leq m$, we obtain

$$\mathbf{E}\#\bar{U}(\bar{Y}_1^m) \leq \mathbf{E}\#U(\bar{X}_1^n) + 2EN + mP(M_n < m) \tag{11.107}$$

By the Chernoff bound (Theorem 11.18), we obtain that $\lim_{n\to\infty} m_n P(M_n < m_n) = 0$ for $m_n = n(L - \epsilon)$ and any $\epsilon > 0$. Thus, plugging in $m = n(L - \epsilon)$ into (11.107) and noticing that $\mathbf{E}N < \infty$, we obtain

$$\underset{m\to\infty}{\text{hilb }} \mathbf{E}\#\bar{U}(\bar{Y}_1^m) \leq \underset{n\to\infty}{\text{hilb }} \mathbf{E}\#U(\bar{X}_1^n) \tag{11.108}$$

Thus, applying Theorem 11.16 and (8.83), we obtain (11.100). □

To the above process $(\bar{Y}_i)_{i\in\mathbb{Z}}$, since it is a process over a finite alphabet, we may apply the theorem about facts and mutual information (Theorem 8.18). By this proposition, the mutual information between blocks for process $(\bar{Y}_i)_{i\in\mathbb{Z}}$ must grow at least as fast as the expected number of inferrable facts,

$$\underset{m\to\infty}{\text{hilb }} I(\bar{Y}_1^m; \bar{Y}_{m+1}^{2m}) \geq \underset{m\to\infty}{\text{hilb }} \mathbf{E}\#\bar{U}(\bar{Y}_1^m) = 1/\alpha \tag{11.109}$$

In the following, we will show that the mutual information does not grow faster. The theorem will be a bit more general since it will concern the stationary coding of Zipfian pseudo Santa Fe processes.

Before we prove the main theorem, we will demonstrate an auxiliary statement. For an event E, we introduce conditional entropy $H(X|E)$ and mutual information $I(X; Y|E)$, which are respectively the entropy of a random variable X and mutual information between random variables X and Y taken with respect to probability measure $P(\cdot|E)$.

Theorem 11.20: *Let $(X_i)_{i\in\mathbb{Z}}$ be the Zipfian pseudo Santa Fe process and consider the variable length code (11.74). If $m \geq L_t^+ n + 1/t$, where $t > 0$, then*

$$P\left(\sum_{i=1}^n L_i \geq m\right) H\left(X_1^n \middle| \sum_{i=1}^n L_i \geq m\right)$$
$$\leq [B_t(m \log 3 + t(m - L_t^+ n)) + C_t] \exp(-t(m - L_t^+ n)) \tag{11.110}$$

for certain coefficients B_t and C_t.

Proof: Write $M_n = \sum_{i=1}^n L_i$. We have

$$H(X_1^n|M_n \geq m) \leq H(X_1^n|M_n, M_n \geq m) + H(M_n|M_n \geq m) \tag{11.111}$$

Moreover, $H(X_1^n | M_n = l) = H(f(X_1) \ldots f(X_n) | M_n = l) \le l \log 3$. Hence,

$$P(M_n \ge m) H(X_1^n | M_n \ge m)$$

$$\le \sum_{l \ge m} P(M_n = l)[H(X_1^n | M_n = l) - \log P(M_n = l | M_n \ge m)]$$

$$\le \sum_{l \ge m} P(M_n = l) \left[l \log 3 - \log P(M_n = l) + \log \left(\sum_{l \ge m} P(M_n = l) \right) \right]$$

$$\le \sum_{l \ge m} P(M_n = l)[l \log 3 - \log P(M_n = l)] \tag{11.112}$$

Function $x \exp(-x)$ is decreasing for $x \ge 1$. Put $A := t(m - L_i^+ n)$. Thus by Theorem 11.18, for $A \ge 1$, we obtain

$$P(M_n \ge m) H(X_1^n | M_n \ge m) \le \sum_{l \ge 0} [(l + m) \log 3 + tl + A] \exp(-tl - A)$$

$$= \exp(-A) \left[(m \log 3 + A) \sum_{l \ge 0} \exp(-tl) + (\log 3 + t) \sum_{l \ge 0} l \exp(-tl) \right]$$

$$\tag{11.113}$$

Putting $B_t := \sum_{l \ge 0} e^{-tl}$ and $C_t := (\log 3 + t) \sum_{l \ge 0} l e^{-tl}$ yields the claim. $\quad \square$

Now let us consider block mutual information for the stationary coding of the Zipfian pseudo Santa Fe process. We will show this fact.

Theorem 11.21: *Let $(\bar{Y}_i)_{i \in \mathbb{Z}}$ be the stationary coding of the Zipfian pseudo Santa Fe process with an exponent α using the variable length code (11.74). The block mutual information $I(\bar{Y}_1^m; \bar{Y}_{m+1}^{2m})$ for process $(\bar{Y}_i)_{i \in \mathbb{Z}}$ satisfies*

$$\limsup_{m \to \infty} \frac{I(\bar{Y}_1^m; \bar{Y}_{m+1}^{2m})}{m^\beta} \le \frac{1}{L^\beta} \frac{(2 - 2^\beta)\Gamma(1 - \beta)}{[\zeta(\beta^{-1})]^\beta}, \quad \beta = 1/\alpha \tag{11.114}$$

The lower limits in particular cases are as follows:

1. *If $p_k \le P(K_i = k)$, then*

$$\liminf_{m \to \infty} \frac{I(\bar{Y}_1^m; \bar{Y}_{m+1}^{2m})}{m^\beta} \ge \frac{1}{L^\beta} \frac{A(\beta)}{[\zeta(\beta^{-1})]^\beta} \tag{11.115}$$

where $A(\beta)$ is defined in (11.52).

2. *If $\lim_{k \to \infty} p_k / P(K_i = k) = 0$, then*

$$\lim_{m \to \infty} \frac{I(\bar{Y}_1^m; \bar{Y}_{m+1}^{2m})}{m^\beta} = \frac{1}{L^\beta} \frac{(2 - 2^\beta)\Gamma(1 - \beta)}{[\zeta(\beta^{-1})]^\beta} \tag{11.116}$$

Proof: For an $\epsilon > 0$ and $l \geq 0$, define events

$$\bar{S}_n^+ := \left(\sum_{i=1}^{n} \bar{L}_i < n(L_t^+ + \epsilon) + 1/t \right) \tag{11.117}$$

$$\bar{S}_n^- := \left(\sum_{i=1}^{n} \bar{L}_i > n(L_t^- - \epsilon) \right) \tag{11.118}$$

$$\bar{T}_n^+ := \left(\sum_{i=-n}^{-1} \bar{L}_i < n(L_t^+ + \epsilon) + 1/t \right) \tag{11.119}$$

$$\bar{T}_n^- := \left(\sum_{i=-n}^{-1} \bar{L}_i > n(L_t^- - \epsilon) \right) \tag{11.120}$$

$$\bar{C}_n^+ := \bar{T}_n^+ \cap \bar{S}_n^+ \tag{11.121}$$

$$\bar{C}_n^- := \bar{T}_n^- \cap \bar{S}_n^- \tag{11.122}$$

$$\bar{B} := (\bar{L}_0 \leq l) \tag{11.123}$$

as well as for each of the above events $\bar{A}$ we define the event A by substituting variables $\bar{L}_i$ with variables L_i.

Let us recall the conditional data processing inequality

$$I(U'; V'|C) \leq I(U; V|C) \tag{11.124}$$

which holds if equalities $U' = g(U)$ and $V' = h(V)$ are satisfied on event C. Hence, by equality (11.80), for $m \geq n(L_t^+ + \epsilon) + 1/t + l$ we have

$$I(\bar{X}_{-n+1}^{-1}; \bar{X}_1^n|\bar{C}_n^+) = I(\bar{X}_{-n+1}^{-1}; \bar{X}_1^n|\bar{B} \cap \bar{C}_n^+)$$

$$\leq I(\bar{Y}_{-m+1}^0; \bar{Y}_1^m|\bar{B} \cap \bar{C}_n^+) \tag{11.125}$$

whereas for $m \leq n(L_t^- - \epsilon)$ we have similarly

$$I(\bar{Y}_{-m+1}^0; \bar{Y}_1^m|\bar{C}_n^-) \leq I(\bar{X}_{-n+1}^0; \bar{X}_0^n|\bar{C}_n^-) \tag{11.126}$$

The above inequalities will be used to bound the block mutual information $I(\bar{Y}_1^m; \bar{Y}_{m+1}^{2m})$ in terms of the block mutual information $I(X_1^n; X_{n+1}^{2n})$.

There is an additional fact that we shall use. Let I_C be the characteristic function of an event C. Observe that

$$P(C)I(X; Y|C) \leq P(C)I(X; Y|C) + P(C^c)I(X; Y|C^c)$$

$$= I(X; Y|I_C) = I(X; Y) - I(X; Y; I_C) \tag{11.127}$$

where $|I(X; Y; I_C)| \leq H(I_C) \leq 1$.

Assume that $m = \lceil n(L_t^+ + \epsilon) + 1/t + l \rceil$. Observe that $H(X_0) < \infty$. Then, applying subsequently (11.127), (11.125), (11.83), and (11.127), we obtain

$$I(\bar{Y}_1^m; \bar{Y}_{m+1}^{2m}) = I(\bar{Y}_{-m+1}^0; \bar{Y}_1^m)$$

$$\geq P(\bar{B} \cap \bar{C}_n^+)I(\bar{Y}_{-m+1}^0; \bar{Y}_1^m | \bar{B} \cap \bar{C}_n^+) - 1$$

$$\geq P(\bar{B})P(\bar{C}_n^+)I(\bar{X}_{-n+1}^{-1}; \bar{X}_1^n | \bar{C}_n^+) - 1$$

$$= P(\bar{B})P(C_n^+)I(X_{-n+1}^{-1}; X_1^n | C_n^+) - 1$$

$$\geq P(\bar{B})I(X_{-n+1}^{-1}; X_1^n) - P(C_n^{+c})I(X_{-n+1}^{-1}; X_1^n | C_n^{+c}) - 2$$

$$\geq P(\bar{B})I(X_1^n; X_{n+1}^{2n}) - P(C_n^{+c})H(X_1^n | C_n^{+c}) - H(X_0) - 2$$

$$\geq P(\bar{B})I(X_1^n; X_{n+1}^{2n}) - 2P(S_n^{+c})H(X_1^n | S_n^{+c}) - H(X_0) - 2 \qquad (11.128)$$

The term $P(S_n^{+c})H(\bar{X}_1^n | S_n^{+c})$ is asymptotically negligible since by Theorem 11.20, we have

$$P(S_n^{+c})H(\bar{X}_1^n | S_n^{+c}) \leq [B_t(m \log 3 + nt\epsilon + tl + 1) + C_t] \exp(-nt\epsilon) \qquad (11.129)$$

Now assume that $m = \lfloor n(L_t^- - \epsilon) \rfloor$. Observe that $H(\bar{X}_0) < \infty$. Then, applying subsequently (11.83), (11.127), (11.126), and (11.127), we obtain

$$I(X_1^n; X_{n+1}^{2n}) \geq I(X_{-n+1}^{-1}; X_1^n)$$

$$= I(\bar{X}_{-n+1}^{-1}; \bar{X}_1^n) \geq I(\bar{X}_{-n+1}^0; \bar{X}_0^n) - 2H(\bar{X}_0)$$

$$\geq P(\bar{C}_n^-)I(\bar{X}_{-n+1}^0; \bar{X}_0^n | \bar{C}_n^-) - 2H(\bar{X}_0) - 1$$

$$\geq P(\bar{C}_n^-)I(\bar{Y}_{-m+1}^0; \bar{Y}_1^m | \bar{C}_n^-) - 2H(\bar{X}_0) - 1$$

$$\geq I(\bar{Y}_1^m; \bar{Y}_{m+1}^{2m}) - P(C_n^{-c})I(\bar{Y}_{-m+1}^0; \bar{Y}_1^m | C_n^{-c}) - 2H(\bar{X}_0) - 2$$

$$\geq I(\bar{Y}_1^m; \bar{Y}_{m+1}^{2m}) - P(C_n^{-c})H(\bar{Y}_1^m | C_n^{-c}) - 2H(\bar{X}_0) - 2$$

$$\geq I(\bar{Y}_1^m; \bar{Y}_{m+1}^{2m}) - 2P(S_n^{-c})H(\bar{Y}_1^m | S_n^{-c}) - 2H(\bar{X}_0) - 2 \qquad (11.130)$$

The term $P(S_n^{-c})H(\bar{Y}_1^m | S_n^{-c})$ is asymptotically negligible since $P(S_n^{-c}) \leq \exp(-nt\epsilon)$ by Theorem 11.18 and $H(\bar{Y}_1^m | S_n^{-c}) \leq m \log 3$.

Write

$$x^+ = \limsup_{n \to \infty} \frac{I(X_1^n; X_{n+1}^{2n})}{n^\beta} \qquad (11.131)$$

$$x^- = \liminf_{n \to \infty} \frac{I(X_1^n; X_{n+1}^{2n})}{n^\beta} \qquad (11.132)$$

From bounds (11.128) and (11.130) we obtain

$$\limsup_{m\to\infty} \frac{I(\bar{Y}_1^m; \bar{Y}_{m+1}^{2m})}{m^\beta} \in \left(\frac{P(\bar{L}_0 \le l)}{[L_t^+ + \epsilon]^\beta} x^+, \frac{1}{[L_t^- - \epsilon]^\beta} x^+ \right) \tag{11.133}$$

$$\liminf_{m\to\infty} \frac{I(\bar{Y}_1^m; \bar{Y}_{m+1}^{2m})}{m^\beta} \in \left(\frac{P(\bar{L}_0 \le l)}{[L_t^+ + \epsilon]^\beta} x^-, \frac{1}{[L_t^- - \epsilon]^\beta} x^- \right) \tag{11.134}$$

If we consider $t \to 0$, $\epsilon \to 0$, and $l \to \infty$, then by Eq. (11.87) we obtain

$$\limsup_{m\to\infty} \frac{I(\bar{Y}_1^m; \bar{Y}_{m+1}^{2m})}{m^\beta} = \frac{x^+}{L^\beta} \tag{11.135}$$

$$\liminf_{m\to\infty} \frac{I(\bar{Y}_1^m; \bar{Y}_{m+1}^{2m})}{m^\beta} = \frac{x^-}{L^\beta} \tag{11.136}$$

Consequently, the requested claims follow by Theorem 11.14. □

Theorem 11.21 was proved by Dębowski (2012a).

In Section 11.1, we have shown that the unencoded Zipfian Santa Fe processes are finite energy, and hence, they do not satisfy a power-law growth of maximal repetition. This property need not translate to the Zipfian Santa Fe processes encoded into a finite alphabet provided we choose some special stationary coding, different to (11.74). We leave the construction of such a code as an exercise for the reader. It would be also interesting to check how enforcing the power-law growth of maximal repetition for the encoded Zipfian Santa Fe process influences the rate of growth of the block mutual information. Are there some nontrivial interactions or inevitable bounds?

11.4 Random Hierarchical Association

In this section, we will construct some processes that satisfy the power-law logarithmic growth of maximal repetition (11.3). The processes, called the RHA processes, were enigmatically mentioned in the introductory Section 1.13. The construction of RHA processes is more general than we need and we will only consider some special case which obeys both the power-law logarithmic growth of maximal repetition (11.3) and the power-law logarithmic growth of mutual information (11.2). Simultaneously, the considered case of an RHA process does not render a strictly positive Shannon entropy rate (11.1). Maybe one could modify the construction a bit to enforce also this condition without breaking conditions (11.2) and (11.3). As an open problem, it would be also interesting to verify whether the power-law decay of mutual information between individual symbols (11.4) holds for some RHA processes.

Differently than elsewhere in this book, in this section, we will denote blocks of consecutive symbols as

$$x_{k:l} := x_k x_{k+1} \dots x_l \tag{11.137}$$

since we need the superscript for additional notations.

Let us recall that in view of Theorems 9.22 and 9.23, a power-law upper bound for the Hartley entropy of an infinite sequence $H_0(n|x_{1:\infty})$ implies a power-law logarithmic lower bound for the maximal repetition $L(x_{1:n})$. Thus, looking for examples of sequences that satisfy the power-law logarithmic growth of maximal repetition (11.3), we may inspect first sequences with a vanishing Hartley entropy rate

$$h_0(x_{1:\infty}) := \lim_{n \to \infty} \frac{H_0(n|x_{1:\infty})}{n} \tag{11.138}$$

Unfortunately, stationary processes consisting of such sequences are asymptotically deterministic, as asserted by the following proposition.

Theorem 11.22: *Stationary processes* $(X_i)_{i \in \mathbb{N}}$ *such that* $h_0(X_{1:\infty}) = 0$ *holds almost surely have the Shannon entropy* $h = 0$.

Proof: The reasoning involves the random ergodic measure $F(A) = P(A|\mathcal{J})$, where $\mathcal{J}$ is the invariant σ-field. By the first ergodic decomposition theorem (Theorem 4.19), Theorem 9.18 and Theorem 9.2, we obtain almost surely

$$H_0(n|X_{1:\infty}) = \log \#\{x_{1:n} : F(X_{1:n} = x_{1:n}) > 0\} \geq H_F(X_{1:n}) \tag{11.139}$$

where $H_F(X_{1:n})$ is the block Shannon entropy of measure F. Thus, the Shannon entropy rate of this measure is $h_F = 0$ by the assumption $h_0(X_{1:\infty}) = 0$, whereas by Theorem 5.35, we obtain $h = \mathbf{E} h_F = 0$. $\qquad\square$

We recall from Section 9.5 that processes with a power-law logarithmic growth of maximal repetition need not have the Shannon entropy rate $h = 0$ in general. Some example of processes with a strictly positive Shannon entropy rate and the maximal repetition growing faster than any sublinear function was exhibited by Shields (1992).

We will present now a simple example of a process with the Hartley entropy $H_0(n|X_{1:\infty})$ upper bounded by a power-law growth. The example will be called a RHA process. The RHA process is parameterized by certain free parameters which we will call perplexities (a name borrowed from computational linguistics, where it denotes a similar concept). Approximately, perplexity k_n is the number of distinct blocks of length 2^n that appear in the process realization. It turns out that controlling perplexities, we can control the value of the Hartley entropy and force the Hartley entropy rate to be zero almost surely.

The RHA processes can be viewed as a stochastic variation on theme of context-free grammars. Whereas some standard stochastic model of context-free grammars is known in computational linguistics as probabilistic context-free grammars (PCFGs) or branching processes (Harris, 1963; Miller and O'Sullivan, 1992; Chi and Geman, 1998), we will develop quite a different approach. PCFGs construct a finite hierarchical structure in a top-down fashion, which results in defining discrete probabilities over finite sequences or trees. In contrast, RHA processes build a hierarchical structure in a bottom-up fashion – with the infinite height – which results in defining a probability measure over infinite sequences.

Before we introduce a stationary RHA process, let us discuss its nonstationary version, which is simpler. The nonstationary RHA process is constructed in two natural steps. In the first step, for each n iteratively, we construct a pool of k_n distinct blocks of length 2^n which are either defined as the natural numbers $1, 2, \dots, k_0$ for $n = 0$ or formed by concatenation of randomly selected pairs chosen from the pool of k_{n-1} distinct blocks of length 2^{n-1} for $n > 0$. In the second step, we obtain an infinite sequence of random numbers by concatenating blocks of lengths $2^0, 2^1, 2^2, \dots$, randomly chosen from the respective pools. As a result there are no more that k_n^2 distinct blocks of length 2^n that appear in the process realization – which provides a simple upper bound for the Hartley entropy. The selection of these blocks is, however, random and we do not know them a priori. To obtain the stationary RHA process, we take the stationary mean of the nonstationary RHA process, where the stationary mean exists since the nonstationary RHA process turns out to be pseudo AMS.

Now we will write down the construction using symbols.

Definition 11.7 *(Nonstationary RHA process)*: Let a sequence of natural numbers $(k_n)_{n \in \{0\} \cup \mathbb{N}}$ satisfy

$$k_{n-1} \le k_n \le k_{n-1}^2 \tag{11.140}$$

The nonstationary RHA process $(X_i)_{i \in \mathbb{N}}$ with perplexities $(k_n)_{n \in \{0\} \cup \mathbb{N}}$ is defined as

$$X_{1:\infty} := Y_{C_0}^0 Y_{C_1}^1 Y_{C_2}^2 \cdots \tag{11.141}$$

where $X_i : \Omega \to \{1, \dots, k_0\}$ and random variables $Y_j^n : \Omega \to \{1, \dots, k_0\}^{2^n}$ and $C_n : \Omega \to \{1, \dots, k_n\}$ are defined as follows.

First, for each $n \in \mathbb{N}$, let $(L_{nj}, R_{nj})_{j \in \{1, \dots, k_n\}}$ be a probabilistically independent random combination of k_n pairs of numbers from the set $\{1, \dots, k_{n-1}\}$ drawn without repetition. That is, we assume that each pair (L_{nj}, R_{nj}) is different, the elements of the pairs may be identical, and the sequence $(L_{nj}, R_{nj})_{j \in \{1, \dots, k_n\}}$ is sorted lexicographically. Formally, we assume that random variables L_{nj} and R_{nj} have the uniform distribution

$$P(L_{n1}, R_{n1}, \dots, L_{nk_n}, R_{nk_n}) = (l_{n1}, r_{n1}, \dots, l_{nk_n}, r_{nk_n})) := \left(\begin{array}{c} k_{n-1}^2 \\ k_n \end{array} \right)^{-1} \tag{11.142}$$

Subsequently, we define random variables

$$Y_j^0 := j, \qquad\qquad j \in \{1, \dots, k_0\} \tag{11.143}$$

$$Y_j^n := Y_{L_{nj}}^{n-1} Y_{R_{nj}}^{n-1}, \qquad j \in \{1, \dots, k_n\}, \quad n \in \mathbb{N} \tag{11.144}$$

Finally, let process $(C_n)_{n \in \{0\} \cup \mathbb{N}}$ be an IID process with the uniform distribution

$$P(C_n = j) := 1/k_n, \qquad j \in \{1, \dots, k_n\} \tag{11.145}$$

probabilistically independent from process $(L_{nj}, R_{nj})_{n \in \mathbb{N}, j \in \{1, \dots, k_n\}}$.

Let us define a few more notations which will be used further. The collection of random variables (L_{nj}, R_{nj}) will be denoted as

$$\mathcal{G} := (L_{nj}, R_{nj})_{n \in \mathbb{N}, j \in \{1, \dots, k_n\}} \tag{11.146}$$

We will also use notations

$$\mathcal{G}_{\leq m} := (L_{nj}, R_{nj})_{n \leq m, j \in \{1, \dots, k_n\}} \tag{11.147}$$

$$\mathcal{G}_{>m} := (L_{nj}, R_{nj})_{n > m, j \in \{1, \dots, k_n\}} \tag{11.148}$$

Let us observe that collection $\mathcal{G}_{\leq m}$ fully determines variables Y_j^m for a fixed m. Moreover, sequence $(X_i)_{i \in \mathbb{N}}$ can be parsed into a sequence of random natural numbers $X_j : \Omega \to \{1, \dots, k_0\}$,

$$X_{1:\infty} = X_1 X_2 \dots \tag{11.149}$$

and we will denote blocks of consecutive random natural numbers as

$$X_{k:l} := X_k X_{k+1} \dots X_l \tag{11.150}$$

Generalizing parsing (11.149), sequence $(X_i)_{i \in \mathbb{N}}$ will be also parsed into a sequence of blocks $X_i^n : \Omega \to \{1, \dots, k_0\}^{2^n}$ of length 2^n, where

$$X_{1:\infty} = Y_{C_0}^0 Y_{C_1}^1 \dots Y_{C_{n-1}}^{n-1} X_1^n X_2^n \dots \tag{11.151}$$

Let us observe that $X_j^0 = X_j$, $X_1^n = Y_{C_n}^n$ and in general there exist unique random variables $K_{nj} : \Omega \to \{1, \dots, k_n\}$ such that

$$X_j^n = Y_{K_{nj}}^n \tag{11.152}$$

Moreover, generalizing notation (11.150), we also denote blocks of length 2^n starting at any position as

$$X_{k:l}^n := X_k^n X_{k+1}^n \dots X_l^n \tag{11.153}$$

Let us prove this useful property.

Theorem 11.23: *Variables K_{nj} are independent from $\mathcal{G}_{\leq n}$ and satisfy*

$$P(K_{nj} = l, K_{n,j+1} = m) = 1/k_n^2, \qquad l, m \in \{1, \dots, k_n\}, j \in \mathbb{N} \qquad (11.154)$$

Proof: Each random number K_{nj} is a function of the random number C_q for some $q \geq n$ and the random pool $\mathcal{G}_{>n}$. Hence, K_{nj} are independent from $\mathcal{G}_{\leq n}$. Now we will show by induction on j that equality (11.154) is satisfied.

The induction begins with $K_{n1} = C_n$ and $K_{n2} = L_{n+1,C_{n+1}}$. These two variables are independent by definition and by definition K_{n1} is uniformly distributed on $\{1, \dots, k_n\}$. It remains to show that so is K_{n2}. Observe that $(L_{n+1,k}, R_{n+1,k})$ are independent of C_{n+1}. Hence, for $l, m \in \{1, \dots, k_n\}$ we obtain

$$P(K_{n2} = l, K_{n3} = m) = \sum_{k=1}^{k_{n+1}} P(L_{n+1,k} = l, R_{n+1,k} = m)P(C_{n+1} = k)$$

$$= \frac{1}{k_{n+1}} \sum_{k=1}^{k_{n+1}} P(L_{n+1,k} = l, R_{n+1,k} = m)$$

$$= \frac{1}{k_{n+1}} \left(\frac{k_n^2}{k_{n+1}} \right)^{-1} \left(\frac{k_n^2 - 1}{k_{n+1} - 1} \right) = \frac{1}{k_{n+1}} \frac{k_{n+1}}{k_n^2} = \frac{1}{k_n^2} \qquad (11.155)$$

so K_{n2} is uniformly distributed on $\{1, \dots, k_n\}$.

The inductive step is as follows: (i) if $K_{n+1,j}$ is uniformly distributed on $\{1, \dots, k_{n+1}\}$ then

$$(K_{n,2j}, K_{n,2j+1}) = (L_{n+1,K_{n+1,j}}, R_{n+1,K_{n+1,j}}) \qquad (11.156)$$

is uniformly distributed on $\{1, \dots, k_n\} \times \{1, \dots, k_n\}$, and (ii) if $(K_{n+1,j}, K_{n+1,j+1})$ is uniformly distributed on $\{1, \dots, k_{n+1}\} \times \{1, \dots, k_{n+1}\}$ then

$$(K_{n,2j+1}, K_{n,2j+2}) = (R_{n+1,K_{n+1,j}}, L_{n+1,K_{n+1,j+1}}) \qquad (11.157)$$

is uniformly distributed on $\{1, \dots, k_n\} \times \{1, \dots, k_n\}$. Now observe that $(L_{n+1,k}, R_{n+1,k})$ are independent of $K_{n+1,j}$. Hence, for $l, m \in \{1, \dots, k_n\}$ we obtain

$$P(K_{n,2j} = l, K_{n,2j+1} = m)$$

$$= \sum_{k=1}^{k_{n+1}} P(L_{n+1,k} = l, R_{n+1,k} = m)P(K_{n+1,j} = k)$$

$$= \frac{1}{k_{n+1}} \sum_{k=1}^{k_{n+1}} P(L_{n+1,k} = l, R_{n+1,k} = m)$$

$$= \frac{1}{k_{n+1}} \left(\frac{k_n^2}{k_{n+1}} \right)^{-1} \left(\frac{k_n^2 - 1}{k_{n+1} - 1} \right) = \frac{1}{k_{n+1}} \frac{k_{n+1}}{k_n^2} = \frac{1}{k_n^2} \qquad (11.158)$$

which proves claim (i). On the other hand, for $l, m \in \{1, \dots, k_n\}$ we obtain

$$P(K_{n,2j+1} = l, K_{n,2j+2} = m)$$

$$= \sum_{p,q=1}^{k_{n+1}} P(R_{n+1,p} = l, L_{n+1,q} = m) P(K_{n+1,j} = p, K_{n+1,j+1} = q)$$

$$= \frac{1}{k_{n+1}^2} \sum_{p,q=1}^{k_{n+1}} P(R_{n+1,p} = l, L_{n+1,q} = m) \tag{11.159}$$

Hence,

$$P(K_{n,2j+1} = l, K_{n,2j+2} = m)$$

$$= \frac{1}{k_{n+1}^2} \sum_{p=1}^{k_{n+1}} P(R_{n+1,p} = l, L_{n+1,p} = m)$$

$$+ \frac{1}{k_{n+1}^2} \sum_{p,q=1,\ p\neq q}^{k_{n+1}} P(R_{n+1,p} = l, L_{n+1,q} = m)$$

$$= \frac{1}{k_{n+1}^2} \left(\frac{k_n^2}{k_{n+1}} \right)^{-1} \left(\left(\frac{k_n^2 - 1}{k_{n+1} - 1} \right) + (k_n^2 - 1) \left(\frac{k_n^2 - 2}{k_{n+1} - 2} \right) \right)$$

$$= \frac{1}{k_{n+1}^2} \left(\frac{k_{n+1}}{k_n^2} + (k_n^2 - 1) \frac{k_{n+1}(k_{n+1} - 1)}{k_n^2(k_n^2 - 1)} \right) = \frac{1}{k_n^2} \tag{11.160}$$

which proves claim (ii). $\qquad\qquad\qquad\qquad\qquad\qquad\qquad\qquad\qquad\qquad\qquad\square$

In view of the above theorem, blocks $X_{k:k+1}^n$ are identically distributed. Thus, although the RHA processes defined in Definition 11.7 are not stationary, as we will see in the next assertion, they have the stationary mean, i.e. they are pseudo AMS processes (see Definitions 10.2 and 10.6). Let

$$X_{k:l} \circ T^j := X_{k+j:l+j} \tag{11.161}$$

be the shift operation.

Theorem 11.24: *RHA processes are pseudo AMS. In particular, for an RHA process $(X_i)_{i\in\mathbb{N}}$, let us denote a stationary process $(\bar{X}_i)_{i\in\mathbb{Z}}$ distributed according to the stationary mean*

$$P(\bar{X}_{1:m} = x_{1:m}) = \lim_{n\to\infty} \frac{1}{n} \sum_{j=0}^{n-1} P(X_{1:m} \circ T^j = x_{1:m}) \tag{11.162}$$

For any $k \in \mathbb{N}$, we have

$$P(\bar{X}_{1:2^n} = x_{1:2^n}) = \frac{1}{2^n} \sum_{j=0}^{2^n-1} P(X_k^n \circ T^j = x_{1:2^n}) \tag{11.163}$$

Proof: Block $X_k^n \circ T^j$ is a substring of block $X_{k:k+1}^n$ for $k \in \mathbb{N}$ and $0 \le j < 2^n$. In particular, there exist functions f_j such that

$$X_k^n \circ T^j = f_j(X_{k:k+1}^n) \tag{11.164}$$

Hence, by Theorem 11.23, probabilities $P(X_{1:2^n} \circ T^j = x_{1:2^n})$ are periodic functions of j with period 2^n. This implies that limits (11.162) exist and satisfy (11.163) for any block $x_{1:2^n}$, so process $(X_i)_{i \in \mathbb{N}}$ is pseudo AMS. $\qquad\square$

Definition 11.8 (*Stationary RHA process*): Process $(\bar{X}_i)_{i \in \mathbb{Z}}$ from Theorem 11.24 will be called the stationary mean of the respective RHA process, or briefly, the stationary RHA process.

It is easier to analyze the behavior of blocks drawn from the RHA process than the behavior of its stationary mean, but we can provide some links between functionals of an RHA process and its stationary mean. First, we will compare the Shannon block entropies for these two processes.

Theorem 11.25: *For any $q \in \mathbb{N}$ we have*

$$H(X_q^{n-1}) \le H(\bar{X}_{1:2^n}) \le H(X_q^{n+1}) + n \tag{11.165}$$

Proof: By the Jensen inequality for function $p \mapsto -p \log p$ and Theorem 11.24, we obtain

$$H(\bar{X}_{1:2^n}) \ge \frac{1}{2^n} \sum_{j=0}^{2^n-1} H(X_k^n \circ T^j) \tag{11.166}$$

Now, we observe that for each $k \ge 1$ and j there exists a q such that X_q^{n-1} is a substring of $X_k^n \circ T^j$. Thus we have $H(X_k^n \circ T^j) \ge H(X_q^{n-1})$. This combined with inequality (11.166) yields $H(X_q^{n-1}) \le H(\bar{X}_{1:2^n})$. On the other hand, using inequality $P(\bar{X}_{1:2^n} = x_{1:2^n}) \ge 2^{-n} P(X_k^n \circ T^j = x_{1:2^n})$ and Theorem 11.24, we obtain

$$H(\bar{X}_{1:2^n}) \le \frac{1}{2^n} \sum_{j=0}^{2^n-1} H(X_k^n \circ T^j) + n \tag{11.167}$$

Now, we observe that for each $k > 1$ and j there exists a q such that $X_k^n \circ T^j$ is a substring of X_q^{n+1}. Thus, we have $H(X_k^n \circ T^j) \le H(X_q^{n+1})$. This combined with inequality (11.167) yields $H(\bar{X}_{1:2^n}) \le H(X_q^{n+1}) + n$. $\qquad\square$

Analogously, we can bound other increasing functions of the stationary mean. We will say that a function $\phi : \mathbb{X}^* \to \mathbb{R}$ is increasing if for u being a substring of

w, we have $\phi(u) \leq \phi(w)$. Examples of increasing functions include the maximal repetition $L(w)$, the Hartley entropy $H_0(m|w)$, and the indicator function $\mathbf{1}\{\phi(w) > k\}$, where ϕ is increasing.

Theorem 11.26: *For an increasing function ϕ and any $q \in \mathbb{N}$, we have*

$$\mathbf{E}\phi(X_q^{n-1}) \leq \mathbf{E}\phi(\bar{X}_{1:2^n}) \leq \mathbf{E}\phi(X_q^{n+1}) \tag{11.168}$$

Proof: By Theorem 11.24,

$$\mathbf{E}\phi(\bar{X}_{1:2^n}) = \frac{1}{2^n} \sum_{j=0}^{2^n-1} \mathbf{E}\phi(X_k^n \circ T^j) \tag{11.169}$$

Now we observe that for each $k \geq 1$ and j there exists a q such that X_q^{n-1} is a substring of $X_k^n \circ T^j$. Thus we have $\phi(X_k^n \circ T^j) \geq \phi(X_q^{n-1})$. This combined with equality (11.169) yields $\mathbf{E}\phi(X_q^{n-1}) \leq \mathbf{E}\phi(\bar{X}_{1:2^n})$. On the other hand, for each $k > 1$ and j there exists a q such that $X_k^n \circ T^j$ is a substring of X_q^{n+1}. Thus we have $\phi(X_k^n \circ T^j) \leq \phi(X_q^{n+1})$. This combined with equality (11.169) yields $\mathbf{E}\phi(\bar{X}_{1:2^n}) \leq \mathbf{E}\phi(X_q^{n+1})$. $\quad\square$

In view of Theorems 11.25 and 11.26, it suffices to investigate the distribution of blocks X_j^n to obtain some bounds for the stationary mean. In the next theorem, we will state a simple upper bound for the Hartley entropy of the nonstationary RHA process. From this bound, we can obtain a lower bound for the maximal repetition via Theorems 9.22 and 9.23.

Theorem 11.27: *For the RHA process,*

$$H_0(2^n|X_{1:\infty}) \leq 2 \log k_n \text{ almost surely} \tag{11.170}$$

Proof: For a given realization of the RHA process (i.e. for fixed blocks Y_j^n), there are at most k_n different values of blocks X_j^n. Therefore, there are at most k_n^2 different values of blocks $X_k^n \circ T^j$ in sequence $(X_i)_{i\in\mathbb{N}}$. $\quad\square$

Similarly, we will bound the Shannon entropies of the nonstationary RHA process. The first result is a corollary of Theorem 11.23, which says that conditional entropy of blocks X_j^n given the entire pool of admissible blocks of the same length $\mathcal{G}_{\leq n}$ is exactly equal to the logarithm of perplexity.

Theorem 11.28: *We have*

$$H(X_j^n|\mathcal{G}_{\leq n}) = \log k_n \tag{11.171}$$

$$I(X_j^n; X_{j+1}^n|\mathcal{G}_{\leq n}) = 0 \tag{11.172}$$

Proof: Given $\mathcal{G}_{\leq n}$, the correspondence between X_j^n and K_{nj} is one-to-one. Hence, $H(X_j^n|\mathcal{G}_{\leq n}) = H(K_{nj}|\mathcal{G}_{\leq n})$. From Theorem 11.23, we further obtain $H(K_{nj}|\mathcal{G}_{\leq n}) = H(K_{nj}) = \log k_n$ and $H(K_{nj}, K_{n,j+1}|\mathcal{G}_{\leq n}) = H(K_{nj}) + H(K_{n,j+1})$. □

The second result is an exact expression for the entropy of the pool of admissible blocks $\mathcal{G}_{\leq n}$, also in term of perplexities.

Theorem 11.29: *We have*

$$H(\mathcal{G}_{\leq n}) = \sum_{l=1}^{n} \log \binom{k_{l-1}^2}{k_l} \qquad (11.173)$$

Proof: The claim follows by the chain rule $H(\mathcal{G}_{\leq n}) = H(\mathcal{G}_{\leq n-1}) + H(\mathcal{G}_{\leq n}|\mathcal{G}_{\leq n-1})$ from $H(\mathcal{G}_{\leq 0}) = 0$ and $H(\mathcal{G}_{\leq n}|\mathcal{G}_{\leq n-1}) = \log \binom{k_{n-1}^2}{k_n}$. □

Combining the above two results, we can provide an upper bound for the unconditional entropy of blocks X_j^n.

Theorem 11.30: *We have*

$$H(X_j^n) \leq \min_{0 \leq l \leq n}(H(\mathcal{G}_{\leq l}) + 2^{n-l}\log k_l) \qquad (11.174)$$

Proof: For $0 \leq l \leq n$ we have $H(X_j^n) \leq H(X_j^n, \mathcal{G}_{\leq l}) = H(X_j^n|\mathcal{G}_{\leq l}) + H(\mathcal{G}_{\leq l})$, whereas $H(X_j^n|\mathcal{G}_{\leq l}) \leq 2^{n-l}H(K_{lj}|\mathcal{G}_{\leq l}) = 2^{n-l}H(K_{lj}) = 2^{n-l}\log k_l$. □

Let us observe that inequality $H(X_j^n) \geq H(X_j^n|\mathcal{G}_{\leq n}) = \log k_n$ gives a certain lower bound for the block entropy of the RHA process. Quite often, this lower bound is orders of magnitude smaller than the upper bound (11.174).

Given Theorems 11.28 and 11.30, we may introduce an important parameter of the RHA process, which we will call the combinatorial entropy rate.

Definition 11.9 *(Combinatorial entropy rate):* The combinatorial entropy rate of the RHA process is defined as

$$h_c := \inf_{l \in \mathbb{N}} 2^{-l}\log k_l \qquad (11.175)$$

Theorem 11.31: *We have*

$$\inf_{n \in \mathbb{N}} 2^{-n}H(X_j^n) = h_c \qquad (11.176)$$

Proof: On the one hand, by Theorem 11.28,

$$\inf_{n\in\mathbb{N}} 2^{-n}H(X_j^n) \geq \inf_{n\in\mathbb{N}} 2^{-n}H(X_j^n|\mathcal{G}_{\leq n}) = \inf_{l\in\mathbb{N}} 2^{-l}\log k_l \tag{11.177}$$

On the other hand, by Theorem 11.30,

$$\inf_{n\in\mathbb{N}} 2^{-n}H(X_j^n) \leq \inf_{l\in\mathbb{N}}\inf_{n\in\mathbb{N}} (2^{-n}H(\mathcal{G}_{\leq l}) + 2^{-l}\log k_l) = \inf_{l\in\mathbb{N}} 2^{-l}\log k_l \tag{11.178}$$

$\square$

Theorem 11.31 combined with Theorem 11.25 yields a bound for the entropy rate of the stationary mean of the RHA process.

Theorem 11.32: *We have*

$$h_c/2 \leq \inf_{m\in\mathbb{N}} \frac{H(\bar{X}_{1:m})}{m} \leq 2h_c \tag{11.179}$$

Proof: Divide the sides of inequality (11.165) by 2^n and take the infimum. $\square$

In particular, the combinatorial entropy rate is zero if and only if the entropy rate of the stationary mean is zero as well.

Finally, we will demonstrate that the stationary RHA process with suitable perplexities constitutes a regular Hilberg process.

Theorem 11.33: *Let $\beta \in (0,1)$. For perplexities*

$$k_n = \lfloor 2^{2^{\beta n}} \rfloor \tag{11.180}$$

the stationary RHA process $(\bar{X}_i)_{i\in\mathbb{Z}}$ satisfies the following:

1. *The Shannon entropy rate is*

$$\lim_{m\to\infty} \frac{H(\bar{X}_{1:m})}{m} = 0 \tag{11.181}$$

2. *The Shannon block entropy is sandwiched by inequalities*

$$C_1 m^\beta \leq H(\bar{X}_{1:m}) \leq C_2 m \left(\frac{\log\log m}{\log m}\right)^{1/\beta - 1} \tag{11.182}$$

3. *The Hartley entropy is upper bounded by inequality*

$$H_0(m|\bar{X}_{1:\infty}) \leq C_3 m^\beta \text{ almost surely} \tag{11.183}$$

4. *The maximal repetition is lower bounded by inequality*

$$L(\bar{X}_{1:m}) \geq C_4 (\log m)^{1/\beta} \text{ for infinitely many } m \text{ almost surely} \tag{11.184}$$

Proof:

1. For perplexities (11.180), the combinatorial entropy rate is $h_c = 0$. Hence, the Shannon entropy rate is zero by Theorem 11.32.

2. By (11.173), entropy $H(\mathcal{G}_{\leq n})$ can be bounded as

$$H(\mathcal{G}_{\leq n}) = \sum_{l=1}^{n} \log \binom{k_{l-1}^2}{k_l} \leq \sum_{l=1}^{n} 2k_l \log k_{l-1} \leq 2nk_n \log k_n \tag{11.185}$$

Hence, from (11.174), for $0 \leq l \leq n$ we obtain an upper bound

$$H(X_j^n) \leq (2lk_l + 2^{n-l}) \log k_l \tag{11.186}$$

If we choose $l = \left\lfloor \beta^{-1} \log \left(\frac{n}{\log n} \right) \right\rfloor$, then for perplexities (11.180) we obtain

$$H(X_j^n) \leq \left[2\beta^{-1} \log \left(\frac{n}{\log n} \right) 2^{n/\log n} + 2^n \left(\frac{n}{\log n} \right)^{-1/\beta} \right] \frac{n}{\log n}$$

$$\leq C_2 2^n \left(\frac{\log n}{n} \right)^{1/\beta - 1} \tag{11.187}$$

On the other hand, by $H(X_j^n) \geq H(X_j^n | \mathcal{G}_{\leq n}) = \log k_n$ we obtain

$$H(X_j^n) \geq 2^{\beta n - 1} \tag{11.188}$$

By (11.187) and (11.188), from Theorem 11.25, we obtain the desired sandwich bound for the entropy of the stationary mean.

3. By Theorems 11.27 and 11.26 we obtain

$$0 = \mathbf{E}\mathbf{1}\{H_0(2^m | X_{1:\infty}) > 2 \log k_m\}$$
$$\geq \mathbf{E}\mathbf{1}\{H_0(2^m | X_j^{n+1}) > 2 \log k_m\}$$
$$\geq \mathbf{E}\mathbf{1}\{H_0(2^m | \bar{X}_{1:2^n}) > 2 \log k_m\} \tag{11.189}$$

Hence almost surely we have $H_0(2^m | \bar{X}_{1:\infty}) \leq 2 \log k_m = 2^{\beta m + 1}$, which implies (11.183) for a certain constant C_3.

4. By Theorems 9.22 and 9.23, inequality (11.184) follows from (11.183). □

The construction of RHA processes and the above results were introduced by Dębowski (2017). The paper discussed also some more precise upper bounds for the maximal repetition and lower bounds for the Shannon entropy but reviewing the paper contents for this book, we have discovered some gaps in the reasoning which we did not manage to fill. For this reason, we have omitted these stronger claims.

For perplexities different from (11.180), RHA processes need not have the vanishing Shannon entropy rate. We find it an interesting open problem to study the rate of growth of the block mutual information and the maximal repetition in such cases and to seek for some interactions and nontrivial bounds between these quantities. We also suppose that the stationary mean of a suitable RHA process satisfies a power-law decay of mutual information between two symbols, as suggested by a result of Lin and Tegmark (2017). Moreover, some other open questions concerning RHA processes are whether RHA processes are AMS in general and whether, for given perplexities, the stationary mean of a RHA process is strongly nonergodic and its random ergodic measure is perigraphic almost surely. We suppose that such claims may be true.

11.5   Toward Better Models

In this final section, we would like to discuss advantages and disadvantages of the toy examples of processes presented in this chapter as possible statistical models of natural language. As we have indicated, there are four statistical phenomena hypothetically satisfied by texts in natural language in general: the strictly positive Shannon entropy rate (11.1), the power-law growth of block mutual information (11.2), the power-law logarithmic growth of maximal repetition (11.3), and possibly the power-law decay of mutual information between individual symbols (11.4). Our assessment of the toy models is mostly based on whether these phenomena hold for the respective models but some bonus points are assigned for other features that make the toy models resemble generation of texts in natural language.

The first class of processes that we have discussed are finite and ultrafinite energy processes. Finite and ultrafinite energy processes have a strictly positive Shannon entropy rate, probably like the natural language. But these processes do not satisfy either a power-law logarithmic growth of maximal repetition or a power-law growth of mutual information, in contrast to the natural language. Since a large subclass of hidden Markov processes are finite energy processes then we should treat the above fact as an indication that texts in natural language cannot be accurately modeled by hidden Markov processes, although these processes were widely used in natural language engineering. Same applies to processes satisfying the Doeblin condition, stronger than the finite energy, although Kontoyiannis (1997) supposed that natural language falls into this class.

The second class of toy examples are variations on the theme of Santa Fe processes. These processes have both a strictly positive Shannon entropy rate and a power-law growing block mutual information. Moreover, the construction of Santa Fe processes can be motivated linguistically. We can imagine that a Santa Fe process is a series of random propositions making assertions about an infinite random extra-textual reality. Consequently, we may split a Santa Fe process into a static description of some time-independent or slowly evolving reality and a narration process, which selects which elementary facts about the reality are being described in the text. Given some simple conditions, Santa Fe processes are strongly nonergodic or perigraphic, and for that reason they satisfy a power-law growing block mutual information, as indicated by theorems about facts and words.

The analyzed Zipfian Santa Fe and pseudo Santa Fe processes are a huge idealization with respect to the natural language. They were proposed as a simple model for the power-law growth of the block mutual information and its possible links to nonergodicity. Yet these processes seem to model, in an idealized form, some phenomena that take place in natural language. Confronted with a text in natural language, we often feel that it consists of a description of some unchanging or slowly evolving reality. The unchanging reality referred to in the texts may concern objects that do not change in time, like mathematical or physical constants. The evolving reality referred to in the texts may consists of objects that evolve with a varied speed, like culture, language, or geography. One should note that the distinction between the unchanging and evolving realities may be a matter of a modeling approximation since some referred objects may evolve so slowly that they may be considered practically constant.

Let us also note that the power-law growth of block mutual information in Santa Fe processes seems to hold only if described objects are referred to at a larger frequency than they evolve. The latter condition may be somewhat unrealistic. It is agreed among linguists that many language constructions which are used less frequently evolve quicker (Pagel et al., 2007). In consequence, block mutual information for natural language might be upper bounded. An interesting open question arises how far the power-law growth of mutual information can be extrapolated for natural language and how large its excess entropy is actually. However, even if the excess entropy of natural language is finite, it may be so large that we may hardly observe boundedness of its estimates in empirical data (Hahn and Futrell, 2019), just like we may hardly observe boundedness of the lexicon (Kornai, 2002).

Santa Fe processes are typically finite energy and for that reason they fail to satisfy the power-law logarithmic growth of maximal repetition. But as we have noted, some finite alphabet encodings of these processes need not be finite energy. An open problem remains whether any possible enforcing of the power-law logarithmic growth of maximal repetition in a Santa Fe process encoding necessarily weakens the power-law growth of block mutual information. We suppose that it need not be the case but this question should be formally researched.

The third class of considered toy examples, the RHA processes, is least understood. We have managed to show that for a simple choice of process parameters, called perplexities, these processes satisfy both a power-law growth of mutual information and a power-law logarithmic bound for maximal repetition. Unfortunately, these processes have a zero Shannon entropy rate for the considered perplexities. A few fundamental questions also remain open: whether RHA processes are AMS, in general, and whether they are strongly nonergodic or perigraphic for the considered perplexities. We also suppose that these RHA processes satisfy the power-law decay of mutual information between individual symbols.

In our opinion, there is a more general open problem whether the power-law logarithmic growth of maximal repetition (11.3) and the power-law decay of mutual information between individual symbols (11.4) imply the power-law growth of block mutual information (11.2). We suppose that laws such as (11.3) and (11.4) may also imply the power-law growth of the number of distinct prediction by partial matching (PPM) words, i.e. a sort of Herdan's law, and the power-law growth of the number of described facts, i.e. the stochastic process being perigraphic. Relating these hypothetical phenomena to natural language, we suppose that the potentially infinitely complex reality repeatedly described in texts created by humans may emerge as a result of our tendency to repeat previously encountered parts of text. It is plausible that the reality described in texts is actively created by texts themselves.

The quest for more accurate models of natural language is open. In this book, searching for statistical models of language, we have restricted our investigations to processes that satisfy the claim of the Birkhoff ergodic theorem, i.e. we have considered only stationary and AMS processes. Honestly, we cannot be certain that the correct probability model of language belongs to these classes. Notwithstanding this, we think that stationary and AMS processes can model a few linguistically relevant phenomena, and they offer a good starting point to investigate what is mathematically and physically possible.

Another delicate question is that we have considered mostly recursive processes. As we have argued in Chapter 1, there is a tension whether we regard the probabilities as frequencies in the empirical data (the frequentist interpretation) or as subjective expectations of intelligent agents who make predictions (the Bayesian interpretation). According to the frequentist view, language should be likely modeled by an ergodic and nonrecursive process, whereas according to the Bayesian view, language should be likely modeled by a recursive but nonergodic process. While doing research in the foundations of intelligence, we should focus on the Bayesian interpretation of probability and seek for a finite description of language within this paradigm. This recursive Bayesian model of language competence should resemble the universal PPM probability with its capacity of approximating any nonrecursive stationary ergodic probability distribution, but it should be more clever in making predictions. The only constraints for its cleverness should be computational tractability and physical scalability.

Statistical language modeling is of a large importance for general artificial intelligence, and we think that mathematicians can also help to construct stochastic processes serving as better generative and predictive models of texts in natural language. We suppose that this interdisciplinary research area will be live in the future. There are a few open mathematical problems that we came across when writing this book. The list of them has been compiled in the next unnumbered chapter. Quite many of these problems can be stated in a relatively simple way, which indicates that there is still much to do for mathematicians at the interface of natural language and stochastic processes.

Problems

11.1 Show that a stationary hidden Markov process $(X_i)_{i \in \mathbb{T}}$ satisfies not only condition (9.31) but also condition (9.32).

11.2 For a stationary process $(X_i)_{i \in \mathbb{Z}}$, we define function

$$\psi(n) = \sup_{i,j \geq 1} \text{ess sup} \frac{P(X_{-j}^0, X_n^{n+i})}{P(X_{-j}^0) P(X_n^{n+i})} \qquad (11.190)$$

The process is called ψ-mixing if $\psi(n) < \infty$ and $\lim_{n \to \infty} \psi(n) = 1$. Show that a ψ-mixing stationary process $(X_i)_{i \in \mathbb{Z}}$ is ultrafinite energy if

$$\lim_{n \to \infty} \text{ess sup} \, P(X_1^n) = 0 \qquad (11.191)$$

This result was shown by Ko (2012, corollary 4.4).

11.3 For a stationary hidden Markov process $(X_i)_{i \in \mathbb{T}}$ denote the underlying Markov process $(Y_i)_{i \in \mathbb{T}}$. Show that process $(X_i)_{i \in \mathbb{T}}$ is finite energy if

$$c := \sup_{x \in \mathbb{X}, y \in \mathbb{Y}} P(X_{i+1} = x | Y_i = y) < 1 \qquad (11.192)$$

Is this condition stronger or weaker than (11.8)?

11.4 Exhibit some examples of hidden Markov processes which are:
(a) not finite energy;
(b) not ergodic;
(c) asymptotically deterministic;
(d) ergodic but not finitary.

Having completed the task, compare your solution with Travers and Crutchfield (2014) and Dębowski (2014).

11.5 Can a finite-state process be not finite energy and not asymptotically deterministic simultaneously?

Future Research

The goal of this book was to develop a theory of discrete stochastic processes so that it could account for a few statistical phenomena exhibited by texts in natural language. In our opinion, the most important of those phenomena take form of power laws concerning the growth or decay of mutual information or the maximal repetition. As promised, we have developed some theory and we have presented a few toy examples of processes which follow some requested statistical phenomena but not all of them simultaneously. However, in spite of a sizeable volume of this book, we feel that in our investigations, we have only scratched the surface of unknown interesting areas.

Rather than offering the ultimate answers, this book was intended to draw interest of the reader to some fundamental mathematical problems of statistical natural language modeling. Here is a handful of open problems that may guide the future research:

1. Is the PPM order a consistent estimator of the Markov order, as conjectured by Csiszar and Shields (2000)? (We guess: yes.)
2. Are there almost sure versions of the theorems about facts and words, i.e. inequalities (8.125) and (8.126) with the expectation operators removed from all terms? (We guess: yes.)
3. Is the Hilberg exponent for the algorithmic mutual information equal to the Hilberg exponent for the PPM mutual information? (We guess: no.)
4. Can a finite-state process be perigraphic? (We guess: no.)
5. Can a finite-state process have a strictly positive Hilberg exponent for the PPM mutual information? (We guess: no.)
6. Can we both upper and lower bound the power-law logarithmic growth of the maximal repetition (11.3) in terms of the power-law growth of the unconditional collision entropy in the general stationary ergodic case? (We guess: no.)
7. Are nonstationary RHA processes AMS? (We guess: yes.)

Information Theory Meets Power Laws: Stochastic Processes and Language Models, First Edition. Łukasz Dębowski.
© 2021 John Wiley & Sons, Inc. Published 2021 by John Wiley & Sons, Inc.

8. Consider a stationary RHA process. In this case, does a power-law decay of mutual information between two symbols (11.4) imply a power-law logarithmic growth of maximal repetition (11.3). (We guess: yes.) Moreover, does condition (11.3) imply that the process is strongly nonergodic and the random ergodic measure is perigraphic almost surely? (We guess: yes.) Do these implications hold true for a general stationary process? (We guess: no.)

9. What is a simple example of a process that has a strictly positive Shannon entropy rate (11.1), is perigraphic, and satisfies a power-law logarithmic growth of maximal repetition (11.3)?

10. Is there a computable function of finite texts drawn from an arbitrary stationary ergodic process which given the text length tending to infinity converges almost surely to 1 if the process is perigraphic and to 0 otherwise? (We guess: no.) Does such a function exist for random ergodic measures of stationary RHA processes? (We guess: yes.)

If the reader solves some of these puzzles, we would like to know of it. Also, if she or he comes up with other related hypotheses.

Bibliography

Aarseth, E.J. (1997). *Cybertext: Perspectives on Ergodic Literature.* Baltimore, MD: Johns Hopkins University Press.

Adams, T.M. and Nobel, A.B. (2010). Uniform convergence of Vapnik-Chervonenkis classes under ergodic sampling. *The Annals of Probability* 38 (4): 1345–1367.

Adger, D. (2019). *Language Unlimited: The Science Behind Our Most Creative Power.* Oxford: Oxford University Press.

Ahlswede, R., Balkenhol, B., and Khachatrian, L.H. (1996). Some properties of fix-free codes. *Proceedings of the 1st INTAS International Seminar on Coding Theory and Combinatorics, 1996,* Thahkadzor, Armenia, pp. 20–33.

Algoet, P.H. and Cover, T.M. (1988). A sandwich proof of the Shannon-McMillan-Breiman theorem. *The Annals of Probability* 16: 899–909.

Allouche, J.-P. and Shallit, J. (2003). *Automatic Sequences. Theory, Applications, Generalizations.* Cambridge: Cambridge University Press.

Altmann, G. (1980). Prolegomena to Menzerath's law. *Glottometrika* 2: 1–10.

Altmann, E.G., Pierrehumbert, J.B., and Motter, A.E. (2009). Beyond word frequency: bursts, lulls, and scaling in the temporal distributions of words. *PLoS ONE* 4: e7678.

Arratia, R. and Waterman, M.S. (1989). The Erdös-Rényi strong law for pattern matching with a given proportion of mismatches. *The Annals of Probability* 17: 1152–1169.

Baayen, R.H. (2001). *Word frequency distributions.* Dordrecht: Kluwer Academic Publishers.

Barron, A.R. (1985a). The strong ergodic theorem for densities: generalized Shannon-McMillan-Breiman theorem. *The Annals of Probability* 13: 1292–1303.

Barron, A.R. (1985b). *Logically smooth density estimation.* PhD thesis. Stanford University.

Behr, F., Fossum, V., Mitzenmacher, M., and Xiao, D. (2003). Estimating and comparing entropy across written natural languages using PPM compression. *Proceedings of Data Compression Conference 2003,* p. 416.

Information Theory Meets Power Laws: Stochastic Processes and Language Models, First Edition.
Łukasz Dębowski.
© 2021 John Wiley & Sons, Inc. Published 2021 by John Wiley & Sons, Inc.

Bentz, C. (2018). *Adaptive Languages – An Information-Theoretic Account of Linguistic Diversity*. Berlin: De Gruyter Mouton.

Bentz, C., Alikaniotis, D., Cysouw, M., and Ferrer-i-Cancho, R. (2017). The entropy of words – learnability and expressivity across more than 1000 languages. *Entropy* 19 (6): 275.

Beran, J. (1994). *Statistics for Long-Memory Processes*. New York: Chapman & Hall.

Beta Writer (2019). *Lithium-Ion Batteries. A Machine-Generated Summary of Current Research*. New York: Springer.

Bialek, W., Nemenman, I., and Tishby, N. (2001a). Predictability, complexity and learning. *Neural Computation* 13: 2409.

Bialek, W., Nemenman, I., and Tishby, N. (2001b). Complexity through nonextensivity. *Physica A* 302: 89–99.

Bienvenu, L., Day, A.R., Hoyrup, M. et al. (2012). A constructive version of Birkhoff's ergodic theorem for Martin-Löf random points. *Information and Computation* 210: 21–30.

Billingsley, P. (1979). *Probability and Measure*. New York: Wiley.

Birch, J.J. (1962). Approximations for the entropy for functions of Markov chains. *The Annals of Mathematical Statistics* 33: 930–938.

Birkhoff, G.D. (1932). Proof of the ergodic theorem. *Proceedings of the National Academy of Sciences of the United States of America* 17: 656–660.

Bloch, W.G. (2008). *The Unimaginable Mathematics of Borges' Library of Babel*. Oxford: Oxford University Press.

Borges, J.L. (1941). *El jardin de senderos que se bifurcan*, chapter La biblioteca de Babel. Editorial Sur.

Boyd, S. and Vandenberghe, L. (2004). *Convex Optimization*. Cambridge: Cambridge University Press.

Bradley, R.C. (1980). On the strong mixing and weak Bernoulli conditions. *Zeitschrift für Wahrscheinlichkeitstheorie und verwandte Gebiete* 50: 49–54.

Bradley, R.C. (1989). An elementary treatment of the Radon-Nikodym derivative. *The American Mathematical Monthly* 96: 437–440.

Bradley, R.C. (2005). Basic properties of strong mixing conditions. A survey and some open questions. *Probability Surveys* 2: 107–144.

Braverman, M., Chen, X., Kakade, S.M. et al. (2019). Calibration, entropy rates, and memory in language models. https://arxiv.org/pdf/1906.05664 (accessed 16 March 2020).

Breiman, L. (1992). *Probability*. Philadelphia, PA: SIAM.

Brown, P.F., Della Pietra, S.A., Della Pietra, V.J. et al. (1992). An estimate of an upper bound for the entropy of English. *Computational Linguistics* 18: 31–40.

Brudno, A.A. (1982). Entropy and the complexity of trajectories of a dynamical system. *Transactions of the Moscovian Mathematical Society* 44: 124–149.

Burger, J.R., George, M.A. Jr., Leadbetter, C., and Shaikh, F. (2019). The allometry of brain size in mammals. *Journal of Mammalogy* 100 (2): 276–283.

Cariolaro, G. and Pierobon, G. (1977). Stationary symbol sequences from variable-length word sequences. *IEEE Transactions on Information Theory* 23: 243–253.

Chaitin, G.J. (1975a). A theory of program size formally identical to information theory. *Journal of the ACM* 22: 329–340.

Chaitin, G.J. (1975b). Randomness and mathematical proof. *Scientific American* 232 (5): 47–52.

Chaitin, G.J. (1982). Gödel's theorem and information. *International Journal of Theoretical Physics* 22: 941–954.

Charikar, M., Lehman, E., Lehman, A. et al. (2005). The smallest grammar problem. *IEEE Transactions on Information Theory* 51: 2554–2576.

Chi, Z. and Geman, S. (1998). Estimation of probabilistic context-free grammars. *Computational Linguistics* 24: 299–305.

Chomsky, N. (1956). Three models for the description of language. *IRE Transactions on Information Theory* 2 (3): 113–124.

Chomsky, N. (1957). *Syntactic Structures*. The Hague: Mouton & Co.

Chomsky, N. (1959). A review of B. F. Skinner's Verbal Behavior. *Language* 35 (1): 26–58.

Chomsky, N. (1986). *Knowledge of Language: Its Nature, Origin, and Use*. Santa Barbara, CA: Greenwood Publishing Group.

Chomsky, N. and Miller, G. (1959). Finite state languages. *Information and Control* 1: 91–112.

Chow, Y.S. and Teicher, H. (1978). *Probability Theory. Independence, Interchangeability, Martingales*. New York: Springer.

Chung, K.L. (1961). A note on the ergodic theorem of information theory. *The Annals of Mathematical Statistics* 32: 612–614.

Cleary, J.G. and Witten, I.H. (1984). Data compression using adaptive coding and partial string matching. *IEEE Transactions on Communications* 32: 396–402.

Cohn, P.M. (2003). *Basic Algebra: Groups, Rings, and Fields*. New York: Springer.

Conrad, B. and Mitzenmacher, M. (2004). Power laws for monkeys typing randomly: the case of unequal probabilities. *IEEE Transactions on Information Theory* 50: 1403–1414.

Cornfeld, I.P., Fomin, S.V., and Sinai, Ya.G. (1982). *Ergodic Theory*. New York: Springer.

Coupé, C., Oh, Y.M., Dediu, D., and Pellegrino, F. (2019). Different languages, similar encoding efficiency: comparable information rates across the human communicative niche. *Science Advances* 5 (9): eaaw2594.

Cover, T.M. and King, R.C. (1978). A convergent gambling estimate of the entropy of English. *IEEE Transactions on Information Theory* 24: 413–421.

Cover, T.M. and Thomas, J.A. (2006). *Elements of Information Theory*, 2e. New York: Wiley.

Cover, T.M., Gacs, P., and Gray, R.M. (1989). Kolmogorov's contributions to information theory and algorithmic complexity. *The Annals of Probability* 17: 840–865.

Crochemore, M. and Ilie, L. (2008). Maximal repetitions in strings. *Journal of Computer and System Sciences* 74: 796–807.

Crutchfield, J.P. and Feldman, D.P. (2003). Regularities unseen, randomness observed: the entropy convergence hierarchy. *Chaos* 15: 25–54.

Crutchfield, J.P. and Marzen, S. (2015). Signatures of infinity: nonergodicity and resource scaling in prediction, complexity, and learning. *Physical Review E* 91: 050106(R).

Csiszár, I. and Körner, J. (2011). *Information Theory: Coding Theorems for Discrete Memoryless Systems*. Cambridge: Cambridge University Press.

Csiszar, I. and Shields, P.C. (2000). The consistency of the BIC Markov order estimator. *The Annals of Statistics* 28: 1601–1619.

Dębowski, L. (2006). On Hilberg's law and its links with Guiraud's law. *Journal of Quantitative Linguistics* 13: 81–109.

Dębowski, L. (2007a). On processes with summable partial autocorrelations. *Statistics and Probability Letters* 77: 752–759.

Dębowski, L. (2007b). Menzerath's law for the smallest grammars. In: *Exact Methods in the Study of Language and Text* (ed. P. Grzybek and R. Kohler), 77–85. Berlin: De Gruyter Mouton.

Dębowski, L. (2009). A general definition of conditional information and its application to ergodic decomposition. *Statistics and Probability Letters* 79: 1260–1268.

Dębowski, L. (2010). Variable-length coding of two-sided asymptotically mean stationary measures. *Journal of Theoretical Probability* 23: 237–256.

Dębowski, L. (2011). On the vocabulary of grammar-based codes and the logical consistency of texts. *IEEE Transactions on Information Theory* 57: 4589–4599.

Dębowski, L. (2012a). Mixing, ergodic, and nonergodic processes with rapidly growing information between blocks. *IEEE Transactions on Information Theory* 58: 3392–3401.

Dębowski, L. (2012b). Maximal lengths of repeat in English prose. In: *Synergetic Linguistics. Text and Language as Dynamic System* (ed. S. Naumann, P. Grzybek, R. Vulanovic, and G. Altmann), 23–30. Wien: Praesens Verlag.

Dębowski, L. (2014). On hidden Markov processes with infinite excess entropy. *Journal of Theoretical Probability* 27: 539–551.

Dębowski, L. (2015a). Hilberg exponents: new measures of long memory in the process. *IEEE Transactions on Information Theory* 61: 5716–5726.

Dębowski, L. (2015b). Maximal repetitions in written texts: finite energy hypothesis vs. strong Hilberg conjecture. *Entropy* 17: 5903–5919.

Dębowski, L. (2016a). Estimation of entropy from subword complexity. In: *Challenges in Computational Statistics and Data Mining* (ed. S. Matwin and J. Mielniczuk), 53–70. New York: Springer.

Dębowski, L. (2016b). Consistency of the plug-in estimator of the entropy rate for ergodic processes. *2016 IEEE International Symposium on Information Theory (ISIT)*, pp. 1651–1655.

Dębowski, L. (2017). Regular Hilberg processes: an example of processes with a vanishing entropy rate. *IEEE Transactions on Information Theory* 63 (10): 6538–6546.

Dębowski, L. (2018a). Is natural language a perigraphic process? The theorem about facts and words revisited. *Entropy* 20 (2): 85.

Dębowski, L. (2018b). Maximal repetition and zero entropy rate. *IEEE Transactions on Information Theory* 64 (4): 2212–2219.

Dębowski, L. (2020). Approximating information measures for fields. *Entropy*. 22 (1): 79.

Dobrushin, R.L. (1959). A general formulation of the fundamental Shannon theorems in information theory. *Uspekhi Matematicheskikh Nauk* 14 (6): 3–104 (in Russian).

Doob, J.L. (1953). *Stochastic Processes*. New York: Wiley.

Downey, R.G. and Hirschfeldt, D.R. (2010). *Algorithmic Randomness and Complexity*. New York: Springer.

Ebeling, W. and Nicolis, G. (1991). Entropy of symbolic sequences: the role of correlations. *Europhysics Letters* 14: 191–196.

Ebeling, W. and Nicolis, G. (1992). Word frequency and entropy of symbolic sequences: a dynamical perspective. *Chaos, Solitons & Fractals* 2: 635–650.

Ebeling, W. and Pöschel, T. (1994). Entropy and long-range correlations in literary English. *Europhysics Letters* 26: 241–246.

Elias, P. (1975). Universal codeword sets and representations of the integers. *IEEE Transactions on Information Theory* 21: 194–203.

Ephraim, Y. and Merhav, N. (2002). Hidden Markov processes. *IEEE Transactions on Information Theory* 48: 1518–1569.

Erdős, P. and Rényi, A. (1970). On a new law of large numbers. *Journal d'Analyse Mathématique* 22: 103–111.

Esposti, Esposti, and Pachet, Pachet (eds) (2016). *Creativity and Universality in Language, Lecture Notes in Morphogenesis*. New York: Springer.

Estoup, J.B. (1916). *Gammes sténographiques*. Paris: Institut Stenographique de France.

Fano, R.M. (1961). *Transmission of Information*. Cambridge, MA: The MIT Press.

Fehr, S. and Berens, S. (2014). On the conditional Rényi entropy. *IEEE Transactions on Information Theory* 60: 6801–6810.

Feistel, R. and Ebeling, W. (2011). chapter self-organization of information and symbols. In: *Physics of Self-Organization and Evolution*. New York: Wiley.

Fekete, M. (1923). Über die Verteilung der Wurzeln bei gewissen algebraischen Gleichungen mit ganzzahligen Koeffizienten. *Mathematische Zeitschriften* 17: 228–249.

Ferenczi, S. (1999). Complexity of sequences and dynamical systems. *Discrete Mathematics* 206: 145–154.

Ferrer-i-Cancho, R. and Solé, R.V. (2001). Two regimes in the frequency of words and the origins of complex lexicons: Zipf's law revisited. *Journal of Quantitative Linguistics* 8 (3): 165–173.

Ferrer-i-Cancho, R. and Solé, R.V. (2003). Least effort and the origins of scaling in human language. *Proceedings of the National Academy of Sciences of the United States of America* 100: 788–791.

de Finetti, B. (1931). Funzione caratteristica di un fenomeno aleatorio. *Atti della Reale Academia Nazionale dei Lincei, Serie 6* 4: 251–299.

Fontana, R., Gray, R., and Kieffer, J. (1981). Asymptotically mean stationary channels. *IEEE Transactions on Information Theory* 27: 308–316.

Franklin, J.N.Y., Greenberg, N., Miller, J.S., and Ng, K.M. (2012). Martin-Löf random points satisfy Birkhoff's ergodic theorem for effectively closed sets. *Proceedings of the American Mathematical Society* 140: 3623–3628.

Gács, P. (1974). On the symmetry of algorithmic information. *Soviet Mathematics Doklady* 15: 1477–1480.

Gács, P. and Körner, J. (1973). Common information is far less than mutual information. *Problems of Control and Information Theory* 2: 119–162.

Gács, P., Tromp, J., and Vitányi, P.M.B. (2001). Algorithmic statistics. *IEEE Transactions on Information Theory* 47: 2443–2463.

Gao, Y., Kontoyiannis, I., and Bienenstock, E. (2008). Estimating the entropy of binary time series: methodology, some theory and a simulation study. *Entropy* 10: 71–99.

Garsia, A.M. (1965). A simple proof of E. Hopf's maximal ergodic theorem. *Journal of Mathematics and Mechanics* 14: 381–382.

Gelfand, I.M., Kolmogorov, A.N., and Yaglom, A.M. (1956). Towards the general definition of the amount of information. *Doklady Akademii Nauk SSSR* 111: 745–748 (in Russian).

Gheorghiciuc, I. and Ward, M.D. (2007). On correlation polynomials and subword complexity. *Discrete Mathematics and Theoretical Computer Science* AH: 1–18.

Gillman, D. and Rivest, R.L. (1995). Complete variable-length "fix-free" codes. *Designs, Codes and Cryptography* 5: 109–114.

Gödel, K. (1931). Über formal unentscheidbare Sätze der Principia Mathematica und verwandter Systeme, I. *Monatshefte für Mathematik und Physik* 38 (1): 173–198.

Grassberger, P. (2002). Data compression and entropy estimates by non-sequential recursive pair substitution. `https://arxiv.org/pdf/physics/0207023` (accessed 16 March 2020).

Gray, R.M. and Davisson, L.D. (1974a). Source coding theorems without the ergodic assumption. *IEEE Transactions on Information Theory* 20: 502–516.

Gray, R.M. and Davisson, L.D. (1974b). The ergodic decomposition of stationary discrete random processses. *IEEE Transactions on Information Theory* 20: 625–636.

Gray, R.M. (1990). *Entropy and Information Theory*. New York: Springer.

Gray, R.M. (2009). *Probability, Random Processes, and Ergodic Properties*. New York: Springer.

Gray, R.M. and Kieffer, J.C. (1980). Asymptotically mean stationary measures. *The Annals of Probability* 8: 962–973.

Guiraud, P. (1954). *Les caractères statistiques du vocabulaire*. Paris: Presses Universitaires de France.

Hahn, M. and Futrell, R. (2019). Estimating predictive rate-distortion curves via neural variational inference. *Entropy* 21: 640.

Harremoës, P. and Topsøe, F. (2001). Maximum entropy fundamentals. *Entropy* 3: 191–226.

Harris, T.E. (1963). *The Theory of Branching Processes*. New York: Springer.

Heaps, H.S. (1978). *Information Retrieval – Computational and Theoretical Aspects*. New York: Academic Press.

Herdan, G. (1964). *Quantitative Linguistics*. London: Butterworths.

Hernández, T. and Ferrer i Cancho, R. (2019). *Lingüística cuantitativa*. EMSE EDAPP, S. L. y Prisanoticias Colecciones.

Hilberg, W. (1990). Der bekannte Grenzwert der redundanzfreien Information in Texten – eine Fehlinterpretation der Shannonschen Experimente? *Frequenz* 44: 243–248.

Hopcroft, J.E. and Ullman, J.D. (1979). *Introduction to Automata Theory, Languages and Computation*. Reading, MA: Addison-Wesley.

Hoyrup, M. (2013). Computability of the ergodic decomposition. *Annals of Pure and Applied Logic* 164: 542–549.

von Humboldt, W. (1836). *Über die Verschiedenheit des menschlichen Sprachbaues und ihren Einfluss auf die geistige Entwicklung des Menschengeschlechts*. Königliche Akademie der Wissenschaften. "On Linguistic Variability and Its Influence upon the Intellectual Development of Mankind".

Ivanko, E.E. (2008). Exact approximation of average subword complexity of finite random words over finite alphabet. *Trudy Instituta Matematiki i Mehaniki UrO RAN* 14 (4): 185–189.

Jacquet, P., Seroussi, G., and Szpankowski, W. (2008). On the entropy of a hidden Markov process. *Theoretical Computer Science* 395 (2–3): 203–219.

Janson, S., Lonardi, S., and Szpankowski, W. (2004). On average sequence complexity. *Theoretical Computer Science* 326: 213–227.

Jelinek, F. (1997). *Statistical Methods for Speech Recognition*. Cambridge, MA: The MIT Press.

Kac, M. (1947). On the notion of recurrence in discrete stochastic processes. *Bulletin of American Mathematical Society* 53: 1002–1010.

Kakihara, Y. (1999). *Abstract Methods in Information Theory*. World Scientific Publishing.

Kallenberg, O. (1997). *Foundations of Modern Probability*. New York: Springer.

Khmaladze, E. (1988). The Statistical Analysis of Large Number of Rare Events. Technical Report MS-R8804. Amsterdam: Centrum voor Wiskunde en Informatica.

Kieffer, J.C. and Yang, E. (2000). Grammar-based codes: a new class of universal lossless source codes. *IEEE Transactions on Information Theory* 46: 737–754.

Knight, F. (1975). A predictive view of continuous time processes. *The Annals of Probability* 3 (4): 573–596.

Knuth, D. (1984). The complexity of songs. *Communications of the ACM* 27: 345–348.

Ko, M.H.F. (2012). Renyi entropy and recurrence. PhD thesis. University of Southern California.

Köhler, Köhler, and Piotrowski, Piotrowski (eds) (2005). *Quantitative Linguistik. Ein internationales Handbuch / Quantitative Linguistics. An International Handbook*. Berlin: Walter de Gruyter.

Kolmogoroff, A. (1933). *Grundbegriffe der Wahrscheinlichkeitsrechnung*. Berlin: Springer-Verlag.

Kolmogorov, A.N. (1965). Three approaches to the quantitative definition of information. *Problems of Information Transmission* 1 (1): 1–7.

Kolpakov, R. and Kucherov, G. (1999a). On maximal repetitions in words. *Journal of Discrete Algorithms* 1: 159–186.

Kolpakov, R. and Kucherov, G. (1999b). Finding maximal repetitions in a word in linear time. *40th Annual Symposium on Foundations of Computer Science, 1999*, pp. 596–604.

Kontoyiannis, I. (1997). The Complexity and Entropy of Literary Styles. Technical Report 97. Department of Statistics, Stanford University.

Kontoyiannis, I. (1998). Asymptotic recurrence and waiting times for stationary processes. *Journal of Theoretical Probability* 11: 795–811.

Kontoyiannis, I. and Suhov, Y. (1994). Prefixes and the entropy rate for long-range sources. In: *Probability, Statistics, and Optimization: A Tribute to Peter Whittle* (ed. F.P. Kelly), 89–98. New York: Wiley.

Kontoyiannis, I., Algoet, P.H., Suhov, Yu.M., and Wyner, A.J. (1998). Nonparametric entropy estimation for stationary processes and random fields, with applications to English text. *IEEE Transactions on Information Theory* 44: 1319–1327.

Kornai, A. (2002). How many words are there? *Glottometrics* 4: 61–86.

Krichevsky, R.E. and Trofimov, V.K. (1981). The performance of universal encoding. *IEEE Transactions on Information Theory* IT-27: 199–207.

Kullback, S. and Leibler, R.A. (1951). On information and sufficiency. *The Annals of Mathematical Statistics* 22: 79–86.

Kuraszkiewicz, W. and Łukaszewicz, J. (1951). The number of different words as a function of text length. *Pamiętnik Literacki* 42 (1): 168–182 (in Polish).

van Lambalgen, M. (1987). *Random sequences*. PhD thesis. Universiteit van Amsterdam.

Lehéricy, L. (2019). Consistent order estimation for nonparametric Hidden Markov Models. *Bernoulli* 25 (1): 464–498.

Lem, S. (1974). *The Cyberiad*. Continuum Publishing Corporation. The Sixth Sally, or How Trurl and Klapaucius Created a Demon of the Second Kind to Defeat the Pirate Pugg.

Li, M. and Vitányi, P.M.B. (2008). *An Introduction to Kolmogorov Complexity and Its Applications*, 3e. New York: Springer.

Lin, H.W. and Tegmark, M. (2017). Critical behavior in physics and probabilistic formal languages. *Entropy* 19: 299.

Löhr, W. (2009). Properties of the statistical complexity functional and partially deterministic HMMs. *Entropy* 11: 385–401.

de Luca, A. (1999). On the combinatorics of finite words. *Theoretical Computer Science* 218: 13–39.

Mandelbrot, B. (1953). An informational theory of the statistical structure of languages. In: *Communication Theory* (ed. W. Jackson), 486–502. London: Butterworths.

Mandelbrot, B. (1954). Structure formelle des textes et communication. *Word* 10: 1–27.

Manin, D. (2019). Running in shackles: the information-theoretic paradoxes of poetry. In: *Handbook of the Mathematics of the Arts and Sciences* (ed. B. Sriraman). New York: Springer 1–14.

Manning, C.D. and Schütze, H. (1999). *Foundations of Statistical Natural Language Processing*. Cambridge, MA: The MIT Press.

de Marcken, C.G. (1996). *Unsupervised language acquisition*. PhD thesis. Massachusetts Institute of Technology.

Markov, A.A. (1913). Essai d'une recherche statistique sur le texte du roman "Eugene Onegin" illustrant la liaison des epreuve en chain. *Bulletin de l'Académie Impériale des Sciences de St.-Pétersbourg* 7: 153–162.

Markov, A.A. (2006). An example of statistical investigation of the text 'Eugene Onegin' concerning the connection of samples in chains. *Science in Context* 19: 591–600.

Martin-Löf, P. (1966). The definition of random sequences. *Information and Control* 9: 602–619.

McMillan, B. (1956). Two inequalities implied by unique decipherability. *IEEE Transactions on Information Theory* IT-2: 115–116.

Menzerath, P. (1928). Über einige phonetische Probleme. In: *Actes du premier Congres international de linguistes*. Leiden: Sijthoff 104–105.

Miller, G.A. (1957). Some effects of intermittent silence. *American Journal of Psychology* 70: 311–314.

Miller, M.I. and O'Sullivan, J.A. (1992). Entropies and combinatorics of random branching processes and context-free languages. *IEEE Transactions on Information Theory* 38: 1292–1310.

Montemurro, M.A. and Zanette, D.H. (2002). New perspectives on Zipf's law in linguistics: from single texts to large corpora. *Glottometrics* 4: 87–99.

Morvai, G. and Weiss, B. (2005). Order estimation of Markov chains. *IEEE Transactions on Information Theory* 51: 1496–1497.

Norris, J.R. (1997). *Markov Chains*. Cambridge: Cambridge University Press.

Ochoa, C. and Navarro, G. (2019). Repair and all irreducible grammars are upper bounded by high-order empirical entropy. *IEEE Transactions on Information Theory* 65: 3160–3164.

Ornstein, D.S. and Weiss, B. (1993). Entropy and data compression schemes. *IEEE Transactions on Information Theory* 39: 78–83.

Pagel, M., Atkinson, Q.D., and Meade, A. (2007). Frequency of word-use predicts rates of lexical evolution throughout Indo-European history. *Nature* 449: 717–720.

Pap, Pap (ed.) (2002). *Handbook of Measure Theory*. Amsterdam: Elsevier Science.

Pereira, F. (2000). Formal grammar and information theory: together again? *Philosophical Transactions of the Royal Society of London, Series A: Mathematical, Physical and Engineering Sciences* 358: 1239–1253.

Pinsker, M.S. (1964). *Information and Information Stability of Random Variables and Processes*. San Francisco, CA: Holden-Day.

Pittel, B. (1985). Asymptotical growth of a class of random trees. *The Annals of Probability* 13 (2): 414–427.

von Plato, J. (1994). *Creating Modern Probability: Its Mathematics, Physics, and Philosophy in Historical Perspective*. Cambridge: Cambridge University Press.

Radford, A., Wu, J., Child, R. et al. (2019). Language models are unsupervised multitask learners. https://openai.com/blog/better-language-models/ (accessed 16 March 2020).

Rényi, A. (1961). On measures of entropy and information. *Proceedings of the 4th Berkeley Symposium on Mathematical Statistics and Probability, Volume 1: Contributions to the Theory of Statistics*, pp. 547–561.

Rączaszek-Leonardi, J. (2012). *Language as a System of Replicable Constraints*, 295–333. New York: Springer.

Rokhlin, V.A. (1962). On the fundamental ideas of measure theory. *American Mathematical Society Translations, Series 1* 10: 1–54.

Rudin, W. (1974). *Real and Complex Analysis.* New York: McGraw-Hill.

Ryabko, B. (1984). Twice-universal coding. *Problems of Information Transmission* 20 (3): 173–177.

Sazonov, V.V. (1964). On a question of R. L. Dobruvsin. *Teorija verojatnostej i ee primenenija* 9 (1): 180–181 (in Russian).

Schönhuth, A. (2008). The ergodic decomposition of asymptotically mean stationary random sources. http://arxiv.org/abs/0804.2487 (accessed 16 March 2020).

Seth, A.K. (2019). Our inner universes. *Scientific American* 321 (3): 40–47.

Shafer, G. and Vovk, V. (2006). The sources of Kolmogorov's "Grundbegriffe". *Statistical Science* 21 (1): 70–98.

Shalizi, C.R. (2001). Causal architecture, complexity and self-organization for time series and cellular automata. PhD thesis. University of Wisconsin-Madison.

Shalizi, C.R. and Crutchfield, J.P. (2001). Computational mechanics: pattern and prediction, structure and simplicity. *Journal of Statistical Physics* 104: 819–881.

Shannon, C. (1948). A mathematical theory of communication. *Bell System Technical Journal* 30: 379–423, 623–656.

Shannon, C. (1951). Prediction and entropy of printed English. *Bell System Technical Journal* 30: 50–64.

Shields, P.C. (1991). Cutting and stacking: a method for constructing stationary processes. *IEEE Transactions on Information Theory* 37: 1605–1617.

Shields, P.C. (1992). String matching: the ergodic case. *The Annals of Probability* 20: 1199–1203.

Shields, P.C. (1993). Universal redundancy rates do not exist. *IEEE Transactions on Information Theory* IT-39: 520–524.

Shields, P.C. (1997). String matching bounds via coding. *The Annals of Probability* 25: 329–336.

Shwartz-Ziv, R. and Tishby, N. (2017). Opening the black box of deep neural networks via information. https://arxiv.org/abs/1703.00810 (accessed 16 March 2020).

Simon, H.A. (1955). On a class of skew distribution functions. *Biometrika* 42: 425–440.

Skinner, B.F. (1957). *Verbal Behavior.* Englewood Cliffs, NJ: Prentice Hall.

Solomonoff, R.J. (1964). A formal theory of inductive inference, part 1 and part 2. *Information and Control* 7: 1–22, 224–254.

Swart, J.M. (1996). A conditional product measure theorem. *Statistics and Probability Letters* 28: 131–135.

Swift, J. (1726). *Travels into Several Remote Nations of the World. In Four Parts. By Lemuel Gulliver, First a Surgeon, and then a Captain of Several Ships.* Benjamin Motte,.

Szpankowski, W. (1993a). Asymptotic properties of data compression and suffix trees. *IEEE Transactions on Information Theory* 39: 1647–1659.

Szpankowski, W. (1993b). A generalized suffix tree and its (un)expected asymptotic behaviors. *SIAM Journal on Computing* 22: 1176–1198.

Takahashi, H. (2008). On a definition of random sequences with respect to conditional probability. *Information and Computation* 206: 1375–1382.

Takahashi, H. (2011). Algorithmic randomness and monotone complexity on product space. *Information and Computation* 209: 183–197.

Takahashi, S. and Tanaka-Ishii, K. (2019). Evaluating computational language models with scaling properties of natural language. *Computational Linguistics* Early Access: https://doi.org/10.1162/coli_a_00355.

Takahira, R., Tanaka-Ishii, K., and Dębowski, L. (2016). Entropy rate estimates for natural language – a new extrapolation of compressed large-scale corpora. *Entropy* 18 (10): 364.

Timo, R., Blackmore, K., and Hanlen, L. (2007). On the entropy rate of word-valued sources. *Proceedings of the Telecommunication Networks and Applications Conference, ATNAC 2007*, pp. 377–382.

Tishby, N. and Zaslavsky, N. (2015). Deep learning and the information bottleneck principle. *2015 IEEE Information Theory Workshop (ITW)*, pp. 1–5.

Tishby, N., Pereira, F.C., and Bialek, W. (1999). The information bottleneck method. *37th Annual Allerton Conference on Communication, Control, and Computing*, pp. 368–377.

Torre, I.G., Luque, B., Lacasa, L. et al. (2019). On the physical origin of linguistic laws and lognormality in speech. *Royal Society Open Science* 6: 191023.

Travers, N.F. and Crutchfield, J.P. (2014). Infinite excess entropy processes with countable-state generators. *Entropy* 16: 1396–1413.

Turing, A.M. (1936a). On computable numbers, with an application to the Entscheidungsproblem. *Proceedings of the London Mathematical Society* 42: 230–265.

Turing, A.M. (1936b). On computable numbers, with an application to the *Entscheidungsproblem*: a correction. *Proceedings of the London Mathematical Society* 43: 544–546.

Upper, D.R. (1997). *Theory and algorithms for hidden Markov models and generalized hidden Markov models*. PhD thesis. University of California.

Vovk, V. (2009). Superefficiency from the vantage point of computability. *Statistical Science* 24: 73–86.

Vovk, V.G. and V'yugin, V.V. (1993). On the empirical validity of the Bayesian method. *Journal of the Royal Statistical Society, Series B* 55: 253–266.

Vovk, V.G. and V'yugin, V.V. (1994). Prequential level of impossibility with some applications. *Journal of the Royal Statistical Society, Series B* 56: 115–123.

V'yugin, V.V. (2007). On empirical meaning of randomness with respect to a real parameter. In: *Computer Science – Theory and Applications*, 387–396. New York: Springer.

West, G. (2017). *Scale: The Universal Laws of Growth, Innovation, Sustainability, and the Pace of Life in Organisms, Cities, Economies, and Companies*. New York: Penguin Press.

Wyner, A.D. (1975). The common information of two dependent random variables. *IEEE Transactions on Information Theory* IT-21: 163–179.

Wyner, A.D. (1978). A definition of conditional mutual information for arbitrary ensembles. *Information and Control* 38: 51–59.

Wyner, A.D. and Ziv, J. (1989). Some asymptotic properties of entropy of a stationary ergodic data source with applications to data compression. *IEEE Transactions on Information Theory* 35: 1250–1258.

Yaglom, A.M. and Yaglom, I.M. (1983). *Probability and Information, Theory and Decision Library*. New York: Springer.

Yeung, R.W. (2002). *First Course in Information Theory*. Dordrecht: Kluwer Academic Publishers.

Zador, A.M. (2019). A critique of pure learning and what artificial neural networks can learn from animal brains. *Nature Communications* 10: 3770.

Zipf, G.K. (1935). *The Psycho-Biology of Language: An Introduction to Dynamic Philology*. Boston, MA: Houghton Mifflin.

Zipf, G.K. (1949). *Human Behavior and the Principle of Least Effort*. Reading, MA: Addison-Wesley.

Ziv, J. and Lempel, A. (1977). A universal algorithm for sequential data compression. *IEEE Transactions on Information Theory* 23: 337–343.

Zvonkin, A.K. and Levin, L.A. (1970). The complexity of finite objects and the development of the concepts of information and randomness by means of the theory of algorithms. *Russian Mathematical Surveys* 25: 83–124.

Index

ε-machine 165
π-system 59
  generated 60
σ-field 59
  Borel 80
  complete 103, 104
  countably generated 61
  extended Borel 113
  generated 60
  invariant 113
  nonatomic 83
  preimage 156
  product 63
  tail 156

a
almost everywhere 85
almost surely 85
argument raising 204

b
binomial coefficient 187
bit 81
  random 81
bound
  Chaitin 209
  Chernoff 328
  excess 232, 234
  Kontoyiannis 270

c
cardinality 87
chain rule 138, 154, 207

code 191
  comma-separated 192
  extension 191
  fixed length 192
  nonsingular 191
  prefix-free 193
  Shannon-Fano 196
  uniquely decodable 192
coding
  quasiperiodic 296
  stationary 325
  synchronizable 299
complexity
  prefix-free Kolmogorov 200, 201
  subword 273
condition
  Doeblin 313
convergence
  almost sure 102
  in probability 102
  of conditioning 151
  to equilibrium 71
criterion
  ergodic 120, 121
  mixing 121, 122

d
disintegration 101
disjoint sets 60
disjunctive normal form 61
distribution 92
  semi 92, 171

Information Theory Meets Power Laws: Stochastic Processes and Language Models, First Edition.
Łukasz Dębowski.
© 2021 John Wiley & Sons, Inc. Published 2021 by John Wiley & Sons, Inc.

distribution (*contd.*)
    stationary 70
    universal semi 172
domain restriction 96
dynamical system 111
    product of 124
    projection of 123

e

entropy
    block 140
    block Rényi 266
    collision 262
    combinatorial rate 342
    conditional 135, 146, 181
    continuity 148
    empirical 177
    excess 143
    Hartley 134, 262, 278
    min- 263
    pointwise 92
    Rényi 261
    Rényi-Arimoto 262
    rate 142
    Shannon 133, 134, 146, 262
event 59
    elementary 59
    negligible 103
expectation 85
    conditional 97

f

field 59
    countably generated 61
    generated 60
filtration 98, 106
function
    characteristic 85
    concave 89
    convex 89
    ergodic set 113
    measurable 66
    real computable 205
    real semicomputable 205
    recursive partial discrete 204

recursively enumerable set 208
stationary set 113
uniquely enumerating 208

h

halting 199, 203
halting problem 205
Hilberg exponent 233

i

independence 136, 152
inequality
    Barron 93
    Bonferroni 74
    Cauchy–Schwarz 91
    Chung-Erdős 106
    conditional Jensen 99
    data-processing 139
    effective inverse Kraft 208
    Fano 159
    Gibbs 93
    inverse Kraft 195
    Jensen 90
    Kraft 195
    Markov 88
    median-mean 91
    on means 264
    Paley–Zygmund 91
    source coding 196
infimum 84
    essential 235
information
    algorithmic mutual 213
    common 164
    conditional mutual
        135, 146
    mutual 134, 146
information symmetry 213
injection 191
invariance of completion 149

J

Jensen gap 274

K

KL divergence   93

L

Lebesgue integral   84
lemma
    Barron   94
    Borel–Cantelli   88
    Fatou   87
    Fekete   141
    Kac   269
Levy laws   98
limit   86
    lower   86
    upper   86
longest match length   280

m

martingale   98
maximal repetition   280
measure
    AMS   294
    ergodic   116
    finite   60
    Lebesgue   80
    mixing   121
    periodic   115
    probability   60
    product   64
    pseudo AMS   294
    random   100
    random ergodic   156
    stationary   116
measure domination   95
median   91
mutual information
    continuity   147

n

norm   264
number
    real computable   205
    real semicomputable   205

o

object encoding   202, 204
order
    Kolmogorov   256
    Markov   181
    PPM   180

p

partition   67
philosophers' stone   219
point   59
    periodic   114
    regular   114
probabilistic fact   243
probability
    algorithmic   210
    conditional   94, 97
    halting   210, 217
    PPM   172
    regular conditional   100
problem
    halting   218
process
    ψ-mixing   348
    k-th order Markov   72
    AMS   301
    asymptotically deterministic   142
    Bernoulli   162
    discrete   67
    ergodic   116
    fair-coin   81
    finitary   144
    finite energy   309
    hidden Markov   72
    IID   68
    infinitary   144
    Markov   69
    Markov aperiodic   70
    Markov irreducible   70
    mixing   121
    nonstationary RHA   336
    perigraphic   247
    prediction   165
    pseudo AMS   301
    pseudo Santa Fe   319

process (*contd.*)
  real 81
  Santa Fe 244, 247, 315
  stationary 116
  stationary RHA 340
  strongly nonergodic 161
  ultrafinite energy 310
  uniformly dithered 312

q

quantile 236

r

random variable 66
  $\mathcal{G}$-measurable 67
  block of 140
  discrete 67
  integrable 84
  real 80
  simple 67
refinement 136, 152
relative frequency 113

s

sequence
  algorithmically incompressible 218
  algorithmically random 224
set
  complete 304
  cylinder 63, 82, 195, 220
  fix-free 304
  nonmeasurable 73
  of inferrable facts 245, 247
  of PPM words 184
  prefix-free 193, 304
  suffix-free 304
set difference xvii
  symmetric 74
shift 112
space
  complete probability 103
  event 59
  finite measure 60
  measurable 60
  nonatomic probability 82

of stationary ergodic measures 125
of stationary measures 125
probability 60
product measurable 63
split of join 150
stationary mean 293
supremum 84
  essential 235

t

test
  Martin-Löf 221, 222
  Solovay 221, 222
theorem
  π-λ 62
  Birkhoff ergodic 117, 128
  Chaitin incompleteness 216
  coding 210, 211
  dominated convergence 87
  ergodic 71
  ergodic decomposition 126, 127, 157
  extension 62
  facts and words 250–253
  invariance 200
  Kolmogorov process 65
  maximal ergodic 118
  monotone convergence 86
  Poincaré recurrence 129
  Radon-Nikodym 95
  Riesz 102
  Schnorr 222
  second order SMB 241
  SMB 170, 178, 303
  stationary process 116
  Vitali-Hahn-Saks 114
  WZ/OW 271
time
  recurrence 269
  waiting 269
transition matrix 69, 72
Turing machine
  prefix-free 198
  universal 200

v

variance 91